Mechanisms and Machine Science

Volume 90

This book series establishes a well-defined forum for monographs, edited Books, and proceedings on mechanical engineering with particular emphasis on MMS (Mechanism and Machine Science). The final goal is the publication of research that shows the development of mechanical engineering and particularly MMS in all technical aspects, even in very recent assessments. Published works share an approach by which technical details and formulation are discussed, and discuss modern formalisms with the aim to circulate research and technical achievements for use in professional, research, academic, and teaching activities.

This technical approach is an essential characteristic of the series. By discussing technical details and formulations in terms of modern formalisms, the possibility is created not only to show technical developments but also to explain achievements for technical teaching and research activity today and for the future.

The book series is intended to collect technical views on developments of the broad field of MMS in a unique frame that can be seen in its totality as an Encyclopaedia of MMS but with the additional purpose of archiving and teaching MMS achievements. Therefore, the book series will be of use not only for researchers and teachers in Mechanical Engineering but also for professionals and students for their formation and future work.

The series is promoted under the auspices of International Federation for the Promotion of Mechanism and Machine Science (IFToMM).

Prospective authors and editors can contact Mr. Pierpaolo Riva (publishing editor, Springer) at: pierpaolo.riva@springer.com

Indexed by SCOPUS and Google Scholar.

More information about this series at http://www.springer.com/series/8779

Jörg Niemann · Adrian Pisla

Life-Cycle Management of Machines and Mechanisms

Springer

Jörg Niemann
Department of Mechanical
and Process Engineering
Fachhochschule Düsseldorf
Düsseldorf, Nordrhein-Westfalen, Germany

Adrian Pisla
Design Engineering and Robotics
Department
Technical University of Cluj-Napoca
Cluj-Napoca, Romania

ISSN 2211-0984 ISSN 2211-0992 (electronic)
Mechanisms and Machine Science
ISBN 978-3-030-56447-6 ISBN 978-3-030-56449-0 (eBook)
https://doi.org/10.1007/978-3-030-56449-0

This Springer imprint is published by the registered company Springer Nature Switzerland AG
The registered company address is: Gewerbestrasse 11, 6330 Cham, Switzerland

Contents

Acronyms/Terms and Abbreviations

A&D	A&D Industry, Aerospace & Defense Industry
AAM	Application Activity Model
ABV	Added Business Value
AEC	Aided Engineering in Construction
AGV	Automated guided vehicle
AI	Artificial intelligence
AICAPA	American Institute of Certified Public Accountants
AIM	Application interpreted model
ANSI	American National Standards Institute
AOV	Average order value
AP	Application protocol
APPTR	Axial piston pump test rig
APUID	Acquisition Program Unique Identification
AR	Augmented reality, to the physical (real) components are added data for connecting characteristics to an image
ARIS	Architecture of Integrated Information Systems
ARM	Application Reference Model
ASCII	American Standard Code for Information Interchange is a character encoding standard for electronic communication. ASCII codes represent text in computers, telecommunications equipment, and other devices (labeling, bar codes, etc.)
ASG	Alternative scenarios generation
ASRS	Automated storage and retrieval systems
AT&T Inc.	American multinational conglomerate holding company headquartered at Whitacre Tower in Downtown Dallas, Texas
ATA	Air Transport Association
ATO	Assemble-to-order
AVM	Architecture virtual machine
BC	Best in the class
BCFTR	Bearing cage friction test rig

BFTM	Bending fatigue testing machine
BMC	Business Model Canvas
BO Model	Business object model
BOM	Bill of materials
BSC	Balanced scorecard
BTO	Build to order
BVC	Best value chain
C	Cheap, not expensive
CAC	Computer-aided control
CAD	Computer-aided design
CAE	Computer-aided engineering
CAM	Computer-aided manufacturing
CAPA	Corrective and preventive action
CAPP	Computer-aided processes planning
CAPP	Computer-aided production planning
CAQ	Computer-aided quality assurance
CAX	Computer aided for X applications
CCSDS	Consultative Committee for Space Data Systems, founded in 1982 for governmental and quasi-governmental space agencies to discuss and develop standards for space data and information systems
CE	Concurrent engineering
CEO	Chief Executive Officer
CHP	Combined heat and power plants (cogeneration plants)
CIL	Correlation with insurance companies and social dedicated legislation
CIM	Computer-integrated manufacturing
CM	Condition monitoring
CM	Configuration management
CMP	Change management profile
CNC	Computer numerical control
COS	Computer operating system
COTS	Commercial items including services
CPG	Consumer packaged goods
CPP	Conventional power plant
CPPS	Cyber-Physical Production System
CPR	Contract provided resources
CPU	Central processing units
CRISP-DA	Cross-industry standard process for data mining
CRM	Customer relationship management
CSI	Customer Satisfaction Index
CSR	Corporate social responsibility
CTC	Company technological culture
CTO	Chief Technical Officer
CUS	Conversions–upgrading–shutdown

D&M	Design and manufacturing
DAC	Digital Age Context
DAG	Directed acyclic graph
DAM	Digital assets management
DAO	Data access objects
DB	Databases
DC	Designated community
DF	Digital factory
DFE	Design for the Environment
DFEA	Design for Environment Alternatives
DHD	Documents handling department
DITA	Darwin Information Typing Architecture
DKP	Direct kinematic problem
DL	Deep learning
DLR	Deutsches Zentrum für Luft- und Raumfahrt
DMS	Data management solution
DMSS	Digital mobile security systems
DMU	Digital mockup
DNC	Direct numerical control
DoD	Department of Defense
DPI	Dots per inch
DPM	Direct part marking
DPS	Digital preservation system/digital preservation strategy
DRP	Disaster recovery plan
DRS	Distributed Resource Scheduler
DSN	Data source name
DSPD	Data sources, processing, and distribution
DT	Digital tool/digital twin
DTO	Data transfer object
DTT	Digital twin technology
EAM	Enterprise asset management
EAS	Engineering analysis and simulation
EBIT	Earnings before interest and taxes
EBITDA	Earnings before interest, taxes, depreciation, and amortization
ECC	Error correcting code memory
ECM	Engineering change management
EDGE	Enhanced Data Rates for Global Evolution network
EDM	Engineering Data Management
EDMS	Electronic data management system
EH&S	Environment, health and safety
EHD	Elastohydrodynamic
EID	Enterprise Identifier
EK	Engineering knowledge
EMD	Electromechanical design
EMK	Engineering and management knowledge

EMS	Energy management system
EOF	End-of-life, management
EOL	End-of-life
EPP	Electronic prototyping platform
ER	Enhanced reality
ERP	Enterprise resource planning
ETO	Engineering to order
FAR	Federal Acquisition Regulation
FASB	Financial Accounting Standards Board
FBS	Function–behavior–structure
FDA	Food and Drug Administration
FGI	Finished goods inventory
FMECA	Failure mode effects and criticality analysis
FMS	Flexible manufacturing systems
FPGA	Field-programmable gate array
FSA	Flexible Spending Account, also known as flexible spending arrangement
FTO	Freedom to Operate
FWTR	Fretting wear test rig
G	Good
GD&T	Geometric dimensioning and tolerancing
GHG	Greenhouse Gas protocol
GRA	Golden Robot Award
GSM/EDGE	The technology Enhanced Data for Global Evolution (EDGE) is a high-speed mobile data standard, intended to enable second-generation Global System for Mobile communication (GSM) and time-division multiple access (TDMA) networks to transmit data at up to 384 kilobits per second (Kbps)
GTO	Game theory optimal
HMI	Human–machine interfaces
Homo Faber	In Latin, means "Man the Smith," "Man the Maker," or "Man the Toolmaker." As used by Max Frisch, it refers to a man who controls his environment through his abilities and tools, to be a maker of things. Including the Internet of things (IoT) or the industrial Internet of things (IIoT)
HVAC	Heating ventilation/air conditioning
I/O	Input/output
I4.0	The Fourth Industrial Revolution, rise from the activities digitalization
ICAM	Integrated Computer-Aided Manufacturing
ICT	Information and Communications Technology
IDEF0	Integration definition for function modeling
IEC	International Electrotechnical Commission

IEEE	Institute of Electrical and Electronics Engineers, a professional association for electronic engineering and electrical engineering with its corporate office in New York City
IFIP	International Federation for Information Processing
IIC	Information item contents, lifecycle management documents
IIoT	Industrial Internet of things
IKP	Inverse kinematic problem
ILS	Integrated logistic support
IMPS	Integrated manufacturing process systems
IoT	Internet of things
IPR	Intellectual property rights
IQ	An intelligence quotient is a total score derived from a set of standardized tests designed to assess human intelligence
IRR	Internal rate of return
ISO	International Standardization Organization
ISV	Independent software vendors
IT	Information Technology
ITAR	International Traffic in Arms Regulations
ITT	Invitation to tender
IUID	Item Unique Identification
JBP	Joint Business Planning
JEDEC	Joint Electron Device Engineering Council, a solid-state technology association independent semiconductor engineering trade organization and standardization body
JTC	Joint Technical Committee
JT™	Data format, for viewing and sharing lightweight 3D product data
KADS	Knowledge analysis and design support
KBS	Knowledge-based systems
KDE SC	K Desktop Environment Software Compilation, founded in 1996 by Matthias Ettrich, a student at the University of Tübingen, troubled at the time by certain aspects of the Unix desktop. Among his qualms was that none of the applications looked, felt, or worked alike. He proposed the formation of not only a set of applications, but, rather, a desktop environment, in which users could expect things to look, feel, and work consistently. He also wanted to make this desktop easy to use; one of his complaints with desktop applications of the time was that his girlfriend could not use them. His initial Usenet post spurred a lot of interest, and the KDE project was born
KPI	Key performance indicators
LAN	Local area network
LCA	Life cycle assessment (or life cycle analysis)
LCC	Life cycle cost
LCM	Lifecycle management
LCP	Life cycle planning

LSF	Logistic storing facilities
LTE	Long-Term Evolution, a 4G communication standard for wireless broadband communication for mobile devices and data terminals, based on the GSM/EDGE and UMTS/HSPA technologies
LTO	Limited time offer
M2M	Machine to machine
MANK	Management knowledge
MBSE	Model-based systems engineering
MDA	Model-driven approach
MDE	Multiple Discipline Engineering
MK	Marketing knowledge
ML	Machine learning
MLM	Machines and mechanisms lifecycle management
MM	Machines and mechanisms
MoU	Memorandum of understanding
MPR	Micro-pitting rig
MR	Mixed reality
MRO	Maintenance, repair, and overhaul
MTO	Made to order/Make to order/Manufacture to order
NC	Numerical control
NDA	Net Digital Age
NEN	Nederlands Normalisatie-instituut
NOPAT	Net operating profit after taxes
NPV	Net present value
OAIS	Open Archival Information System
OASIS	Open standards opens source
OEM	Original Equipment Manufacturer
OMG	Object Management Group
ORC	Organic Rankine cycle is a steam generator that uses an organic, high molecular mass fluid with a liquid–vapor phase change, or boiling point, occurring at a lower temperature than the water–steam phase change
OSLC	Open Services for Lifecycle Collaboration
PAS	Publicly Available Specification
PASM	Product accompanying service model
PBM	Policy-based management
PC	Personal computer
PCB	Printed circuit boards
PCH	Project change histories
PCPR	Previous contract provided resources
PD	Product data
PDM	Product data management
PESTEL	Political, economic, social, technology, environmental, legal
PiP	Picture in picture
PIV	Particle image velocimetry

PLC	Programmable logic controllers
PLCS	Application Protocol for Product Life Cycle Support
PLM	Product lifecycle management
PLMP	Product lifecycle management platform
PM	Parts management
PMBOK	Product management book of knowledge
PMT	Project management triangle
POD	Pin on disk
POF	Problem optimization formulation
POS	Point of sale
PP	Product process
PPC	Production planning and control
PROSA	Product sustainability assessment
ProSeCo	Product and service co-design
PSH	Process Simulate Human
PSM	Product structure management
PSS	Product-service system
PTAB	Primary Trustworthy Digital Repository Authorisation Body
PTO	Paid time off
PTR	Pump test rig
PVM	Process virtual machine
R&D	Research and development
RACI	Responsible, Accountable, Consulted, Informed
RCF	Relative centrifugal force
REACH	Registration, Evaluation, Authorisation and Restriction of Chemicals, an European Union regulation dating from 18 December 2006
RFID	Radio-frequency identification
RGB	Red Green Blue, is an additive color model in which red, green, and blue light are added together in various ways to reproduce a broad array of colors. The name of the model comes from the initials of the three additive primary colors
RIO	CompactRIO is a real-time embedded industrial controller made by National Instruments for industrial control systems. The CompactRIO is a combination of a real-time controller, reconfigurable IO modules, FPGA module, and an Ethernet expansion chassis
RMA	Records Management Applications
RMMD	Refined mechanism-machine design
RMT	Reliable Memory Technology
ROCE	Return on capital employed
RoHS	Restriction of Hazardous Substances Directive 2002/95/EC
ROI	Return on investment
RPA	Robotics process automation
RPM	Rotations per minute

RPUID	Real Property Unique Identification
RQ	Robotics quotient is a scoring way of a company or individual's about the ability to work effectively with robots
RTO	Recovery Time Objective
RTPV	Real-time project visibility
S	Swift
SADT	Structured analysis and design technique
SCM	Supply chain management
SE	Sale Engineer
SEC	Securities and Exchange Commission, an independent agency of the United States federal government
SEM	Systems engineering methodologies
SEO	Systems Engineering Organization
SHRM	Society for Human Resource Management
SIPOC	Suppliers, inputs, process, outputs, customers
SLM	Simulation lifecycle management/service lifecycle management
SMBO	Sequential model-based optimization
SME	Small and medium-sized enterprises
SMRL	Semantic Markup Rule Language
SOP	Standard operating procedure
SPDM	Simulation and Process Data Management
SSCC	Serial Shipping Containers Code
ST	Synchronous technology
STEP	Standard for the Exchange of Product Model Data
STEP	Is an open-source two-dimensional physics simulation engine that is included in the KDE SC
SVM	System virtual machine/support vector machine, a supervised machine learning model that uses classification algorithms for two-group classification problems.
SWOT	Strengths, weaknesses, opportunities, threats
SysML	Systems Modeling Language
TaT	Turnaround time
TCD	Total customer demand
TCO	Total cost of ownership
TCP	Tool Center Point
TLCM	Total Life Cycle Management
TOC	Total ownership cost
TPM	Trusted Platform Module
TRL	Technology readiness level
TTM	Time to market
TTR	Tiltrotor test rig
TTR	Turbocharger test rig
TWTR	Thrust washer test rig
UA	User agent
UI	User interface

UID	Unique Identification
UII	Unique item identifier
UML	Unified Modeling Language
UMTS/HSPA	Technologies, where UMTS provides a clear evolutionary path to high-speed packet access (HSPA). HSPA refers to the combination of high-speed downlink packet access (HSDPA) and high-speed uplink packet access (HSUPA) that allows data rates up to 14.4 Mbit/s in the downlink
VC	Variant configurator
VM	Virtual machine
VMC	Virtual machine controls
VMM	Virtual machine monitor (hypervisor)
VMT	Virtual machine templates
VP	Vice President
VR	Virtual reality
VSM	Viable system model
VSOE	Vendor-specific objective evidence
VTR	Valve test rig
W3C	World Wide Web Consortium
WACC	Weighted average cost of capital
WAN	Wide area network
WBS	Work breakdown structure
WEEE	Waste Electrical and Electronic Equipment Directive (2002/96/EC)
WIPO	World Intellectual Property Organization
WPDM	Web-based PDM
XML	Extensible Markup Language

Part I
Life Cycle System Modeling: Factors of PLM Design

Chapter 1
Product Life Cycle and Services Management

1.1 The New Paradigm

Industrial manufacturing and the consumption of technical products have led to a dramatic depletion of natural resources and an increasing strain on the environment due to emissions. Society's heightened ecological awareness is taking effect, with the result that more and more companies are publicly committing themselves to environmental protection. In the process, laws and requirements are bringing about a change in the management of resources. Many companies now recognise the fact that they can make cost savings by encapsulating critical technical processes and handling problem materials more frugally. Today, this development is leading to a rediscovery of the product life cycle. In consequence, this development is also strengthening the sustainability sought by politics and society with regard to responsible commercial trading. Commercial sustainability means that all trade should be orientated towards maintaining all resources. The question at the core of manufacturing is how to achieve overall value creation with one product over its entire lifetime by taking life cycle management into account. Consequently, a change in strategies has taken place which not only takes economical aims but also ecological and societal aspects into consideration in the design and utilisation of technical products (Fig. 1.1).

Manufacturers have to accept more and more responsibility for the usability of their technical products and for the consequences of usage. However, many companies only follow statutory general conditions in pre-sales and after-sales in order to avoid losing their markets. There is a general impression that the cost-benefit ratio, especially in after-sales business, is insufficient. This also applies to industrial recycling.One main factor is the availability of actual information about the products and a lack of synergy between final assembly and after-sales operations [1].

The development of modern products is being decisively influenced by the application of technologies contributing towards increased efficiency. Products are becoming complex highly-integrated systems with internal technical intelligence enabling the user to implement them reliably, economically and successfully even in the fringe

J. Niemann and A. Pisla, *Life-Cycle Management of Machines and Mechanisms*, Mechanisms and Machine Science 90,
https://doi.org/10.1007/978-3-030-56449-0_1

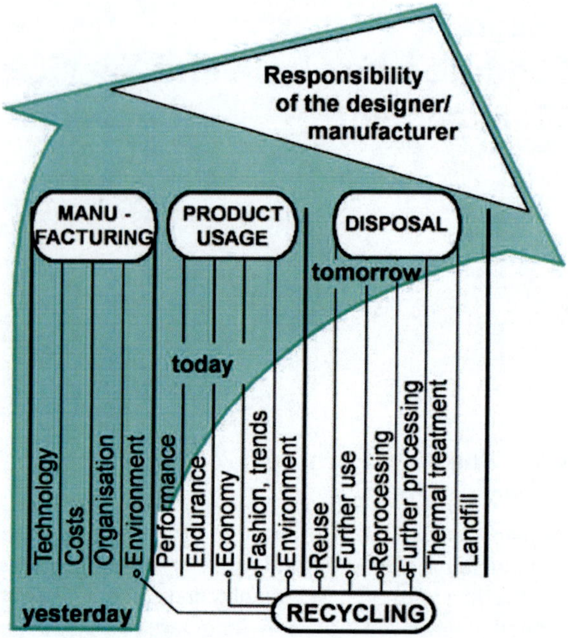

Fig. 1.1 Increased responsibility over the entire product life cycle

ranges of technology. As a result, business strategies are aiming more and more at perfecting technical systems, optimising product usage and maximising added value over the entire lifetime of a product. In this context, the total management of product life cycles coupled with the integration of information and communications systems is becoming a key success factor for industrial companies.

When manufacturing technical products, industrial companies generally direct their strategies towards economic targets. Their main business lies in developing, producing and operating products either for individual customers or for complete sectors of the market. Service and maintenance are considered by many companies to be necessary in order to achieve lasting business relationships with customers.

Several studies indicate that the role of these services will change from being a product-accompanying service to becoming the main *revenue driver*. This means that the original product itself will turn into a vehicle (platform) to sell such services as *main business* [1].

Consequently, industrial manufacturing companies are increasingly concentrating their businesses on engineering, assembly and services. They are following new paradigms in order to add value through customer orientation, system management and services during the lifetime of their products. The machine manufacturing industry and other industrial fields such as the automobile industry have reduced their own capacities down to the main or core technologies and final assembly. Parts and components are manufactured by suppliers or specialised companies. Profit is increasingly becoming a result of business operations in design, engineering, final assembly and service. These phases of production are the core competencies of companies which produce strong market or customer-orientated products and add value during a product's life cycle [2].

The functionalities of products are defined in the processes of design and engineering. The functionality of products and their specific or characteristic properties for usage are determined (as built) or altered by assembling, maintaining and disassembling real configurations. In the usage phase, special know-how regarding design and characteristic properties is required, such as specific process knowledge for optimising utilisation and performance. Increasing technical complexity is promoting product-near services and manufacturer assistance. This brings about new business models for marketing only the functionality of capital-intensive products rather than selling the products themselves.

There is a new paradigm behind these tendencies: in order to add value and maximise utilisation, products are linked in the manufacturer's network from the beginning right up to the end (Fig. 1.2). In order to realise this paradigm, manufacturers need life cycle management (LCM) systems, tools and technologies. The concentration of all processes into the total life cycle of a product and the optimisation of usage of each single technical product can be described as a new paradigm. Seen from a global point or macro-economical point of view, this is only logical. Seen from an operational or micro-economical point of view, it is proving difficult to initiate such strategies. This is because fundamental structural changes are required in products as well as in organisations and production technologies and also that the economic benefits involved are either uncertain or associated with risks [1].

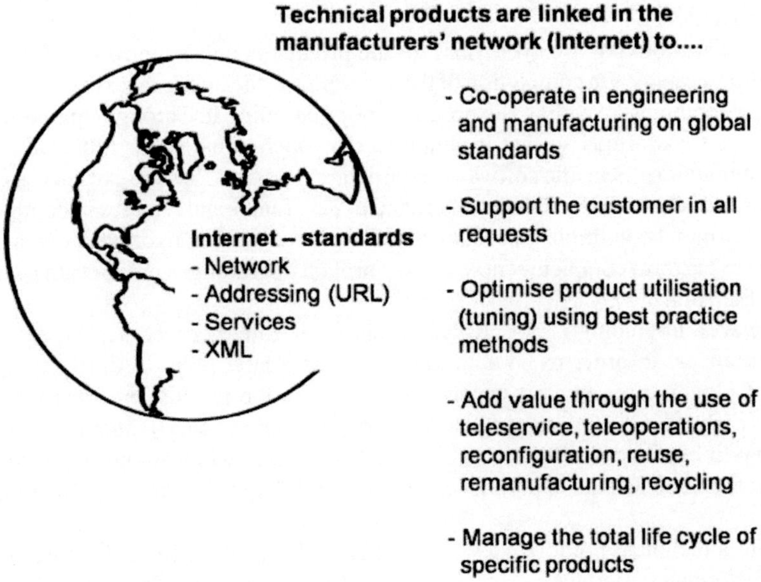

Fig. 1.2 The vision of life cycle management

Additionally, locally optimised product life cycles (i.e. optimisation of individual processes) may not exhibit superior performance globally from multi-objective perspectives. Therefore, the performance of product life cycles needs to be evaluated from holistic and multi-objective perspectives.

However, there is a futuristic vision in the life cycle management of optimising the total exploitation of each product and reducing environmental impact to a minimum. In reality, the different types of products need to be taken into account individually. For some products, it makes economic sense to link them to the manufacturer's network. If the futuristic vision is followed that all machines and high-quality technical products remain in the manufacturer's information network, the Internet will attain a central importance in total life cycle management [3–5].

The strategies followed by companies are significantly dependent upon the type of product involved. In a preliminary classification, three categories with varying time scales and strategies may be defined. The first category is goods with a short lifetime and a low product value or complexity. Such non-durable technical consumer goods are usually mass-produced and manufactured in large series. Here the main emphasis of life cycle management is placed on the rational organisation of services, marketing and product recycling techniques. Robust techniques can be used for recycling due to the fact that the added value profit is low in relation to the value of the product. The second category is assigned to series products with a limited number of variants. Life cycle management for these products includes services and maintenance as well as industrial recycling and the partial reuse of parts and components. The third category is reserved for high-quality capital goods. The main emphases here are on maximum utilisation strategies, maintaining performance and additional added value in the field of after-sales. Industrial recycling only plays a minor economic role in this category of products.

A forward-looking life cycle plan for the product is one example of a maximum utilisation strategy. On completion of the usage phase, the owner faces the alternatives of either scrapping/recycling the product or of upgrading it. Through upgrading, the product is transformed so that it obtains a new operational status reflected in new product functions. Specific software or hardware modifications are carried out on the product to equip it with advanced, extended or new functional features in comparison with its original condition. Consequently, the product can be improved, extended or utilised to perform completely new tasks. Through upgrading, a product almost starts a new life (Fig. 1.3).

However, upgrading is not always possible due to either technical or economic circumstances. In order to be able to upgrade at a later point in time, far-sighted product planning is required which commences in the product engineering stage. In this early phase of development, the fundamental product features—including later modification possibilities—are fixed. Numerous technical and organisational measures decide whether a product can be successfully transformed to attain another level.

From a technical point of view, the modular design of a product's construction is of particular importance. Modular product design in accordance with the laws of system technology enables the variable and economically-viable re-design of a

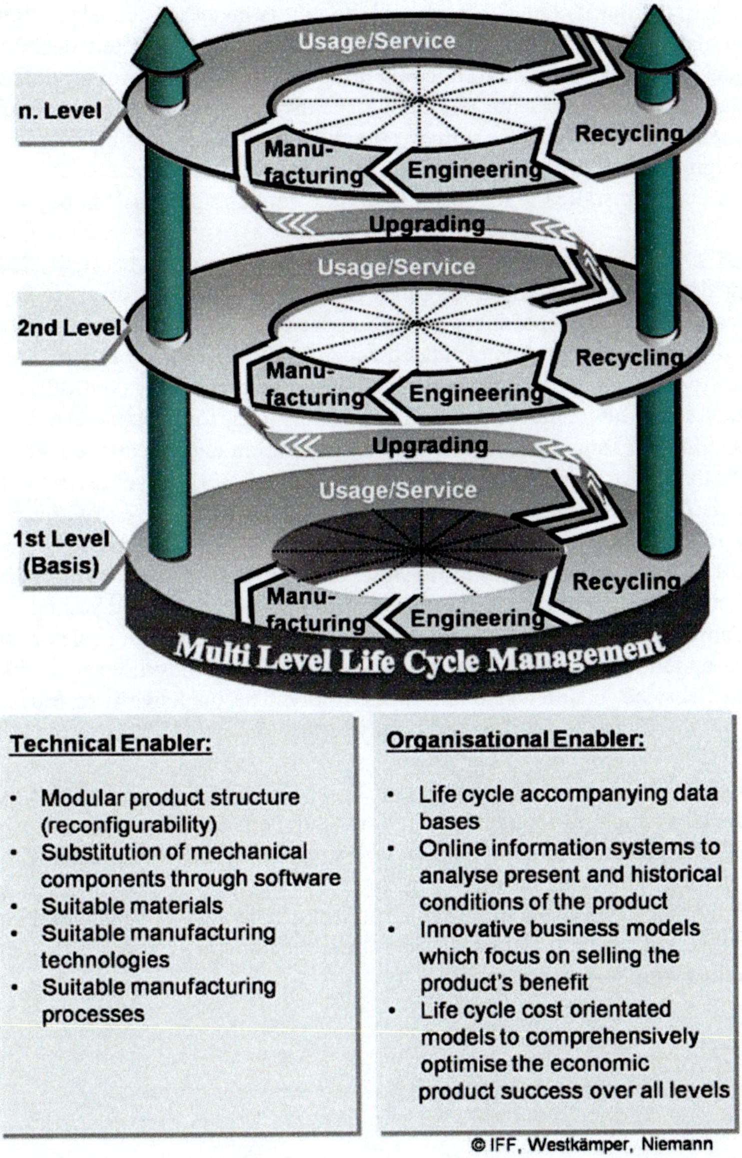

Fig. 1.3 Products have several lives [2]

product throughout its entire lifetime. If the fact is taken into consideration that a product may be modified many times over or even altered completely during its lifetime, such product constructions not only bring about advantages for product maintenance but also create enormous potentials. The increasing substitution of mechanical components with software also supports the short-term usage of a product for variable task assignments. Retrofitting times can be shortened due to the fact that modified software can be installed much faster than hardware components can be exchanged [1].

From a technical point of view, product optimisation can be supported using lifelong data acquisition. Data-logging enables the behaviour of a product to be statistically analysed or products and processes to be monitoring online. The data obtained using this method is evaluated according to specific criteria and discloses optimisation potentials. This permits machines to be completely controlled with the result that, in the future, not only will it be possible to perform technical optimisation but also to take economical factors into consideration and to carry out far-sighted planning thanks to the availability of "real" machine data. Life cycle simulation techniques also enable us to predict product behaviour even in the early phases of the design process. Such real machine data dynamically improve the life cycle model used in life cycle simulation. Up till now, conventional manufacturing paradigms have focused on profit aspects associated with manufacturing and selling products to the end-customer. The new paradigm takes into account the life cycle of technical products and the optimisation of value and benefits during the phases of engineering, assembly, service, maintenance and disassembly. The objective is to reduce environmental losses and to fulfil public or governmental restrictions over the life cycle [5, 6].

Following the new paradigm of optimisation and added value over the total life of products, a structural change in the relationship between the manufacturer and the user will take place. Both have different views regarding the same business processes in the life of products, as shown in Fig. 1.4.

Different views held about the same product are the result of industrial developments in the twenty-first century. In the future, the holistic view will offer new ages of manufacturing.

1.2 Manufacturer's Viewpoint

In general, the life cycle of products can be divided into the phases of design and engineering, manufacturing, assembly, usage, service, disassembly and recycling.

The main objective is to fulfil markets and customer requirements to ensure the efficient utilisation of manufacturing resources. The new view adds value in the usage and recycling phases as a result of customer-related services including maintenance and disassembly for reconfiguration, reuse and recycling. More than ever before, this view of the usage and recycling phases makes it indispensable to take into account

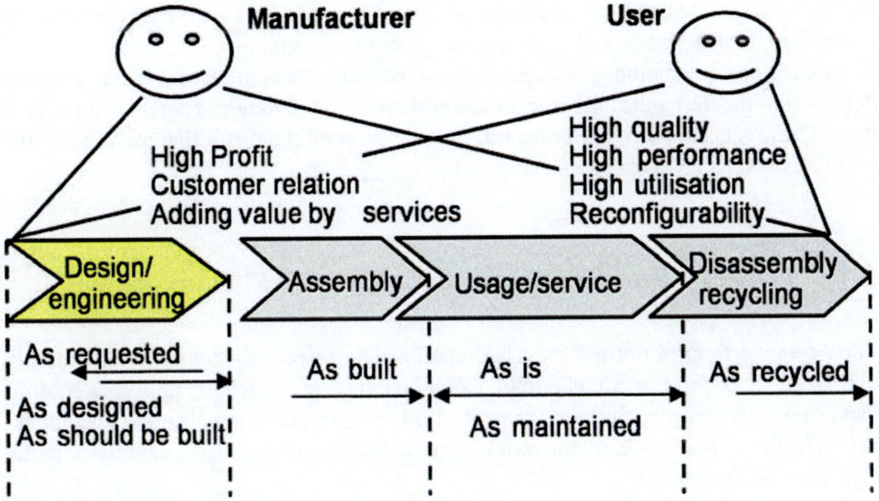

Fig. 1.4 Views of manufacturers and users on the life cycle of technical products

the various aspects of life cycle design and engineering or the capability of systems to be assembled, disassembled and diagnosed in all phases, especially in that of usage.

It is also necessary to describe the architecture of a product which, in effect, is a mixture of goods and services. Using a model of the integrated architecture, interdependencies between goods and services can be managed more easily because it clarifies how various parts contribute to realising a function. As mentioned before, the early phases of the manufacturing process are mostly outsourced to suppliers. Therefore, it is necessary to consider the relationship between manufacturers and suppliers from an economical and environmental perspective. This creates profitable product-orientated services throughout all operations by supporting the diagnostics of actual features, as well as the partial disassembly and assembly for reconfiguration or upgrading and the final disassembly for recycling [7, 8].

1.3 Customer's Viewpoint

Customers are generally interested in achieving high product utilisation in the usage phase at the lowest cost, even if this demands that manufacturing processes need to be changed. This requires flexible manufacturing systems which provide guaranteed process performance and require minimal set-up times and costs. The high efficiency of the usage of complex technical products depends on specific skills and know-how concerned with the details of machines, mechatronic components, software and process optimisation. These costs can be overcome by using specific skilled services and assistance or support provided by manufacturers. Users prefer buying specialised services to reduce the fixed costs of products as well as the costs of inspection, maintenance and reconfiguration or upgrading.

The economic efficiency of capital-intensive products in industrial manufacturing depends on the demands and profiles of products, technical requirements and capacities. These requirements are constantly changing with the result that manufacturing systems need to be permanently adapted [7].

1.4 Goals of a Sustainable Product Life Cycle Management

The new paradigm of optimising a technical product's cost–benefit is orientated not only towards economic but also towards environmental aspects by applying ecological criteria. It assumes that the concentration on core competencies and specialisation offers new potentials to add value or reduce the cost of usage by industrialising services and disassembly.

A common understanding between manufacturers and users is a prerequisite for activating potentials in order to obtain the maximum benefit from each technical product during its life cycle and to fulfil economic and environmental objectives (Fig. 1.5).

Common sense and active optimisation demand technical solutions that link products at any point in the time of their entire life cycle to the information networks of manufacturers and users. This can be achieved by integrating technical products into global IT networks and electronic services. It is evident today that we have the technologies to do this and also to follow the technical trend for developing intelligent machines connected up to communications systems [9–11].

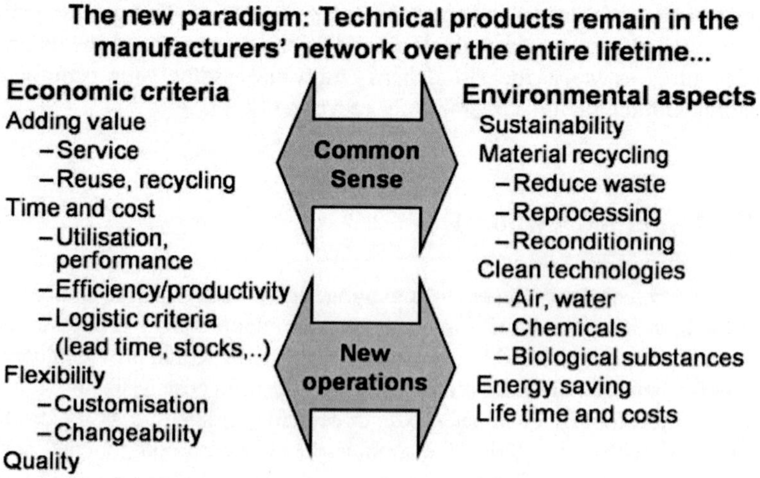

Fig. 1.5 Objectives of life cycle management

Following this new paradigm, the vision is to permanently link products to manufacturers' networks. Communication is the platform for any product-orientated service with the aim of achieving maximum benefit over a product's lifetime. First the basics of the product life cycle management fields and service management are explained.

1.5 Definitions of Terms in Life Cycle

1.5.1 Product

By the term product basically all kinds of products can be understood, as for example the machine, the transportation, the production plant, the used good or the service [5]. In this paper it is referred to a product, when talking about a machine, a capital good or a production plant.

1.5.2 Product Life Cycle

The life cycle of a product contains all phases that a product is going through, that means "*[...] from the idea over the development and construction, production as well as distribution and service to the decommissioning of a product [...]*". [1, 2]

In the following the product life cycle is divided into the classical threefold division:

- **Design Phase** (acquisition phase)—Conception and definition, design and development, production, installation
- **Usage Phase**—Operation and maintenance of products
- **End-of-Life Phase** (continued use and disposal phase)—modernization, recycling, disposal [7]

1.5.3 Product Life Cycle Management

The product life cycle management can be defined as follows:

> Product life cycle management (PLM) is the business activity of managing, in the most effective way, a company's products all the way across their life cycles; from the very first idea for a product all the way through until it is retired and disposed of. PLM is the management system for a company's products. It doesn't just manage one of its products. It manages, in an integrated way, all of its parts and products, and the product portfolio. PLM manages the whole range, from individual part through individual product to the entire portfolio of products. At the highest level, the objective of PLM is to increase product revenues, reduce

product-related costs, maximise the value of the product portfolio, and maximise the value of current and future products for both customers and shareholders. [8]

Logically this means: Effective management of a product along the whole lifecycle means from the first idea to the disposal of the product. It is about a management system for products of companies. All parts, products as well as the whole product portfolio of a company are managed in an integrated way. The goal at the highest level of the product life cycle management is to increase product turnover, reduce product-related costs, and maximize the value of the product portfolio as well as the products' current and future value from a customer and shareholder view [8].

1.6 Services Management

The term service is difficult to narrow down as there is no clear existing definition. However there are different approaches to define the term [9].

Services in contrast to material goods are not tangible, consequently they are immaterial goods and from this follows that it is impossible to store services. At the moment when the service is delivered it is also consumed. The term "Industrial service" is a service that is closely linked to a capital good and is provided by a company. An industrial service can enable or improve the benefit of a product [12].

1.6.1 Definition of Terms

The service business can generally be subdivided into the following fields [2, 3].

- Pre-Sales-Services
- At-Sales-Services
- After-Sales-Services
- Independent-Services

First services are differentiated according to the demand. Then the services are subdivided according to their connectedness. Regarding Pre-Sales-Services, At-Sales-Services and After-Sales-Services it is differentiated according to the date of the service provision. Those services are connected with own products. Independent services on the contrary comprise all services connected with external products [3, 4].

Figure 1.6 illustrates the individual differentiations.

The term pre-sales service describes services that are delivered before the product is being sold. The usage of the product is enabled through the pre-sales service [11, 12].

At-sales services comprise all services that are delivered directly when purchasing the product [13].

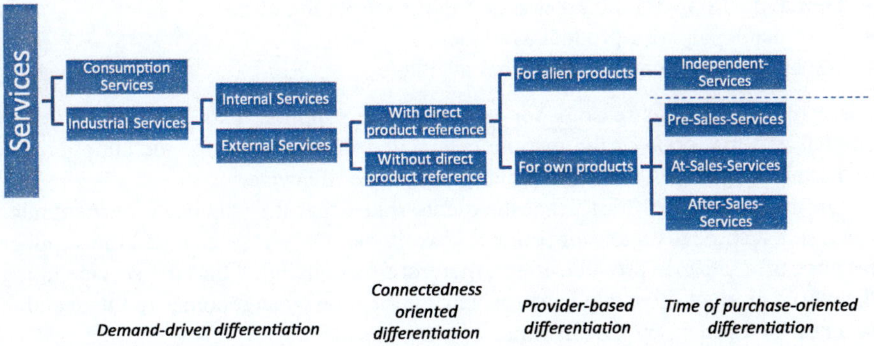

Fig. 1.6 Differentiation of services (Modified acc. to [10])

The term after-sales service comprises all technical or commercial services, which are delivered after purchasing the product. Through the use of after-sales services the value of the product can be maintained or improved. In the field of capital goods the after-sales service is of enormous significance. It is the basis for the replacement part and maintenance business [12].

Independent services comprise all services that a company offers but that are associated with products of external companies.

1.6.2 Importance of Service Business

The significance of the service business is constantly growing. Figure 1.7 shows a customer survey [13] that displays the reasons for a change of supplier.

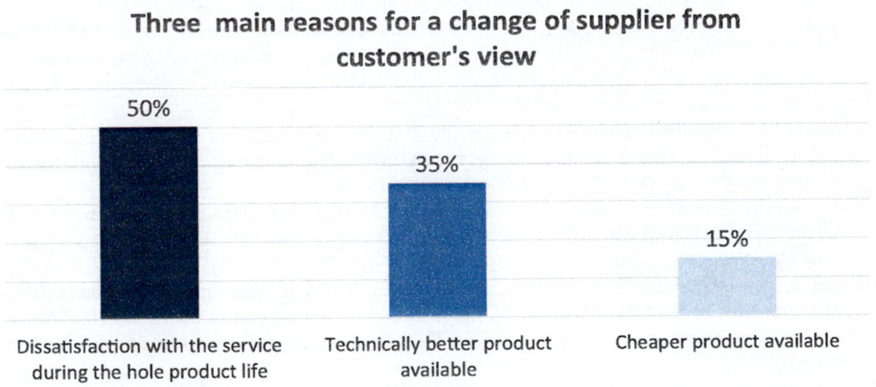

Fig. 1.7 Main reasons for a supplier change [13]

- Dissatisfaction with the service during the whole life cycle
- Technically superior product availabe
- Cheaper/more economical product available

...are the three main reasons for changing the supplier. Though it needs to be emphasized that 50% of the respondents stated that they change the supplier due to dissatisfaction with the service along the whole life cycle.

On the contrary 20% of the respondents stated that they choose a new supplier because a technically superior product is available. Only 15% changed the supplier because of a cheaper product. It can therefore be concluded that the service along the whole product life cycle is of special significance when it comes to the decision of a new investment by the customer.

References

1. Niemann, J., Tichkiewitch, S., Westkämper, E.: Design of Sustainable Product Life Cycles. Springer Verlag, Heidelberg, Berlin (2009)
2. Niemann, J.: Life Cycle Management—das Paradigma der ganzheitlichen Produktlebenslaufbetrachtung. In: Spath, D. et al. (Hrsg.): Neue Entwicklungen in der Unternehmensorganisation. Springer-Vieweg, VDI Buch, Berlin (2017)
3. Niemann, J.: Ökonomische Bewertung von Produktlebensläufen- Life Cycle Controlling. In: Spath, D. et al. (Hrsg.): Neue Entwicklungen in der Unternehmensorganisation. Springer-Vieweg, VDI Buch, Berlin (2017)
4. Westkämper, E., Niemann, J.: Digitale Produktion—Herausforderung und Nutzen. In: Spath, D. et al. (Hrsg.): Neue Entwicklungen in der Unternehmensorganisation. Springer-Vieweg, VDI Buch, Berlin (2017)
5. Sendler, U.: Das PLM-Kompendium. Referenzbuch des Produkt-Lebenszyklus-Managements. Springer (Xpert.press), Berlin, Heidelberg (2009). Online: https://www.springerlink.com/content/n3k448
6. Arnold, V.; Dettmering, H.; Engel, T.; Karcher, A.: Product Life Cycle Management beherrschen. Ein Anwenderhandbuch F R Den Mittelstand. Springer, Dordrecht (2011). Online: https://gbv.eblib.com/patron/FullRecord.aspx?p=769901. Checked 25 May 2019
7. Kuhrke, B., Dervisopoulos, M., Abele, E.: Bedeutung und Anwendung von Lebenszyklusanalysen beiWerkzeugmaschinen. In: Schweiger, S. (Hg.) Lebenszykluskosten optimieren. Paradigmenwechsel für Anbieter und Nutzer von Investitionsgütern. 1. Aufl. Gabler, Wiesbaden (2009), S. 51–80
8. Stark, J.: Product Life Cycle Management. In: 21st Century Paradigm for Product Realisation, vol. 1. Springer International Publishing, Cham (2015). Online: https://gbv.eblib.com/patron/FullRecord.aspx?p=3109724. Checked 25 May 2019
9. Pepels, W.: Servicemanagement. 1. Aufl. Rinteln: Merkur-Verl. (Das Kompendium) (2005). Online: https://deposit.ddb.de/cgi-bin/dokserv?id=2662124&prov=M&dok_var=1&dok_ext=htm. Checked 25 May 2019
10. Seiter, M.: Industrielle Dienstleistungen. Springer Fachmedien Wiesbaden, Wiesbaden (2013)
11. Niemann, J.: Die Services-Manufaktur, Industrielle Services planen –entwickeln – einführen. Shaker Verlag, Ein Praxishandbuch Schritt für Schritt mit Übungen und Lösungen. Aachen (2016)
12. Kenning, P., Markgraf, D.: After-Sales-Services. Hg. v. Springer Gabler Verlag. Gabler Wirtschaftslexikon. Online: https://wirtschaftslexikon.gabler.de/Archiv/55435/after-sales-service-v6.html. Checked 25 May 2019

13. Service. Hauptgründe für Lieferantenwechsel. In: Absatzwirtschaft 2003, 03/2003. Online: https://www.impuls-consulting.de/impuls/progof/datadocs/absatzwirschaft_72003.pdf?PHSESSID=1b28317225387bcfb080ab8350132bc5. 25 May 2019

Chapter 2
Life Cycle Design Phase

The life cycle design phase is of great importance regarding the integrated consideration of the product life cycle. Empiric studies showed that 70–85% of the total costs are determined in this phase. In this phase it is possible to actively influence the life cycle costs of a product at minimal effort. Configuration and the system's range of functions as well as the utilized materials, purchased parts and production processes are determined in this early phase. The costs resulting out of those decisions predominantly accrue in the usage phase. Like this an optimized machine or plant regarding the total costs can only be developed if the long-term expectable costs and performances are estimated and taken into account. Furthermore in this life cycle phase the ecological as well as social dimensions of a product are determined and like that the greatest part of the environmental and social impacts over the whole life cycle is defined [1, 2].

2.1 Generic Approach in Product Development

The goal of the product development is to create a technical solution which fulfills the required functions for the planned lifetime. Decisions about the product structure, connection and joining technologies as well as materials are made. The general problem solving cycle is used as a basis for a general procedure. It is comprised of four phases:

1. Analysis of problem and situation—gathering information
2. Formulation of problem and goal—ormulation and specification of problems
3. Synthesis of system—determination of characteristics
4. Evaluation and decision—assessment of the solution.

A subdivision of the product development process in different phases increases the clarity [1].

J. Niemann and A. Pisla, *Life-Cycle Management of Machines and Mechanisms*, Mechanisms and Machine Science 90,
https://doi.org/10.1007/978-3-030-56449-0_2

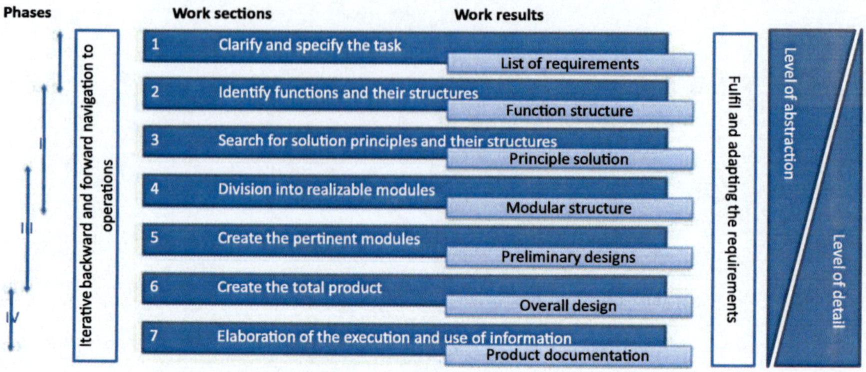

Fig. 2.1 General procedure for development and construction [1, 3]

Figure 2.1 illustrates the general procedure model regarding construction and design according to VDI 2221.

Here the overall approach is subdivided into seven work steps out of which seven working results emerge. The working steps are being run through iteratively several times. Further those working steps are assigned to phases [1, 3].

2.2 Interdisciplinary Collaboration

Figure 2.2 shows the life cycle costs of a product. Further the conventional product manufacturing is compared to the interdisciplinary product manufacturing and cost

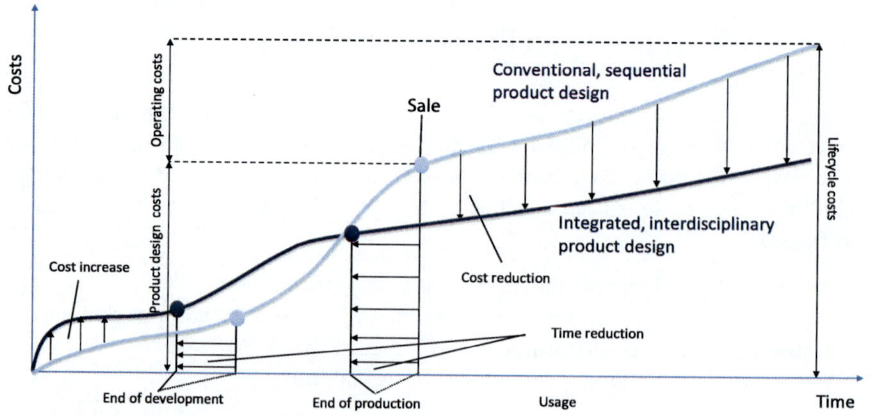

Fig. 2.2 Cost and time effects [5]

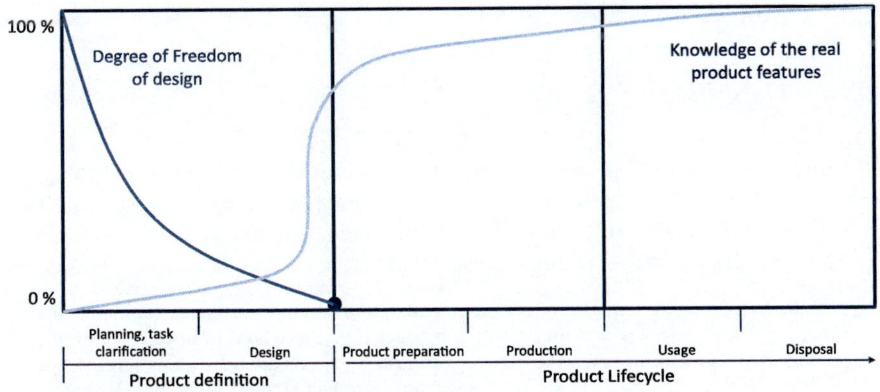

Fig. 2.3 Dilemma of product development [6]

as well as time effects are presented. Due to an often not coordinated, consecutive approach in the conventional product manufacturing, frequently expensive and suboptimal products occur. Through an integrated and interdisciplinary product manufacturing in a project team shorter development times, a faster product manufacturing, cost reduction as well as an improvement of quality are possible. Like this the so called dilemma of construction can be reduced [4–6].

Figure 2.3 illustrates this dilemma of the product development. In the initial phase the costs can be influenced to the greatest extent, though at that time the least about future costs is known [4]. A project team out of all areas that are participating in the product development process is put together.

The team is working independently led by a project manager. Like this departmental boundaries can be overcome.

Initially a small core team is formed, which is made up of experts of the areas construction, work preparation, marketing and sales. If needed this core team is complemented by experts of the areas quality assurance, assembly, control and regulation, recycling and environment. Like this in such a project team also insights and knowledge bases of neighboring disciplines are implied. In order to lead such a team, the project manager needs professional as well as social competences like for example:

- Ability to plan complex processes
- Ability to judge complex processes
- Awareness and early detection
- Ability to concentrate statements
- Abiltiy to continuously observe as is and taget parameter
- Leadership skills in a team [6].

2.3 Life Cycle Design

2.3.1 *Definition of Terms*

In a life cycle-oriented product development technical, economic and ecological requirements of all phases of the product life cycle are taken into account. This constitutes an extension of the so called simultaneous engineering.

The term life cycle design can be defined as the

> [...] Process of the systematic consideration and optimization of technical, economic and ecological characteristics and impacts of a product over the whole life cycle within the product development process. The goal is, by making use of the decision-making scope in the product development, to satisfy the extended product responsibility, which enables the maximum product value for customers and producers over the life cycle at the lowest possible economic, ecological and social expense and risk. [6, 7]

2.3.2 *Cost Allocation and Their Targeted Manipulation*

For many purchasers the main focus is the procurement price of a machine. Though more and more customers decide to consider and calculate the whole life cycle costs. In the future the consideration of the life cycle costs will play an always bigger role regarding purchase decisions. Figure 2.4 displays a possible cost distribution for a product regarding technology and lifetime.

In case of a procurement-based decision-making procedure the buyer would decide for supplier B, though he has the highest operating, maintenance and disposal costs.

On the contrary regarding a life cycle-oriented decision supplier C performs best [8, 9]. Figure 2.5 once again shows the suppliers. In this illustration the compensation of costs can be recognized. Regarding an interdisciplinary life cycle-oriented product

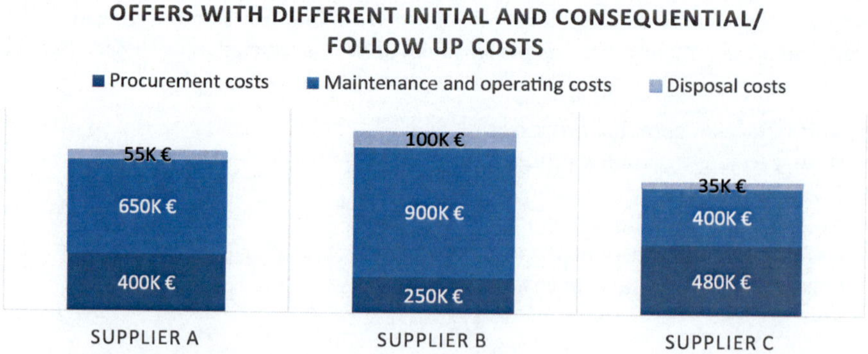

Fig. 2.4 Exemplary offers with different one-off costs and follow-up costs (Modified acc. [8])

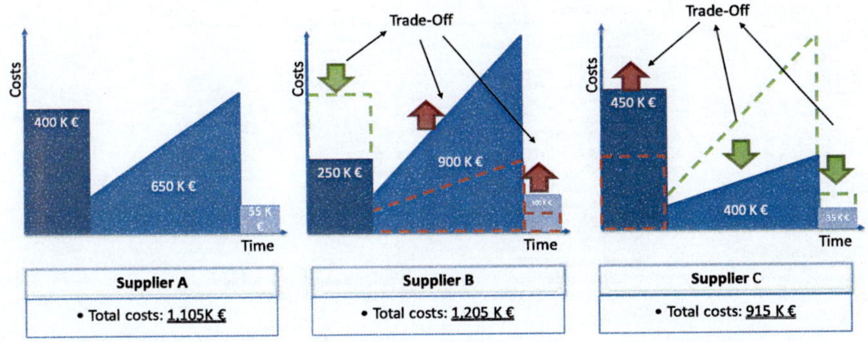

Fig. 2.5 Comensation of costs—trade-offs (Modified acc. [9])

development higher costs occur, which though can be compensated in the usage and disposal phase. On the contrary within the traditional product development other phases are not taken into consideration. This can lead to increased operating, maintenance and disposal costs.[10] A life cycle-oriented product development integrates other phases of the life cycle. One-time costs like transport costs can be reduced by the direct consideration of packaging guidelines. With regard to operating costs energy can be saved and losses as well as costs for auxiliary and operating materials can be reduced. This can be reached for example by avoiding energy conversion, reducing friction losses or the use of standard lubrications. Maintenance costs can be lowered for example by an appropriate arrangement of components and assembly and an accompanying good access to replace positions as well as an avoidance of special tools. With regard to the selection of materials a low variety of materials should be paid attention to in order to further utilize materials at a low loss in value [4].

2.3.3 Examples of Design Changes and Its Impact on Life Cycle

In the design phase it is possible to actively influence the individual phases of the product life cycle. Subsequently three examples of adaptive constructions and the related effects on the product life cycle are shown.

2.3.3.1 Adaptive Constructions in Order to Reduce Manufacturing Costs

The example displayed in Fig. 2.6 shows a potential for cost-savings, which occurs in the design life cycle phase. A rocker arm of a diesel engine of the company MTU

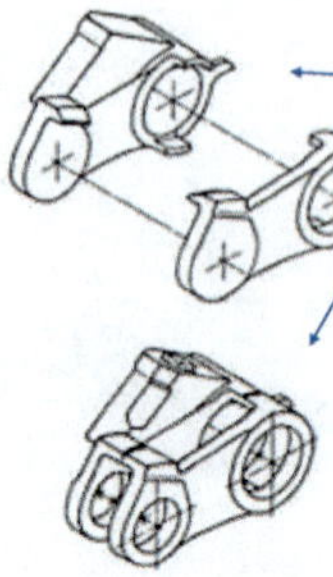

Before - 2 parts	After - 1 part
2 forged parts Ck 15	1 forged part 16 MnCr 5
Finished part	Finished part
Production process: 1. Prepare both halves 2. Welding 3. Straightening 4. Mechanical processing 5. Re-welding of the pressure surface 6. Mech. proc. after re-welding 7. Heat treatment (Ck15 bad) 8. Straightening after heat treatment 9. Grinding	Production process: 1. Mechanical processing 2. Heat treatment (16 MnCr 5 well) 3. Grinding
Material costs 15.00 € Production time / piece: 161 min Production costs 320.00 € (100%)	Material costs 15.60€ Production time / piece: 106 min Production costs 210.00 € (67%)

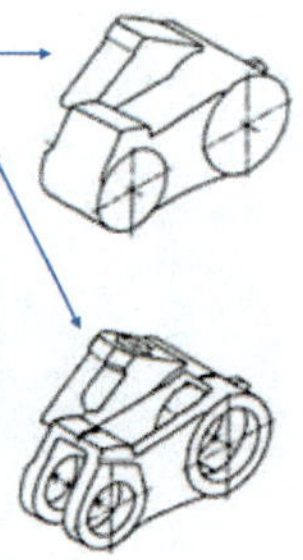

Fig. 2.6 Cost reduction through integral [4]

running at medium-speed is shown. Here a cost reduction of 33% of the production costs could be reached by a revision of the component's construction. Developers should not only consider the pure function during the construction but also develop a cost-awareness [4].

2.3.3.2 Adjustment Construction Optimized for Usage

Figure 2.7 shows labeling machines, which are used in filling and packaging machines. The illustration shows an integrated construction, which was not common before. After extensive market research and customer surveys, a claim for increased flexibility in the processing of different tank constructions was determined.

Therefore a labeling machine in a modular design was developed. Through this modular design changeover times are reduced to a considerable extent. Like this

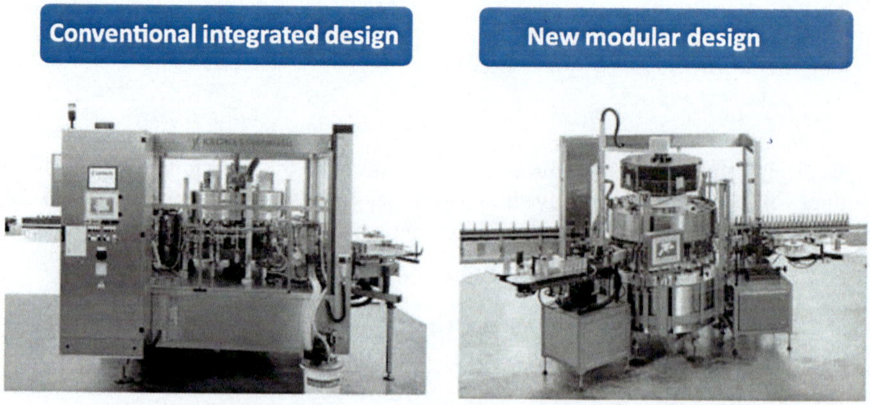

Fig. 2.7 Conventional "integrated construction" before–after [4]

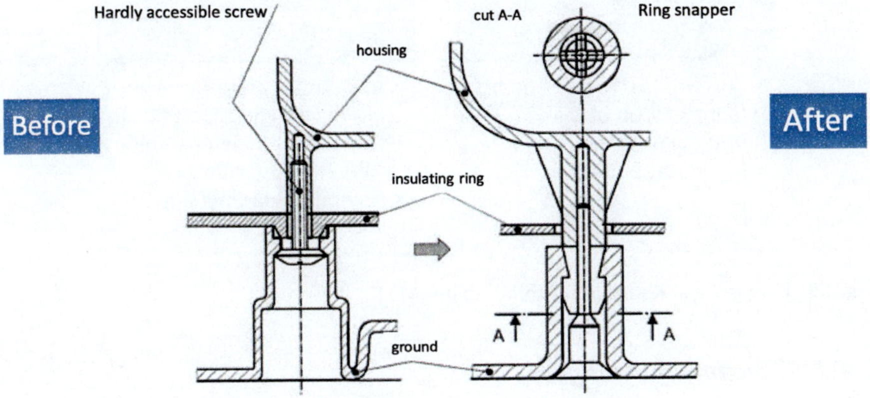

Fig. 2.8 Adaptive construction of the bond ground/case [4]

the machine is quickly convertible to different labeling systems. A good access of the aggregates is given. They can be exchanged and upgraded quickly and are transportable for maintenance and configuration works [4].

2.3.3.3 Adjustment Construction with Economical Disposal

In this example a Siemens coffee machine TC 22 was developed under the condition of a good decomposability and reduction of used materials. The goal of this product development was that the machine splits into its usable parts by one precisely focused hammer stroke.

The screw connection was identified as the weak point (Fig. 2.8, left). As a solution the screw connection was replaced by a ringsnapper (Fig. 2.8, right). Like this no more unscrewing is needed and the connection can be unfixed by a focused hammer stroke on the heating plate. The disposal costs like this can be lowered from 0.29 to 0.17 €/device [4].

2.4 Services in the Design Phase

Already in the design phase services are offered. Though it cannot always clearly be differentiated between a service and an activity in the product business. In the following exemplary services in the design phase are presented. Services in the design phase can be classified in the categories pre-sales-services, at-sales-services or independent-services.

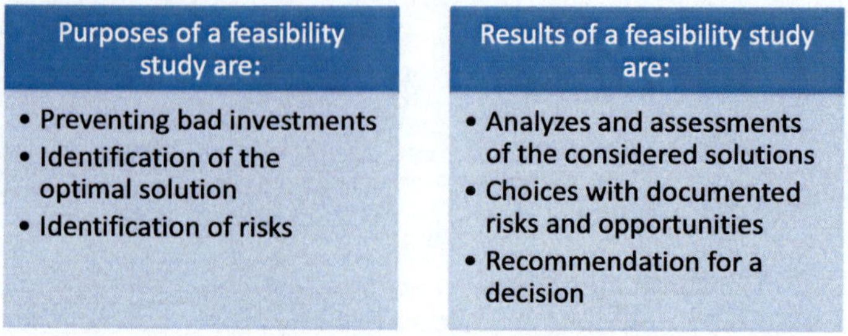

Fig. 2.9 Purposes and results of feasibility studies [11]

2.4.1 Feasibility Analyses

Within a practicability analysis possible solution approaches for a determined project are examined regarding their practicability. Hereby solution approaches are analyzed and judged, risks are identified and prospects forecasted. It is questioned if with the given framework conditions and considered solution approaches the agreed project solutions can be realized [11].

Figure 2.9 illustrates the purposes and results of feasibility studies.

Feasibility studies as a service can be categorized into pre-sales, because such services will be executed close before the buying decision.

2.4.2 Financial Services

Some companies have own financial divisions and offer their own financial services to the customers playing the role of a financial service provider. One example is machine leasing. Financial service can be categorized in at-sales services as the customer at the time of the service delivery has already decided for a product. In the case that the customer decides for an external leasing company, this service could be categorized as independent service. A business relation between the three parties (customer), supplier (producer) and lessor (leasing company) results. The lessee uses and owns a capital good but is not the proprietor. Thus no capital sum needs to be put on, but a monthly rate needs to be agreed upon [12–17].

References

1. Herrmann, C.: Ganzheitliches Life Cycle Management. Nachhaltigkeit und Lebenszyklusorientierung in Unternehmen. Springer Verlag, Heidelberg (2010)
2. VDI 2884: Beschaffung, Betrieb und Instandhaltung von Produktionsmitteln unter Anwendung von Life Cycle Costing (LCC), Dec 2005

3. VDI 2221: Methodik zum Entwickleln und Konstruieren technischer Systeme und Produkte, May 1993
4. Ehrlenspiel, K., Kiewert, A., Lindemann, U., Mörtl, M.: Kostengünstig Entwickeln und Konstruieren. Kostenmanagement bei der integrierten Produktentwicklung. 7. Aufl. Springer Vieweg (VDI-Buch), Berlin (2014). Online verfügbar unter https://dx.doi.org/10.1007/978-3-642-41959-1
5. Bhagwati, M. (Hg.): Konstruktion. Begriff und Bedeutung der Konstruktion (2016). Online: https://www.daswirtschaftslexikon.com/d/konstruktion/konstruktion.htm. Checked 25 May 2019
6. Feldhusen, J., Grote, K. H., Kochan, D., Beyer, C., Vajna, S., Lashin, G. et al.: Die PEP-begleitenden Prozesse. In: Jörg Feldhusen und Karl-Heinrich Grote (Hg.) Pahl/Beitz Konstruktionslehre. Methoden und Anwendung erfolgreicher Produktentwicklung. 8. Aufl. Berlin Heidelberg: Springer Vieweg, 2013, S. 25–236
7. Mansour, M.: Informations- und Wissensbereitstellung für die lebenszyklusorientierte Produktentwicklung. Techn. Univ., Diss.—Braunschweig (2006). Essen: Vulkan-Verl. (Schriftenreihe des Instituts für Werkzeugmaschinen und Fertigungstechnik der TU Braunschweig)
8. Noske, H., Kalogerakis, C.: Design-to-Life cycle-Cost bei Investitionsgütern am Beispiel von Werkzeugmaschinen. Einleitung. In: Stefan Schweiger (Hg.) Lebenszykluskosten optimieren. Paradigmenwechsel für Anbieter und Nutzer von Investitionsgütern. 1. Aufl. Gabler, Wiesbaden (2009), S. 135–152
9. Niemann, J.: Ökonomische Bewertung von Produktlebensläufen-Life Cycle Controlling. In: Spath, D. et al. (Hrsg.): Neue Entwicklungen in der Unternehmensorganisation. Springer-Vieweg, Berlin, VDI Buch (2017)
10. Manja, R.: Life Cycle Costing. Wikipedia. Online: https://bit.ly/1RSy6D6. Checked 25 May 2019
11. Angermeier, G.: Machbarkeitsstudie. Berleb Media GmbH. Taufkirchen (2015). Online: https://www.projektmagazin.de/glossarterm/machbarkeitsstudie. Checked 25 May 2019
12. Siemens Financing: Vom Kauf zum Leasingvertrag. Online verfügbar unter https://finance.siemens.de/financialservices/ger/produkte_loesungen/leasing/leasing_know-how/seiten/vom_kauf_zum_leasingvertrag.aspx. Checked 25 May 2019
13. Niemann, J., Pisla, A.: Sustainable potentials and risks assess in automation and robotization using the life cycle management index tool—LY-MIT. Sustainability **10**, 4638 (2018)
14. Niemann, J., Schemann, T., Erkens, J.: Servitization—pathway of transformation from product manufacturer towards a service provider. In: 2018 International Conference on Production Research—Africa, Europe, Middle East 5th International Conference on Quality and Innovation in Engineering and Management, 25–26 July 2018, Cluj-Napoca, Romania
15. Stöhr, C., Janssen, M., Niemann, J.: Smart services. In: 14th International Symposium in Management, Challenges and Innovation in Management and Entrepreneurship, 27–28 Oct 2017, Timisoara, Romania
16. Niemann, J.: Eine Methodik zum dynamischen Life Cycle Controlling von Produktionssystemen. Jost-Jetter Verlag, Heimsheim, 2007IPA-IAO Forschung und Praxis 459). Stuttgart, Univ., Fak. Maschinenbau, Inst. für Industrielle Fertigung und Fabrikbetrieb, Diss. (2007)
17. Niemann, J.: Life Cycle Management-das Paradigma der ganzheitlichen Produktlebenslaufbetrachtung. In: Spath, D., Westkämper, E., Bullinger, H.-J., Warnecke, H.-J. (Hrsg.) Neue Entwicklungen in der Unternehmensorganisation. Springer-Vieweg, VDI Buch, Berlin (2017)

Chapter 3
Life Cycle Usage Phase

In the life cycle usage phase the after-sales services mentioned in the previous chapter are employed. After-sales services can generate additional turnover and margins through longt-term service contracts. Further will the customer loyalty be increased by regular interactions, whereby information is generated of the product use, which can lead to product improvements. In addition this constant contact generates information about product statuses and current customer needs, which play a significant role for follow-up businesses [1].

In the following maintenance and its disciplines as well as the parts business are discussed.

3.1 Maintenance

The fundamental goal of maintenance from the economical point of view is to secure a high availability by increasing the maintenance intervals or reducing the maintenance times. Thereby the available maintenance downtime costs are taken into account [2].

3.1.1 Definition of Terms

The term maintenance according to DIN 31051 is defined as follows:

> Combination of all technical and administrative measures as well as measures by the management during the life cycle of a unit, which serve the preservation or recovery of a functioning state, so that the required function can be fulfilled. [3]

J. Niemann and A. Pisla, *Life-Cycle Management of Machines and Mechanisms*, Mechanisms and Machine Science 90,
https://doi.org/10.1007/978-3-030-56449-0_3

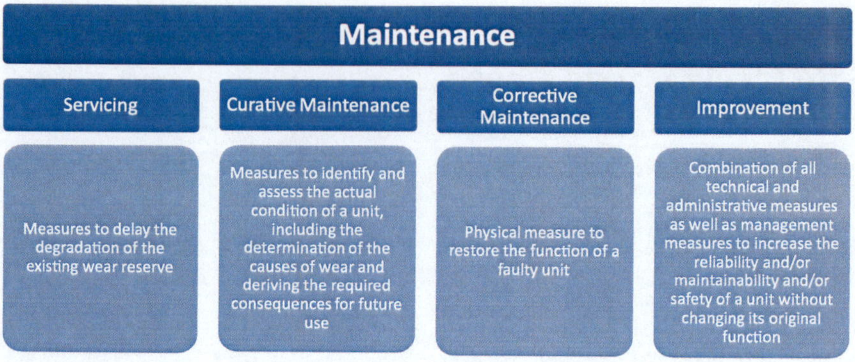

Fig. 3.1 Subdivision and definition of maintenance [3]

The maintenance is subdivided into the general measures maintenance, inspection, repair and improvement. Figure 3.1 illustrates this subdivision as well as the particular definitions [3].

Furthermore the important terms for this elaboration in connection with the wear are needed. They are displayed in Fig. 3.2.

After manufacturing a certain "wear reserve" exits, which over time will be "consumed" by utilization. When reaching the wear limit, this should be rebuild by determined repair or improvement works.

Figure 3.3 shows the degradation of the wear reserve as well as its "re"-creation.

Fig. 3.2 Explanation of the terms wear, wear reserve and wear limit [3]

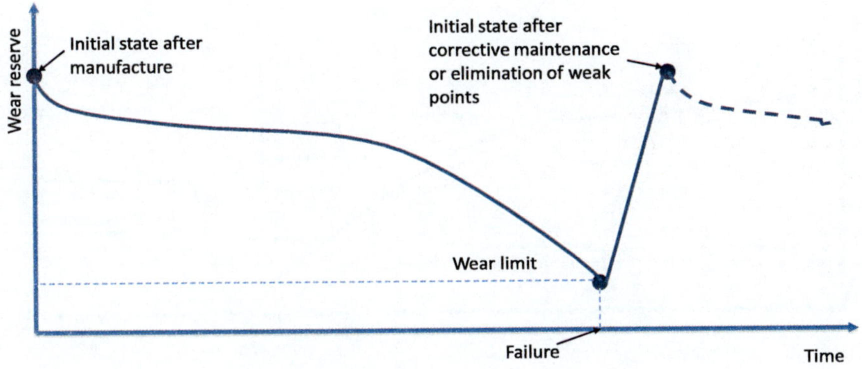

Fig. 3.3 Degradation of wear reserve and its recreation [3]

3.1.2 *Maintenance Strategies*

The term maintenance strategy can be defined as follows:

> A maninenance strategy strives for the realization of an overall concept, defined by the management according to technically possible and economically beneficial aspects, in order to maintain the availability of existing plants. [4]

In the following three significant maintenance strategies are presented and explained.

3.1.2.1 Failure-Based Maintenance (Failure Strategy)

Considering failure-based maintenance the failure is first awaited before action is taken. The operator does not influence the machine failure, consequently no relevant effort for maintenance and strategy is needed. As soon as during a failure consequential losses or damages to health can be expected, the choice of this strategy is not possible. The complete consumption of the wear reserve is characteristic for this strategy. The process of the wear reserve in combination with this strategy is seen in Fig. 3.4.

3.1.2.2 Preventive Maintenance (Preventive Strategy)

The preventive maintenance is a maintenance strategy in which, based on fixed intervals, the point of maintenance is determined.

Such intervals can for example be determined in the following ways:

- *performance-related* after the delivery of a determined production output

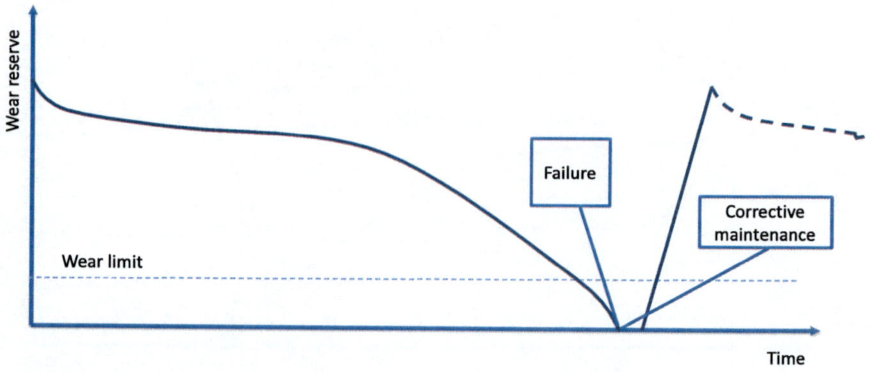

Fig. 3.4 Process of wear reserve at failure-based maintenance (Modified acc. [3, 5])

- *output-dependent* after the delivery of a determined amount of production

This strategy is suitable if in case of a failure high costs or a danger for people can be expected. In order to execute this strategy, knowledge about failure behavior, use intensity and lifetime of the machine is necessary. The risk of a failure can indeed be minimized but not eliminated by applying this strategy. The repair is done independently of the current state, that's why a waste of wear occurs as components are exchanged too early. This can lead to increased costs of spare parts and repair. Figure 3.5 displays the process of the wear reserve at preventive maintenance and illustrates the waste of wear. As the repair is executed after a determined interval, a different consumption of wear reserve can be found [4].

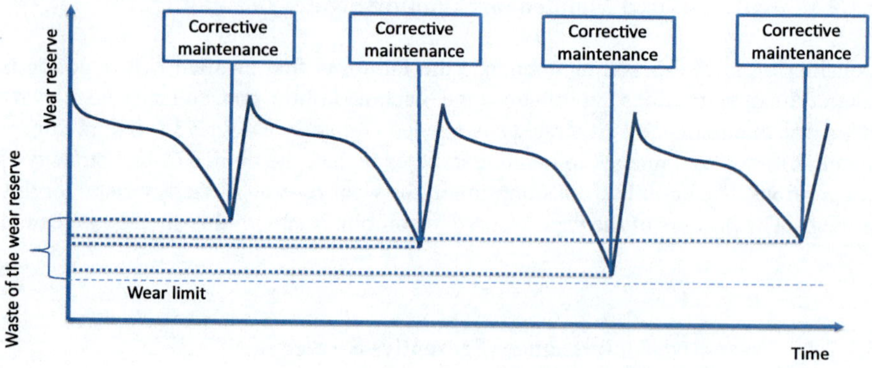

Fig. 3.5 Process of wear reserve during preventive maintenance (Modified acc. [3])

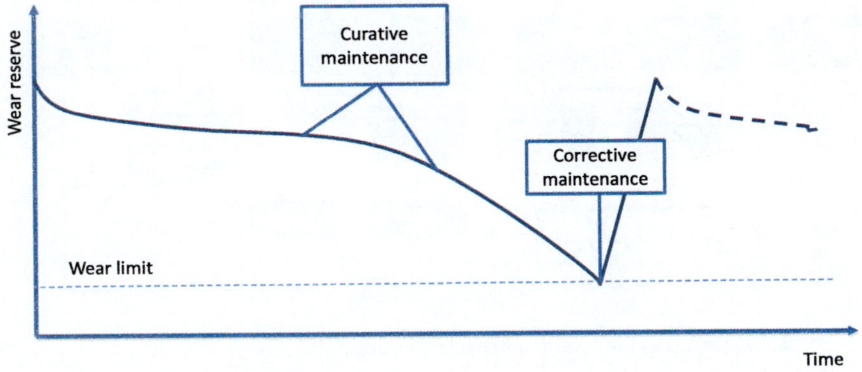

Fig. 3.6 Process of wear reserve during condition-based maintenance (Modified acc. [3])

3.1.2.3 Condition-Based Maintenance (Curative Strategy)

The condition-based maintenance is oriented towards a determined state. The repair is done when reaching a determined limit of wear. Thereby the wear reserve is consumed to a great extent.

Figure 3.6 shows the wear process during the inspection strategy [4, 6]. By regular inspections the actual state is determined, whereby the development of the wear reserve can be assessed (forcecast). Thus mistakes can be found earlier and failure progressions can be better diagnosed. Consequently failures can be avoided to a great extent and machine running times as well as operational safety can be increased. This strategy needs modern measure and testing techniques, qualified personnel and good knowledge about the wear process of a machine.

3.1.3 Introduction of a State-Oriented Maintenance

Figure 3.7 shows the reference model of the introduction of a state-based maintenance. Considering the description of the plant structure a functional model is set up.

By means of this model operating modes and states for selected parts are described and in case of failure complemented by information about the behavior. In the following the state describing target characteristics are determined. In order to measure the actual states, suitable measurement techniques and instruments need to be chosen. After each measurement procedure the generated measurement values are archived in order to determine possible trends. The determined actual states are then compared to the target states to judge the current state of the plant. If an impermissible value is attained with this comparison, a diagnosis starts and a connection

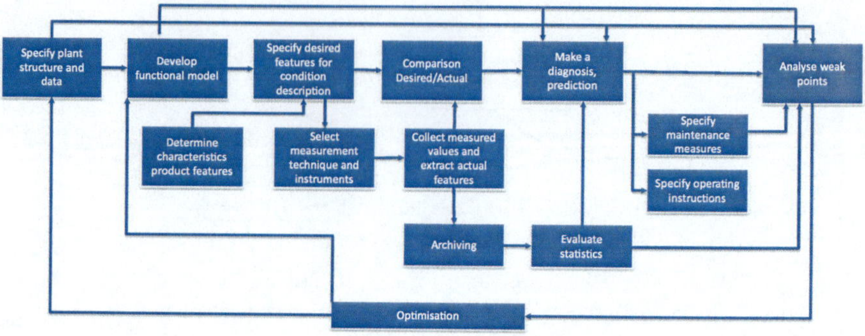

Fig. 3.7 Reference model for the state-oriented maintenance [6]

between type of failure, cause and location is generated. Out of that operating instructions and maintenance measures can be derived. With those information, which are complemented by continuously archived measurement values, weaknesses can be identified and subsequently analyzed. In the following an optimization of the detected weaknesses can be achieved [6].

The demand for contractually agreed maintenance services is increasing with the technical and economic development at a national as well as at an international level. Therefore it is important to exactly structure and write the maintenance contracts. DIN EN 13269—maintenance—instruction for the creation of maintenance contracts contains proposals for the structure and the content of a contract. The following contract elements should be taken into account when making a maintenance contract:

- Contract header
- Target
- Contract-relevant definitions
- Scope of tasks
- Technical agreements
- Incoterms/agreements
- Organizational agreements
- Legal agreements [7].

3.2 Spare Parts

Spare parts are interchangeable units of a product. They are a central part of the maintenance process. Spare parts maintain the functionality of a product or recreate it [4].

A spare part according to DIN 13306 is defined as follows:

> Unit as a replacement for an equivalent unit in order to maintain the originally required function of the device [8].

The goal of the spare parts management is to provide the correct amount of the right spare part in the correct location at the right point in time. The stocking and availability of spare parts are in conflict of objectives as the stocking constitutes a cost problem and the availability of spare parts is connected to the availability of plants. The task of the spare parts management is to secure a balance between low stocks and a high availability. In order to manage this task, the following subtasks need to be in the foreground.

- Optimal planning of spare parts in order to ensure an optimal stock level matching the chosen maintenance strategy
- Safe storage and fast retrievability of spare parts or an optimal procurement of spare parts
- Demand-based povision of spare parts
- Continuous control and adjustment of stocks [4].

The main task of the spare parts logistic is to optimally manage the final stock and thus achieve an optimum in maintenance business. Thereby the focus should be on

> […] the cost-minimizing comparison of the opposing costs for a (potential) parts shortage and stock costs […]. [9]

The spare parts are classified in A, B and C parts according to their value and in X, Y and Z according to the demand. Depending on the classification of the spare part and which maintenance strategy is applied, a changed stocking strategy is applied. The quality of the spare parts logistic, which is characterized by a high reliability or high competence, plays an important role. The costs of the spare parts logistic should be as low as possible, though a high delivery service with flexible delivery times and a high delivery loyalty are expected (i.e. 24 h, 7 days/week) as well as a high flexibility (selective processing of the orders according to priority). The flexibility plays a significant role as a failure cannot always be foreseen and can be related to high failure costs [9].

Figure 3.8 summarizes the requirements of industrial suppliers to meet customers demand in spare part logistics. In order to meet/master the challenges of the spare parts logistic, companies exist that put together so called spare parts packages. Like this the customer is able to react to failures or other necessary maintenance measures on-site as quick as possible [10–13].

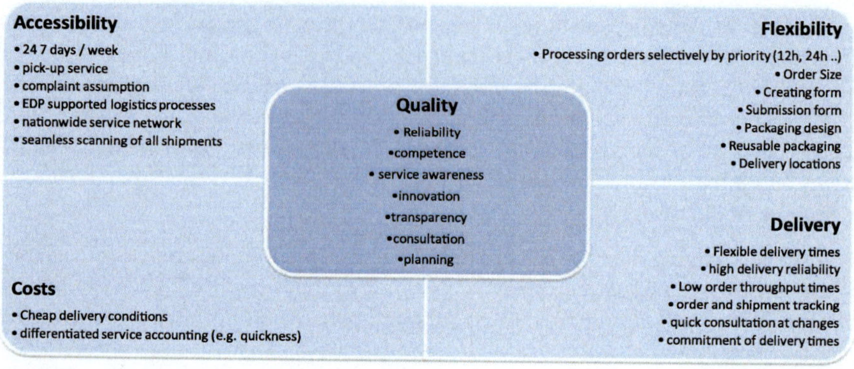

Fig. 3.8 Requirements of the spare parts logistic [9]

References

1. Helbling Management Consulting GmbH: After-Sales-Services. Kunden binden, Umsatz und Erträge steigern. Online: https://www.helbling.de/hol/aktuelles/after-sales-services-kunden-binden-umsatz-und-ertraege-steigern. Checked 25 May 2019
2. Herrmann, C.: Ganzheitliches Life Cycle Management. Nachhaltigkeit und Lebenszyklusorientierung in Unternehmen. Springer Verlag, Heidelberg (2010)
3. DIN 31051: Grundlagen der Instandhaltung, Dec 2010
4. Strunz, M.: Instandhaltung. Springer, Berlin, Heidelberg (2012)
5. Leidinger, B.: Wertorientierte Instandhaltung. Kosten senken, Verfügbarkeit erhalten. Springer Gabler, Wiesbaden (2014)
6. VDI 2888: Zustandsorientierte Instandhaltung, Dec 1999
7. DIN EN 13269: Anleitung zur Erstellung von Instandhaltungsverträgen (2006)
8. DIN 13306: Instandhaltung - Begriffe der Instandhaltung, Dec 2010
9. Schuh, G., Stich, V., Wienholdt, H.: Ersatzteillogistik. In: Günther Schuh und Volker Stich (Hg.): Logistikmanagement. Handbuch Produktion und Management 6. 2., vollst. neu bearb. und erw. Aufl. 2013. Springer (VDI-Buch, 6), Berlin, Heidelberg (2013), S. 165–208
10. Bühler, A. G. U.: Switzerland. Online: https://www.buhlergroup.com/global/de/downloads/Brochure_PITSTOP_. Checked 25 May 2019
11. Niemann, J., Tichkiewitch, S., Westkämper, E.: Design of Sustainable Product Life Cycles. Springer Verlag, Heidelberg Berlin (2009)
12. Niemann, J.: Life Cycle Management-das Paradigma der ganzheitlichen Produktlebenslaufbetrachtung. In: Spath, D. et al. (Hrsg.) Neue Entwicklungen in der Unternehmensorganisation. Springer-Vieweg, VDI Buch, Berlin (2017)
13. Niemann, J.: Ökonomische Bewertung von Produktlebensläufen-Life Cycle Controlling. In: Spath, D. et al. (Hrsg.) Neue Entwicklungen in der Unternehmensorganisation. Springer-Vieweg, VDI Buch, Berlin (2017)

Chapter 4
End-of-Life Phase

In this chapter a closer look is taken at the topics recycling and disposal, which occur at the end of a life cycle.

4.1 End-of-Life Stage in Product Life Cycles

In order to create the basic prerequisites for the sensitization of the topic and the end-of-life strategies, in this chapter a definition of the basic building block is provided in advance. This researched finding will be added with the goals and challenges as well with an explanation of how the strategies are to be managed.

4.1.1 Definition and Explanation

The concept of a product life cycle assumes that products are subject to a process of becoming and decay just as it is assumed in biology for living beings. Furthermore, it is assumed that a product undergoes different life phases as part of this process [1]. The life cycle of a product can be contemplated through different point of views, though. A common approach is the market cycle model which considers one single product and describes its value from the sales perspective. It covers the period from the launch on the market through to its withdrawal from the market. The individual phases of the life cycle are differentiated to varying degrees depending on the literature. Very often the five phases of introduction, growth, maturity, saturation and decline are distinguished [1].

Nevertheless, the life cycle of a product or rather of its materials can be described from an ecological perspective, as well. As can be seen in Fig. 4.1, this approach focuses only on the material flow itself and does not take the market or rather sales

J. Niemann and A. Pisla, *Life-Cycle Management of Machines and Mechanisms*, Mechanisms and Machine Science 90,
https://doi.org/10.1007/978-3-030-56449-0_4

perspective into account. It describes the different stages of added value from material degradation to the manufacture of the product, its use and disposal [2].

In contrast to the market life cycle, this model considers the stage after the use of a product. The corresponding phase is the so-called "end-of-life" stage. However, the exact definition about when this stage is reached is handled very differently in the literature. Rose [4] defines the end-of-life stage as the "point in time, when a product no longer satisfies the initial purchaser or first user" whereas de Britto and Decker [5] differentiate the entry into the end-of-life stage at the end of the physical useful life. The European Commission [6], on the other hand, has a slightly alternative approach. According to one of their publications, the end-of-life stage is reached when a product is perceived as "waste". However, there are different ways to handle end-of-life products according to its type, characteristics and current condition. The possible approaches of handling these products are the so-called end-of-life strategies.

4.1.2 Goals and Challanges

Depending on their goals, end-of-life strategies may diverge more or less. Nevertheless, almost every approach essentially aims at the implementation of a circular economy. Within the principle of a circular economy, resources are valued more efficiently, that is, the evolvement from a unidirectional source-to-sink (a.k.a. cradle-to-grave) economy, where materials are extracted, manufactured, used and discarded, to closed material cycles (cradle-to-cradle) as can be observed [7].

As mentioned previously, the classifications of end-of-life strategies may vary. ISO has classified the approaches on the basis of their potential environmental gains: prevention, reuse, recycling, energy recovery and disposal [8]. The strategies focused within this elaboration, however, are based on an approach by Bauer [2]. Since energy recovery and landfilling do not support a circular economy because they are not sustainable, Bauer's strategies deal with recovering as many elements of a product as possible so that no added-value and material gets destroyed. In contrast to that the strategies presented in the following sections aim at saving as much added-value as possible. In addition to the traditional disposal, end-of-life products can also be recycled, directly reused or reprocessed. Figure 4.1 emphasizes how these strategies can be implemented into the life cycle so that the concept of a circular economy gets realized.

Beside from environmental benefits, the reason why companies should establish an end-of-life systems can also be out of self-interest. Other motivations are e.g. financial benefits. Some commonly known companies such as IBM aim at gaining precious information about the usage and wear of a product to enhance the product design [4].

According to Rose [9], managing the end-of-life of a product is also characterized by two major challenges. These challenges can be distinguished between technical

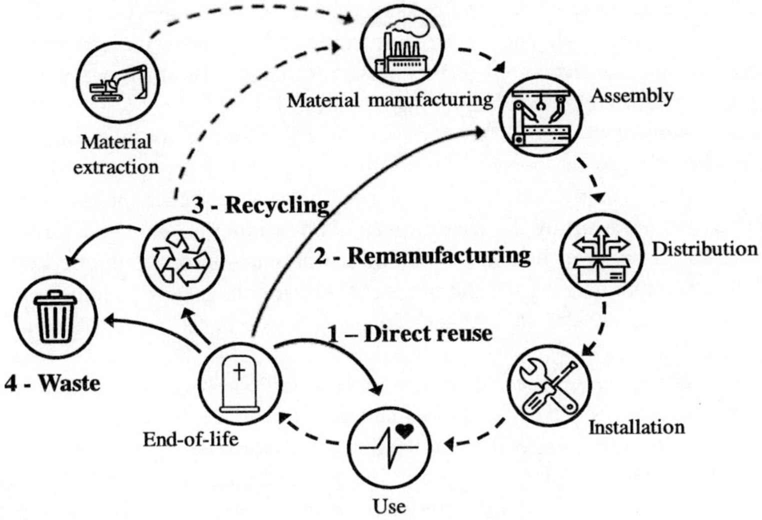

Fig. 4.1 The concept of a circular economy including the 4 main end-of-life strategies [3]

and non-technical. The non-technical challenges deal with problems such as economical profitability or the opportunities to properly arrange the supply chain. The technical difficulties relate to uncertainty about the condition of products collected at the end of their life.

Each alternative strategy is driven by individual challenges, though. The different challenges will be addressed during this elaboration. At this point, however, it can already be stated, that for achieving a sustainable product life cycle management it is inevitable to consider the after-usage treatment even before the product is designed.

The different added-value strategies will be presented more detailed during the course of the following sections.

4.2 End-of-Life Strategies

In this section, various end-of-life strategies are discussed in more detail as they have proven to be particularly relevant during literature research and are applied in practice in most manufacturing companies.

4.2.1 Direct Reuse

The first environmentally motivated strategy for a product after its originally intended life cycle is the “reuse” strategy. This strategy aims at returning a product or rather its components directly back into the usage phase "for the same purpose for which they were conceived" [6]. The "reuse"—strategy is designed to increase the lifetime of a product and therefor avoid further waste. Reusing a product as a whole is the ideal end-of-life strategy in terms of sustainability [4].

An extension of the use of products by means of reuse can take place through active intervention of the owner. The product remains in the possession of the user and is further utilized by conversion (=reuse). In this case, reuse means a change of use which is carried out by the user himself. The product is "changed" in whole or in parts. Another way to extend the use of a product is through maintenance. The user cleans, maintains or repairs the product for further usage [10]. Non-commercial upcycling of used cans to lampshades for private use or as a gift to friends is an example of this.

Furthermore, there are specific design approaches to support the circular usage of a product. "Design for Ease of Maintenance and Repair" e.g. intends to simplify the maintenance and repair capability of products so that users can extend the lifetime themselves. A rather similar concept is "Design for Disassembly and Reassembly" [10]. The benefits of this design approach become clear when considering that a product's end-of-life is normally determined by only one subassembly or component while other parts of the product, however, may obtain a residual life. In this case, the remaining lifetime of the residual parts can be extended through partial disassembly from the original product and reusing it in other assemblies [11].

Another related approach is to only disassemble those parts of products, which reached their individual end-of-life. Products designed to enable this strategy are so-called "Multi-Life-Products". More generally described, these products are designed to enable substitution of modules and/or whose interfaces in order to adapt to future market developments by modifying, replacing or adding individual modules without redesigning the product [12].

Generally, the costs of disassembly exceed those for assembly because the impact of economies of scale is attenuated. Disassembling of products needs much higher effort since their individual condition (after its end-of-life) diverge. The major issue of the component reuse concept, however, is the design of durable interfaces that enable subassemblies to last through multiple generations of products [11]. A famous example of a successful module-based strategy is the company Dell. It managed to develop a computer with modular design so that key components (e.g. disk drive, motherboard) can be exchanged allowing the user to remain up-to-date. Nevertheless, some companies intentionally prevent module reuse in order to maintain revenue sources. A famous example is Apple, which regularly redesign the power-supply connectors of their products.

4.2.2 *Remanufacturing*

The remanufacturing strategy describes a process in which used products get refurbished after their initial lifetime in order to go through another life cycle. The European Commission explains this approach as the sequencing of certain manufacturing activities with the intention to "*return it [a part or the whole product -author's note] to like-new or better performance*" [6]. However, one important characteristic of this approach is that remanufacturing of worn-out products inevitably requires manufacturing processes because otherwise the original state cannot be restored. Additionally, the products may be complemented by new parts if necessary. Hence, remanufacturing aims at restoring as much added-value of the products as possible. It is based on the idea of recovering the lasting friction of the products in order to promote multiple uses of materials [13]. After accomplishing the remanufacturing process, the products are given a warranty which is at least as high as a newly manufactured product [14].

One of the first studies regarding remanufacturing by Lund [15] has shown the tremendous benefits of remanufacturing for the environment. Having an equivalent quality compared to new products, remanufactured products require 50–80% less energy. Beside from environmental aspects, remanufactured products can also safe up to 80% of production costs. Additionally, manufacturers are capable of upgrading functions and improve the quality of their products regularly [13].

Compared to the reusing strategy, remanufacturing is slightly more intricate. According to Sundin and Bras [13] seven different steps are required within the process of remanufacturing in order to restore the quality of a product. The individual remanufacturing steps accomplished in-between the end-of-life and the new distribution stage can be seen in Fig. 4.2 depending on type or condition of a product [13].

Similar with the presented reuse strategy, the process or rather the above-mentioned steps of remanufacturing can also be performed by the owner itself. However, this non-commercial approach requires special design conditions that allow an easy handling of disassembly and assembly [10]. Remanufacturing by private persons for subsequent sale is also called "Base Remanufacturing". On the other

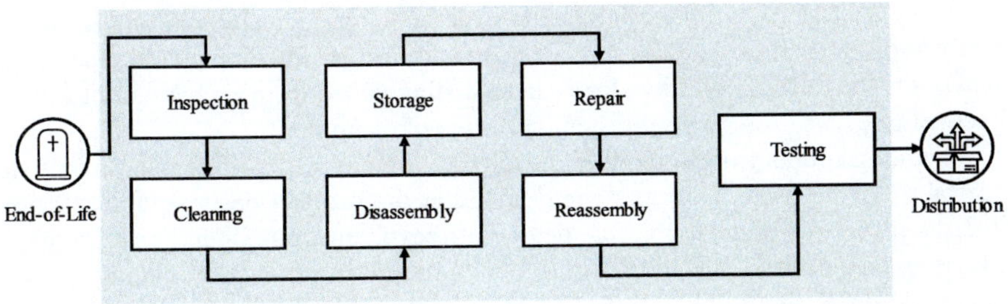

Fig. 4.2 The seven steps of remanufacturing

hand, “Mass Remanufacturing” refers to industrial processing and larger volumes. The commercial approach of mass remanufacturing must deal with the difficulties of processing used products. That is, the individual condition of the product or its components is consequently different. Hence, not only the order of steps can vary but also the intensity of each individual process (e.g. disassembly is more labor intensive). As a result, some processes cannot be automated and have to be performed manually. Compared to the original manufacturing of a product, remanufacturing is less able to capture economics of scale [16, 17].

Compared to each other, remanufacturing and reuse are almost similar strategies. Both strategies aim to create a product that is similar to the original one and is manufactured from the initial materials. Both strategies base on the same concept though can be distinguished by the degree of required process operations. Whereas reuse just usually only requires decent cleaning, remanufacturing is characterized through the use of machinery and further processing [2]. Furthermore remanufacturing can be differentiated according to the phase of the life cycle into which the product is reintegrated after its processing [10].

4.2.3 Recycling

The word recycling first refers to the idea of the economic cycle. The raw materials, products or product and machine parts used are returned to the usage or production cycle from which they originate. Any quantity of recovered and reused raw materials does not have to be purchased on the global world market and represents a source of raw materials within the sphere of influence of the affected regions and/or companies. Depending on the recycling rate, the recycled materials flow back into the usage or production cycle or are removed by the landfill or thermal influences. Figure 4.3 shows the various forms of recycling and disposal and illustrates the relationships that arise [18]. The area framed in blue indicates the procedures that fall within the scope of recycling and are dealt with in this section.

Depending on reconditioning or preparation, recycling objects usually undergo treatment processes that are attributed to the utilization or processing. In the case of reconditioning, technical production processes predominate, while preparation is process-related [19]. A further subdivision can be made by differentiating between reutilization, further utilization, reprocessing and further processing [20]. Table 4.1 shows how the different subdivisions can be distinguished from each other, in which practical relevance is also established by means of examples.

Whether a product is completely recycled at the end of its life cycle or only at a certain recycling rate is already determined at the product design stage [18]. In order to ensure the highest possible degree of recyclable components in a product, it is necessary to think ahead both in production and in production planning [21]. Ideally, products should be designed to be reusable, technically more durable, easy to repair and low in pollutants. The more recyclable products are designed, the easier it is to return the resources gained to the production process. The use of regionally

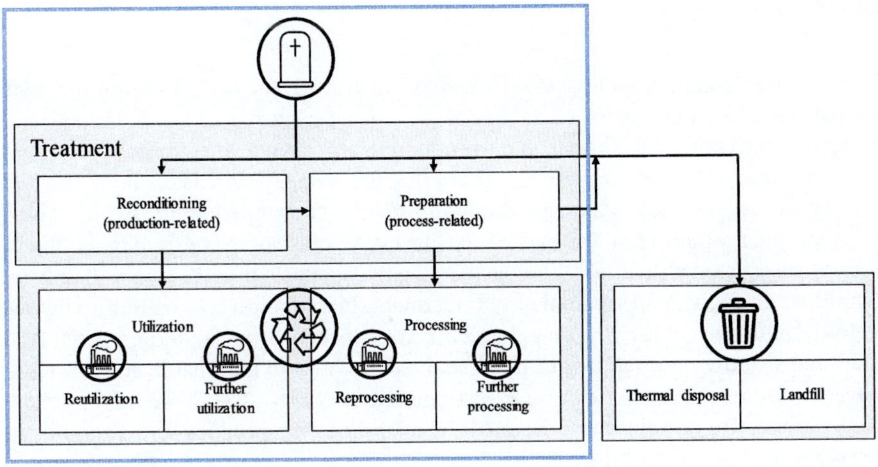

Fig. 4.3 Presentation of recycling processes

Table 4.1 Practical relevance to recycling processes

Treatment	Recycling	Explanation	Example
Reconditioning	Reutilization	Used products are reused for the same purpose	Reusable bottles or packaging
	Further utilization	Used products are reused for a purpose other than the original one	A mustard jar becomes a drinking jar
Preparation	Reprocessing	Renewed use of used materials and production waste in a similar production process	The remelting of glass
	Further processing	Use of waste materials and production waste in a production process not yet completed by them, resulting in other materials or products with different properties	Manufacture of paperboard from waste paper

available raw materials and resources as well as a modular structure of products and machines would allow regions and the companies operating in them to carry out technological developments from within themselves and thereby achieve a certain degree of precautionary sovereignty and economic independence [22].

4.2.4 *Waste*

The just mentioned recycling rate in the end-of-life strategy of recycling includes on the one hand the part of reusable components/resources and on the other hand materials that cannot be returned to the production process and are therefore classified in the category of waste. Figure 4.4 shoes that the strategy of waste can be divided into product categories, which are thermally treated or dumped on the landfill site.

A product which is at the end of its life cycle and cannot be remanufactured, reused or recycled will only undergo one of the two disposal processes or will only be landfilled through an upstream thermal treatment. Incineration is essentially a detour before landfilling, which must either be chosen because the materials cannot be landfiled without pretreatment or because the materials are flammable and therefore have a calorific value that can be used to generate energy [23]. Mineral materials, for example, are not directly deposited. In relation to the product level, glass is an example.

All products that are disposed of thermally or disposed of in landfills instead of being recycled cause not only even more waste to be incinerated and higher costs for the general public, but also lead to an undeniable (enormous) loss of natural resources. Due to the higher costs of the disposal processes compared to the recycling processes, a recycling-friendly construction of products is here to be appealed as well [24].

4.3 An End-of-Life Phenomenon

The warranty has just expired, and the TV or mobile phone purchased a few years ago is "suddenly and unexpectedly" defective and must be sent in for repair. This type of

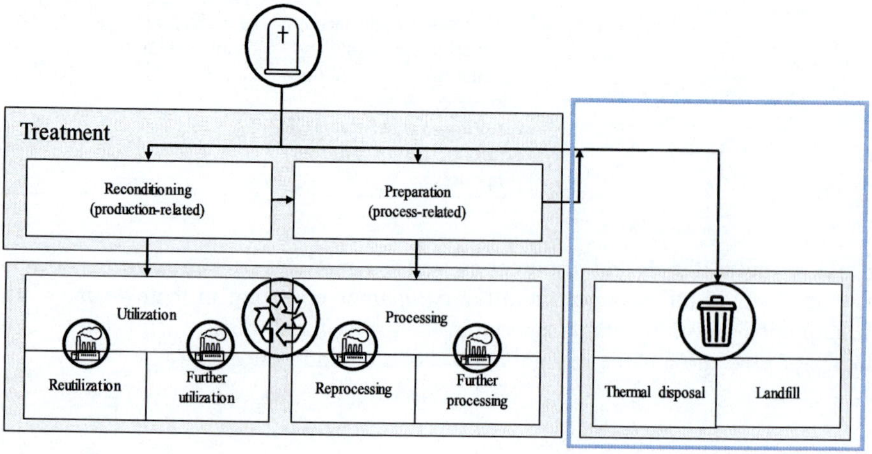

Fig. 4.4 Presentation of waste processes

procedure is typical for special consumer goods and is supposedly increasingly used in practice. It is about the obsolescence of a product. In general, obsolescence can be divided into natural and artificial/planned product ageing. The following section focuses primarily on the planned obsolescence.

4.3.1 Demarcation of Predetermined Breaking Points

A predetermined breaking point is a defined safety feature of a product which is designed in such a way that in the event of damage a break occurs only at the planned point. The installation of a predetermined breaking point is planned, part of the safety concept and documented [25]. If the predetermined breaking point is triggered during use, the product can either no longer be used or was designed from the outset in such a way that the device can be operated again after a repair. An ideally developed predetermined breaking point does not reduce the shelf life or usability of a product but serves exclusively product safety. With the planned obsolescence, on the other hand, the so-called predetermined breaking points are misused not to guarantee the safety of a product, but to cause premature ageing of the product. An example of this are monitors or LCD televisions in which heat-sensitive electrolytic capacitors are located directly next to power components that reach temperatures of over 100 °C. In contrast to the predetermined breaking point, these special points are neither documented nor made public by the manufacturer.

4.3.2 Planned Obsolescence

The term “planned obsolescence” is used when consumer goods are deliberately designed with vulnerabilities that cause premature or artificial product aging to shorten the life of a product [26]. This means that a product manufactured with an aging character will age, become obsolete or unusable [27]. This makes certain product functions weaker, the product no longer fulfils the desired purpose or the product is only compatible with a new additional product or no longer corresponds to the desired design. The area in the life cycle of a product in which the planned obsolescence is applied is illustrated in Fig. 4.5.

Most of the sources used for literature research divide this type of path to the end of product life into technical and non-technical obsolescence which can be subdivided further [28–30]. After elaboration, each obsolescence type is established based on examples.

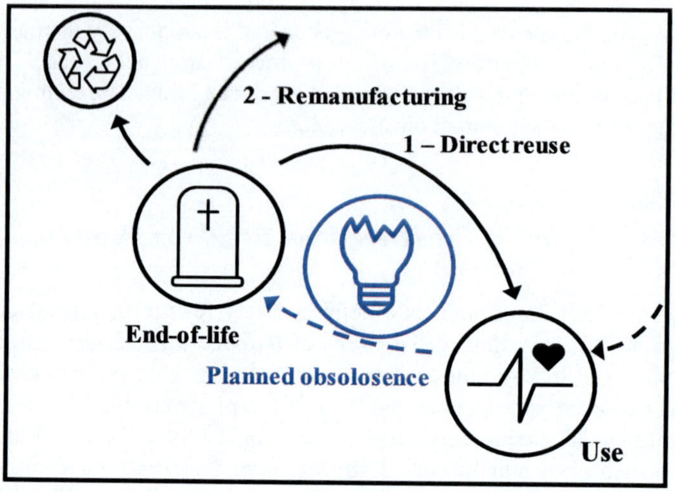

Fig. 4.5 Positioning of planned obsolescence in the product life cycle

4.3.3 Forms of Planned Obsolesence

For overview the different types of planned obsolescence, publications from the last 5–10 years were used to highlight the continuous presence as well as the current debate about the topic [31–34].

The various forms of obsolescence shown in the Table 4.2 as well as the practical examples found in the literature search indicate a clear use of this phenomenon. Nevertheless, the deliberate and planned use of planned obsolescence is difficult to prove today, since on the one hand the fact of premature product ageing and on the other hand the intention behind it must be proven. Thus, the question results, how is it possible to measures purposeful premature product ageing?

4.4 Recycling

The term recycling signifies:

> [...] The return of by-products and residues occurring during production and consumption back into the production/consumption cycle. This includes the collection, transport and the actual recycling of waste as well as the needed energy and resources for the disposal of waste. [35]

Figure 4.6 shows and differentiates the different types of recycling.

The term recycling can be subdivided into product and material recycling. The product recycling can also be subdivided

Table 4.2 Forms of planned obsolescence and their practical relevance

Obsolescence	Form	Explanation	Example
Technical	Physical	Inadequate materials or constructions can mean premature shutdown of equipment	The drawer drive with belts from DVD players, due to the aging and hardening of the belt only work 2–3 years. The worm gear drive principle could help here
	Functional	Here the product itself remains functional but can no longer be used to its full extent due to new requirements	Strong emergence in the computer industry, device only works with newer operating system
	Economic	For cost reasons, maintenance or repair of the device is waived The repair or the spare parts exceed here the costs of a new purchase of the equipment	The repair of washing machines by a technician in most cases exceeds the price of a new one
Non-technical	Psychological	The product is still fully functional in terms of quality and performance but is considered obsolete or outdated by the consumer because it seems less desirable for fashion reasons or of other changes	The fashion industry had an average of two collections a year. Nowadays, the collections of international fashion groups such as H&M and Zara have multiplied

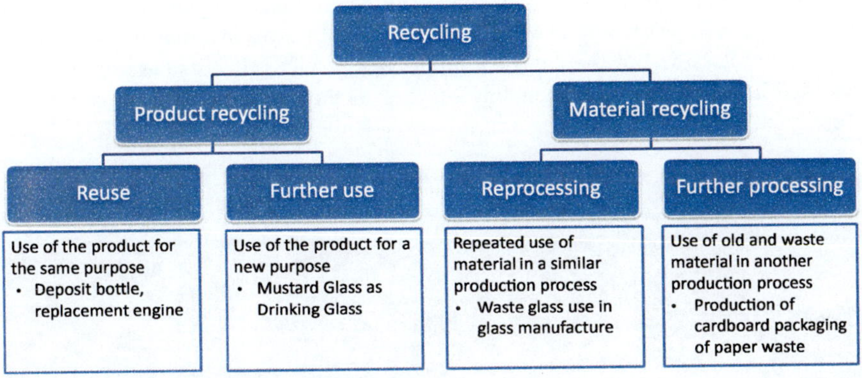

Fig. 4.6 Differentiation of types of recycling [36]

- **Reuse**—The product is reused for the same purpose—examples are the reuse of deposit bottles or exchange engines.
- **Further use**—The product is reused with a new purpose. An example is the reuse of a mustard glass as a drinking glass.

Regarding material recycling also a subdivision is done:

- **Reprocessing**—Materials are used in a similar production process in a repetitive way. An example is scrap glass when producing glass
- **Further processing**—Materials are used in a not yet run through product process—an example of further utilization is the use of paper waste to produce cartons [35, 36].

4.5 Modernization

Complex products after their use phase can have a second use phase due to modernization works [37]. Regarding a machine modernization it is about product recycling, well said reuse. The product after a modernization is used for the same purpose.

Figure 4.7 shows a portal milling machine type FP 2000 SA. It was modernized by the company Rottler machine tools GmbH [38].

In the course of this machine modernization the following steps have been done by the company Rottler machine tools Ltd.:

- Geometrically updated
- Equipped with a Siemens 840 D control system
- Extension of the tool magazine from 40 to 60 tool places.
- Update of the CNC-fork type
- Newly designed KSS-supply, central lubrication system and hydraulic system [38].

Through this modernization an extension of the product life cycle is reached. Well said, the machine gets a second use phase. A service like this machine modernization can also be classified in the category after-sales services. In this case though it is classified in the category independent services, as this modernization, was done for a foreign product.

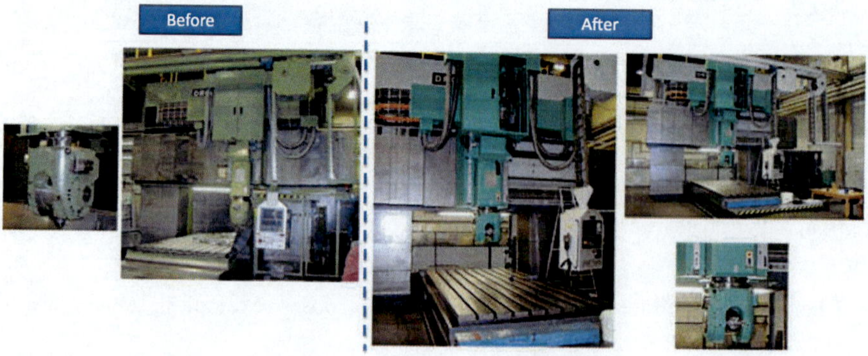

Fig. 4.7 Modernization of a portal milling machine type FP 2000 SA [38]

4.6 Disposal

At the end of the product life cycle an avoidance of waste according to the waste hierarchy is not feasible any more.

(a) Avoidance
(b) Preparation for reuse
(c) Recycling
(d) Other forms of energetic recycling
(e) Disposal [39].

An exclusion of the economic cycle is the logical consequence. A disposal of waste can for example happen in the form of collection, combustion or depositing [37].

4.7 Conclusion

Through literature research, the end-of-life strategies of remanufacturing, reuse, recycling and waste just discussed have proven to be particularly relevant in practice, as they are applied in the majority of manufacturing companies. The aspects considered, the high-value addition of remanufacturing and reuse as well as the strategies influenced by the recycling quota, thus covered the main routes in or across the production process of all material and product flows. Nevertheless, due to the ever-increasing scarcity of resources and environmental awareness, the two aspects of value addition and recycling quota for life cycle products must be developed sustainably. A forward-looking and promising response to end-of-life strategies is to increase value creation and recycling rates [40]. This poses new challenges for the planners and technologies of a used product, because the rethinking extends from the design, through the production to the use of the product. Overcoming the challenge with the help of the cradle-to-grave approach could help to develop new approaches. The increasingly frequent planned obsolescence in the product life cycle counteracts the two aspects mentioned above and will occur more frequently due to the shorter innovation cycles. This could be helped by a method that helps to find out how a planned product aging is to be measured in order to sanction the companies practicing it.

References

1. Raubold, U.: Lebenszyklusmanagement in der Automobilindustrie (2011)
2. Bauer, T., Brissaud, D., Zwolinski, P.: Design for High Added-Value End-of-Life Strategies (2017)
3. Zhang, F.: Intégration Des Considérations Environnementales En Entreprise: Une Approche Systémique Pour La Mise En Place de Feuilles de Routes, Grenoble, France Universite de Grenoble, www.theses.fr/s95811 (2014)

4. Rose, C.: Design for Environment: A Method for Formulating Product End-of-Life Strategies (2000)
5. Brito, M. P., Decker, R.: Modelling product returns in inventory control—exploring the validity of general assumptions. Int. J. Prod. Econ (2003)
6. European Commission: Directive 2008/98/EC of the European Parliament and of the Council of 19 November 2008 on Waste and Repealing Certain Directives. https://eur-lex.europa.eu/legal-content/EN/TXT/?uri=celex:32008L0098.
7. Zinck, S., Ayed, A.C., Niero, M., Head, M., Wellmer, F., Scholz, R., Morel, F.: Life Cycle Management Approaches to Support Circular Economy. In: Benetto E. (ed.) (2018)
8. ISO 2002: Integrating Environmental Aspects into Product Design and Development. ISO/TR 14062:2002. https://www.iso.org/standard/33020.html
9. Rose, C., Ishii, K., Stevels, A.: ELDA and EVCA: tools for building product end-of-life strategy. J. Sustain. Prod. Des. **1** (2002)
10. Walcher D., Leube M.: Kreislaufwirtschaft in Design und Produktmanagement Co-Creation im Zentrum der zirkulären Wertschöpfung. Springer Fachmedien Wiesbaden GmbH (2017)
11. Allwood, J. M.: Squaring the Circular Economy: The role of recycling within a hierarchy of material management strategies. In: Handbook of Recycling State-of-the-art for Practitioners, Analysts, and Scientists (2014)
12. Feldhusen, J., Gebhardt, B.: Product Life cycle Management für die Praxis. Umsetzung und Anwendung. Springer-Verlag, Berlin Heidelberg, Ein Leitfaden zur modularen Einführung (2008)
13. Sundin, E., Bras B.: Making functional sales environmentally and economically beneficial through product remanufacturing. J. Clean. Prod. **13**(9), 913–925 (2005)
14. Ijomah, W.: Addressing decision making for remanufacturing operations and design-for-remanufacture. Int. J. Sustain. Eng. https://dx.doi.org (2009)
15. Lund, R.T.: Remanufacturing: The experience of the U.S.A. and implications for the Developing Countries. World Bank Technical Paper No. 3
16. Ferrer, G., Ayres, R.U.: The impact of remanufacturing in the economy. In: Ecological Economics (1984)
17. Gallo, M. et al.: A perspective on remanufacturing business: issues and opportunities. Int. Trade Econ. Policy Perspect. (2012)
18. Rost, N.: Recycling: Rohstoffquellen des 21. Jahrhunderts, www.reginalentwicklung.de (2007)
19. Friedel, A.: Einfluss der Produktgestalt auf den Energieaufwand beim Recycling mechanischer Bauteile und Baugruppen. Springer-Verlag, Berlin, Heidelberg (1999)
20. Bruns, K.: Analyse und Beurteilung von Logistikentsorgungssystemen, deutscher Universitätsverlag, Wiesbaden (1997)
21. Bode, E.: Konstruktionsatlas. Hoppenstedt Verlag, Darmstadt (1996)
22. Beamon, B.: Designing the green supply chain. Logist Inf. Manage. **12**(4), 332–342 (1991)
23. Schneider, U. et al.: recyclingfähig kostruieren, Wien (2011)
24. Rochat, N.: Schluss mit Wegwerfen, www.aquava.ch (2013)
25. Kessler, T., Brendel, J.: Planned Obsolescence and Product-Service Systems: Linking Two Contradictory Business Models. Rainer Hampp Verlag, Bayreuth (2016)
26. Bulow, J.: An economic theory of planned obsolosence. Q. J. Econ. (1986)
27. Pope, K.: Understanding Planned Obsolosence. London (2017)
28. Stuyck, J.: Production Differentiation in Terms of Packaging Presentation, Advertising, Trade Marks, etc. Deventer (1983)
29. Elias, J. A., Burgers, J.: Time—A Vocabulary of the Present. New York (2016)
30. Akyurek, K. B., Ciravoglu, E. G.: Developing a quantitative research method on planned obsolosence in architecture. In: Product Lifetimes and the Environment—Conference Proceedings (2017)
31. Packard, U.: The Hidden Persuaders. Published by Penguin, London (1961)
32. Schneider, J.: Werkzeuge für den Bestand. Published by epubil, Berlin (2011)
33. Cooper, T.: Longer Lasting Products: Alternatives to the Throwaway Society (2010)
34. Giles, S.: Made to Break-Technology and Obsolescence in America (2006)

35. Förstner, U.: Umweltschutztechnik. 7, vollst. bearb. und aktualisierte Aufl. Berlin, Heidelberg: Springer (VDI-Buch), 2008. Online: https://dx.doi.org/10.1007/978-3-540-77883-7
36. Janorschke, B.: Rebel, Birgit; Kott. Recycling rückgebauter industrieller Bausubstanz, Matthias (2009)
37. Walther, G.: Nachhaltige Wertschöpfungsnetzwerke. Überbetriebliche Planung und Steuerung von Stoffströmen entlang des Produktlebenszyklus. Techn. Univ., Habil.-Schr.—Braunschweig, 2009. 1. Aufl. Wiesbaden: Gabler (Produktion und Logistik) (2010). Online: https://dx.doi.org/10.1007/978-3-8349-8643-6
38. Rottler Werkzeugmaschinen GmbH. Mudersbach. Online: https://www.rottler-maschinenbau.de/fileadmin/images/pdf-presseberichte/ueberholungen_d.pdf. Checked 25 May 2019
39. Europäisches Parlament: RICHTLINIE 2008/98/EG DES EUROPÄISCHEN PARLAMENTS UND DES RATES, 19.11.2008 über Abfälle und zur Aufhebung bestimmter Richtlinien. In: Amtsblatt der Europäischen Union, S. L 312/3 -L 312/30. Online: https://eur-lex.europa.eu/LexUriServ/LexUriServ.do?uri=OJ:L:2008:312:0003:0030:de:PDF. Checked 25 May 2019.
40. Niemann, J., Tichkiewitch, S., Westkämper, E.: Design of Sustainable Product Life Cycles. Springer Verlag, Heidelberg, Berlin (2009)

Chapter 5
Life Cycle Evaluation

A sustainable development through a holistic product life cycle management requires an appropriate analysis of the effects of decisions along the whole product life cycle. This analysis is done based on the three fields of sustainability. The fields with the corresponding overall phase of life analysis methods can be taken from Fig. 5.1 [1].

This chapter deals with the subsections of the Method PROSA (Product Sustainability Assessment) as well as with existing standards, guidelines and standard sheets.

5.1 Life Cycle Assessments

The life cycle of a product is connected with different influences and effects on the environment. Those can emerge from the raw material extraction, production, distribution, use to the recycling and disposal. To fulfill the model of a sustainable product life cycle, a future-oriented environment protection in order to minimize risks for human beings and the environment is necessary and the motto is:

> [...] the follow-up care is to be replaced by precaution. [1]

The society has an increasing environmental awareness. Therefore the customers want to be informed about the impacts on the environment related with the product. Consequently the method life cycle assessment has been developed. The definition of the life cycle assessment is the following:

> Combining and judging the input and output flows and the potential environmental effects of a product system throughout its life cycle. [3]

This can help for example to show improvements of the environmental characteristics of a product in the individual phases of a life cycle [1, 3].

A life cycle assessment contains four phases. Those can be taken from Fig. 5.2.

J. Niemann and A. Pisla, *Life-Cycle Management of Machines and Mechanisms*, Mechanisms and Machine Science 90,
https://doi.org/10.1007/978-3-030-56449-0_5

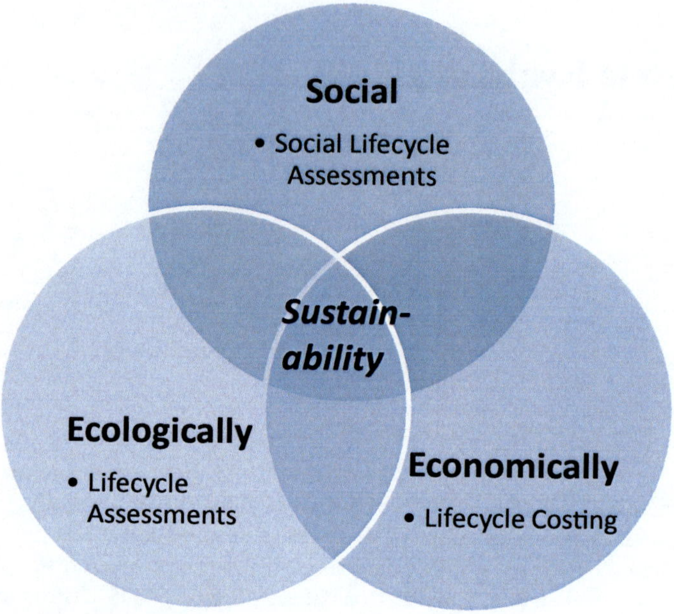

Fig. 5.1 Three fields of sustainability (Own source modified acc. to [2])

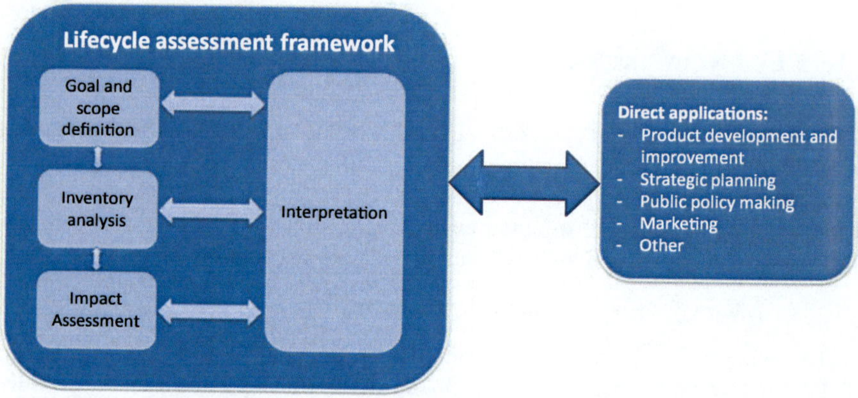

Fig. 5.2 Phases of a life cycle assessment [3]

In the first phase of the life cycle assessment the goals as well as the framework of studies are determined. The framework depends on the object of investigation and on the application of the life cycle assessment [3].

The second phase of the life cycle assessment, the creation of a factual balance sheet, serves as a survey of input and output data. Data are collected that are important to reach the set goals [3].

The third phase of the life cycle assessment is about the impact assessment. During this phase additional information in order to support the valuation of the factual balance sheets' results are provided. Like this the environmental relevance can be assessed in a better way [3].

The final phase of the life cycle assessment is the evaluation phase. The results of the preceding phases are judged and the resulting conclusions, recommendations and supports are discussed and documented [3].

5.2 Social Life Cycle Assessments

Regarding the social life cycle assessment the social aspects along the product life cycle are investigated. They are of great importance for the analysis and the improvement of the sustainability of products. Social aspects are a big challenge regarding the product evaluation as they are very diverse and rated in different ways depending on different groups of interest, countries or regions. Every phase of the product life cycle is connected with one or more geographic locations where process steps are carried out. Depending on the country or the region, it can be calculated with different social standards [1, 4].

The methodic procedure regarding the social balance can be subdivided into four process steps as the life cycle assessment:

1. Determination of the goal and the framework of investigation
2. Factual balance sheet
3. Estimation of impacts
4. Evaluation of results [5].

Most common social aspects arise from the following fields:

- Hotspots during the production, usage as well as during the disposal (for example child labor, wages below subsistence level)
- Impacts during the product use (for example lopsided posture when operating a machine)
- Indirect impacts on the society (e.g. use of mobile phones) [6].

The PROSA list of social indicators can help when chosing the aspects that need to be considered in greater detail. The list was created out of several dozen indicator lists with over 3000 social indicators and sorted according to four steakholder groups:

- Employees
- Neighboring or regional population
- Society
- Private, commercial and state users [7].

5.3 Life Cycle Costing

The goal of the life cycle costing (life cycle cost analysis) is a holistic and systematic consideration of the costs along the whole product life cycle.

The term "total cost of ownership" is applied in the field of information technique as a synonym for the term life cycle costing. This analysis serves for example customers as a support when making decisions regarding investments, as not only the acquisition costs but all costs along the product life cycle are considered.

The early information about the costs in the phases of the product life cycle help the producer optimize the product [1].

Figure 5.3 shows an iceberg as an illustration of the life cycle costs. At first glance only the acquisition costs of the product can be seen as the follow-up costs are not directly visible, though those costs are significant as they account for the greatest part of the costs [8].

This is shown by Fig. 5.4. It shows the cost distribution of a machine tool over a life cycle of 10 years.

In the following, a closer look is taken at the VDI 2884 guideline as well as at the VDMA 34,160 standard sheet.

5.3.1 VDI 2884 Standard

VDI 2884 acquisition, operation and maintenance of production, mean applying the life cycle costing (LCC). One goal of the guideline is to support operators regarding the choice of different production means. Therefore a methodical framework was made available in order to take a decision regarding the acquisition based on the resulting life cycle costs. Further the goal of this guideline is to support producers by means of a methodical framework regarding the development of products against the background of a life cycle cost consideration [11, 12].

Fig. 5.3 Life cycle costs illustration (Own source modified acc. to [8, 9])

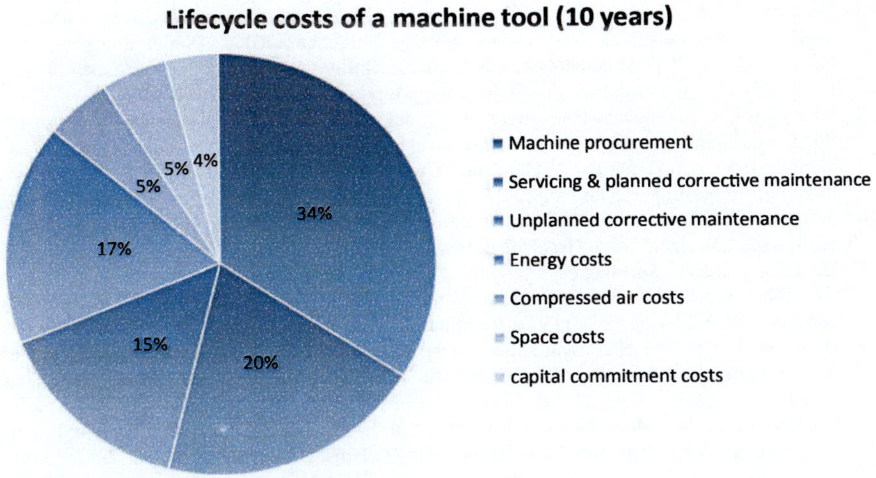

Fig. 5.4 Life cycle costs of a machine tool (10 years) [10]

5.3.2 VDMA 34160 Standard

VDMA 34160 Forecasting model for the life cycle costs of machines and plants [13].

In this standard sheet the structured definiton and prognosis of the life cycle costs for machines, plants and components are described. In the described forecasting model for the determination of life cycle costs no price effects like financing or capital costs are considered. The model determines cost blocks for every phase as well as calculation rules. Furthermore the VDMA provides an Excel tool for the calculation of life cycle costs, which is based on the standard sheet VDMA 34160 [9, 11–15].

References

1. Herrmann, C.: Ganzheitliches Life Cycle Management. Nachhaltigkeit und Lebenszyklusorientierung in Unternehmen. Springer Verlag, Heidelberg (2010)
2. Aachener Stiftung Kathy Beys: Drei Säulen Modell. Aachen (2015). Online: https://www.nachhaltigkeit.info/artikel/1_3_a_drei_saeulen_modell_1531.htm. Checked 25 May 2019
3. DIN EN ISO 14040: Umweltmanagement – Ökobilanz – Grundsätze und Rahmenbedingungen, Oct 2006
4. Grießhammer, R., Droste, A.: Die Sozialbilanz bei PROSA. Hg. v. Öko-Institut e.V. Freiburg (2019). Online: https://www.prosa.org/index.php?id=204. Checked 25 May 2019
5. Grießhammer, R., Droste, A.: Liste sozialer Indikatoren. Hg. v. Öko-Institut e.V. Freiburg (2019). Online: https://www.prosa.org/index.php?id=202. Checked 25 May 2019
6. Grießhammer, R., Droste, A.: Sozialbilanz und SocioGrade. Hg. v. Öko-Institut e.V. Freiburg (2019). Online: https://www.prosa.org/index.php?id=180. Checked 25 May 2019
7. Grießhammer, R., Droste, A.: Soziale Indikatoren. Hg. v. Öko-Institut e.V. Freiburg (2019). Online: https://www.prosa.org/index.php?id=203. Checked 25 May 2019

8. Deutges, D.: Wer nur auf den Kaufpreis schaut, handelt mit Zitronen. Die echten Maschinenkosten durchleuchten. Carl Hanser Verlag. München (2012). (Sonderdruck aus der Fachzeitschrift WB Werkstatt+Betrieb 9/2012). Online: https://monforts-wzm.de/uploads/media/2012-09_wb_Monforts_TCO_Internet_pdf.pdf. Checked 25 May 2019
9. Niemann, J.: Ökonomische Bewertung von Produktlebensläufen- Life Cycle Controlling. . In: Spath, D. (Hrsg.) u.a.: Neue Organisationsformen im Unternehmen : In: Spath, Dieter, Westkämper, Engelbert (Hrsg.) u.a.: Handbuch Unternehmensorganisation: Springer, Strategien, Planung, Umsetzung. Berlin U.A (2016)
10. Schweiger, S., Dressel, K., Pfeiffer, B. (eds.): Serviceinnovationen in Industrieunternehmen erfolgreich umsetzen. Neue Geschäftspotenziale gezielt durch Dienstleistungen ausschöpfen. Heidelberg, Berlin, Springer (2011)
11. VDI 2884: Beschaffung, Betrieb und Instandhaltung von Produktionsmitteln unter Anwendung von Life Cycle Costing (LCC) (VDI-Richtlinie) (2005)
12. Niemann, J.: Die Services-Manufaktur, Industrielle Services planen –entwickeln – einführen. Shaker Verlag, Ein Praxishandbuch Schritt für Schritt mit Übungen und Lösungen. Aachen (2016)
13. Verband Deutscher Maschinen- und Anlagenbau e.V. (VDMA): Handbuch für das Excel-Berechnungs-Werkzeug zur Berechnung von Lebenszykluskosten in der Investitionsgüterindustrie. Online verfügbar unter https://www.vdma.org/article/-/articleview/1180530. Checked 25 May 2019
14. Niemann, J., Tichkiewitch, S.: Westkämper Engelbert: Design of Sustainable Product Life Cycles. Springer Verlag, Heidelberg, Berlin (2009)
15. Niemann, J.: Life Cycle Management-das Paradigma der ganzheitlichen Produktlebenslaufbetrachtung. In: Spath, D., Westkämper, E., Bullinger, H., Warnecke, H.-J. (Hrsg.) Neue Entwicklungen in der Unternehmensorganisation. Springer-Vieweg, VDI Buch, Berlin (2017)

Chapter 6
Life Cycle Information Support

In holistic product life cycle management information appear in every phase of the product. The goal is to give an information feedback to the initial life cycle design phase in order to develop new, better and more innovative products. Like this a continuous improvement process is possible.

Figure 6.1 shows the information flows that occur in a holistic product life cycle management. In order to support the collection, processing and administration of information, it makes sense to apply a life cycle management software (PLM software). Some companies today work with different systems describing data in their different departments. The development for example works with a CAD-Software, which applies specific parts lists. In the production an enterprise resource planning software is applied and in sales special data processing programs are used. The goal of using PLM software is to standardize and stabilize descriptive information of every department. In all phases of the product life cycle product-related data result, from the development up to service measures, which are subject to constant changes by different users. That's why it is no longer sufficient to only grant the development, but all responsible departments need access to all product-related data [1].

The application of PLM software brings various advantages. Decisions can be made faster and more precise due to current and complete information. Changes can be done faster and costs can be lowered. The customer satisfaction can be increased due to an excellent maintenance service, which has access to all relevant data like parts lists.

6.1 Analytics of Life Cycle Data

Life cycle management is divided into many thematic fields and structures. In order to be able to present a state of the art conscientiously, it is necessary to begin with

J. Niemann and A. Pisla, *Life-Cycle Management of Machines and Mechanisms*, Mechanisms and Machine Science 90,
https://doi.org/10.1007/978-3-030-56449-0_6

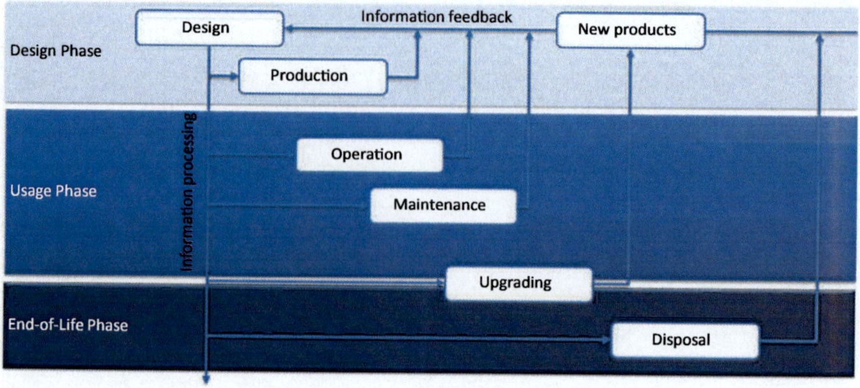

Fig. 6.1 Information flows of the product life cycle management [2–5]

some definitions. In the topic of this report *life cycle data support*, data is a very essential part. So first of all there will be a closer look on what are data.

Freely referred to Merriam-Webster dictionary: Data are facts which describe statistical or measurable properties in a way and which can be used as a fundamental basis for scientific statements and offer possibilities of calculating, reasoning and discussing themselves or the context in which they have been taken [6–8].

While there are many data within PLM concepts, it is important to point out which possibilities exist to draw insight into the ways how data is dealt within these processes. The content in particular plays a decisive role in the handling of life cycle processes. Therefore the focus will be shifted from the pure consideration of the data to the way the data is collected and dealt with.

From a scientific point of view there would exists many possibilities to deal with data in an appropriate way, but in the context of life cycle management these possibilities can be concentrated on the most efficient ones. Currently there are four ways which can be preferred for drawing insight into product data. These categories are descriptive analytics, diagnostic analytics, predictive analytics and prescriptive analytics. They will be presented in the following Fig. 6.2 and afterwards compared with each other.

- **Descriptive Analytics** is of all these 4 possibilities the easiest to present data in a usable way. Results can be seen directly after they are measured and it's possible to use these data for the trends prognosis and further developments.
- **Diagnostics Analytics** gives fundamental information about source and courses and can present conventional developments based on historical data.
- **Predictive Analytics** Enables a huge data complexity and allows this way computerised learning. In addition, it can improve the results.
- **Prescriptive Analytics** This is the most advanced technique of analytics and can make prognoses combined with optimisations of outcomes. It also has the possibility to give supporting recommendations during the analytic process [9].

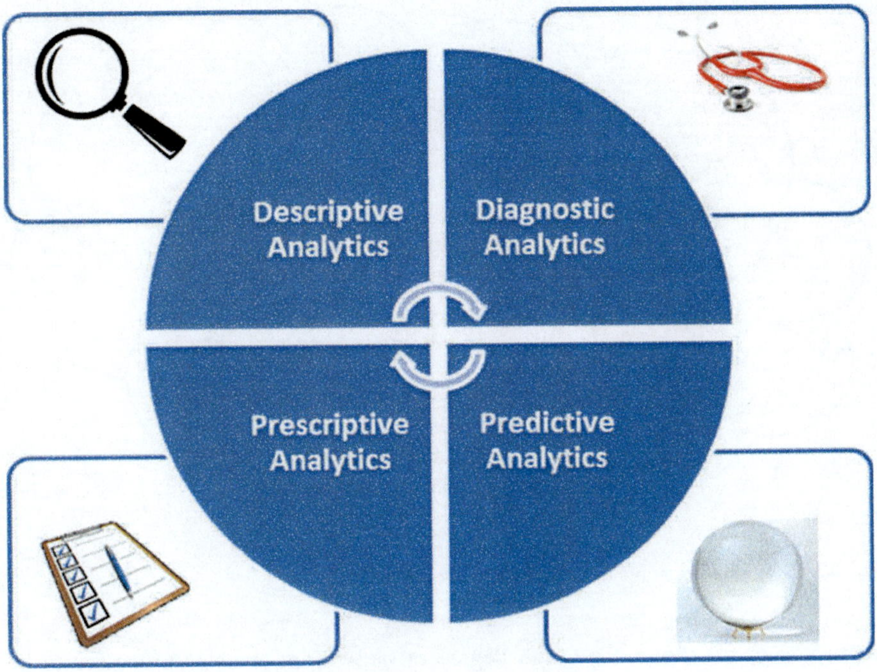

Fig. 6.2 Insight drawing analytics (*Source* Own graph acc. to [9])

So there are several different concepts of how analytics can deal with data. The type of data that results from these processes is diverse. There are also many software products that deal with this data. Nowadays, software products are developed and improved very quickly, so it is not easy to say how many products are available on the world market. But it is a fact that there are many of them and that their number is increasing at the same time as the possibilities increase of the ways how life cycle concepts are connected to economy [9, 10].

Because of the huge amount of subareas within life cycle processes, it is important not to focus only on single part solutions for these areas, but to see the whole concept from an abstract point of view.

A thoroughly designed solution is a helpful choice for this purpose. A basic software solution in this case is not a good alternative for the versatility of the application areas. In this context a cloud system can be seen as a possible solution to get many tools for part-solution aspects combined in one holistic package. In this context there would be a possible way to integrate an application by hosting it into a cloud [11].

There are many potential solutions within a data cloud as can be seen in Fig. 6.3. A core component of the data cloud is the accessibility of product data (e.g. relationships, component, assembly, and other product information) that can be accessed by other cloud components. These additional components may include applications

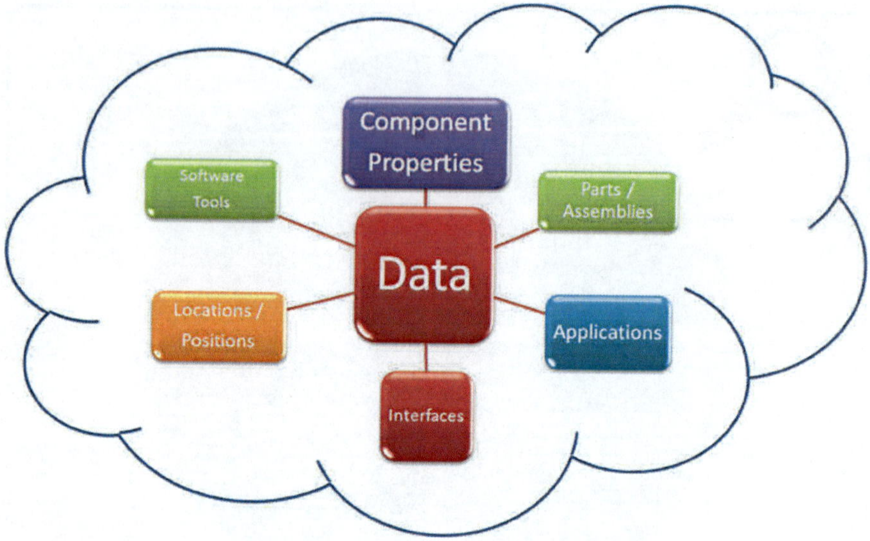

Fig. 6.3 Data cloud example (*Source* Own figure acc. to [12])

connected to the cloud. The high degree of networking enables processes to be processed flexibly and quickly [12]. Within the cloud all aspects of PLM data should be combined. Therefore it is necessary to get an overview of the separate points or topics of data which can be reviewed in a life cycle process. The following Fig. 6.4 will present several common aspects which are included in a life cycle process and it will show consequently, why there always is data as a core aspect at the beginning of life cycle.

In Fig. 6.4 there are several important steps connected to possible influence factors or relevant kinds of data presented. Data can be seen at the beginning of this chain, because it is one of the main influence factors for all other parts of the life cycle. With a closer look at different steps it is obvious that each step is instructed on data.

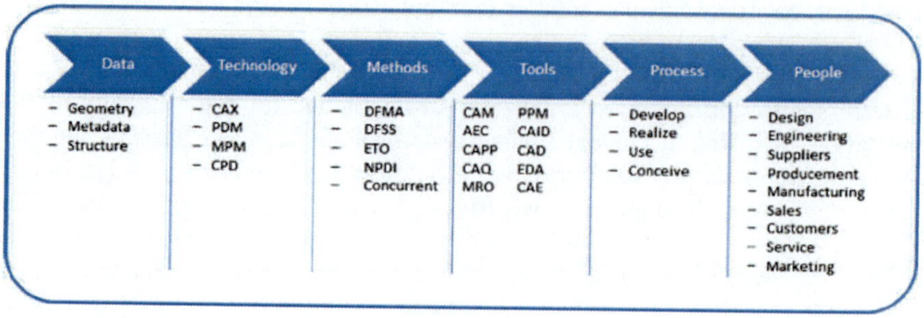

Fig. 6.4 Holistic life cycle data (*Source* Modified acc. to [13])

6.2 Support Function of Life Cycle Data

According to the linking status of data to almost all other aspects of a life cycle it can be seen which high degree of importance this information has for the life cycle management. So it is easily imaginable that there are some other qualities that are based on data and can support life cycle processes. In the following, the advantages that data can offer for a life cycle concept are examined in more detail.

As mentioned above, there are some types of analyses that enable people involved in life cycle management to analyze historical data developments and make predictions about future developments. Alternatively, it is also possible to check the current status of the life cycle processes during their use [9].

By a closer look at these analytics there are many possibilities in how far they can support a life cycle process. Among other things, from the obtained data, it would be possible to generate tem-plates for the database, templates that can be used in forecasting processes to sup-plement the missing data or to plan a life cycle. This could include standardized val-ues derived from analytics.

From each year the performance of databases with all their solutions increases. It is therefore very important to focus on these developments. The development of databases is closely linked to the development of information technology (abbreviated hereafter as IT) [14].

In this context the "International Federation for Information Processing" (abbreviated hereafter as IFIP) plays an important role. It is a global organisation that consists out of many members of research areas and professionals highly linked to IT technologies with the aim to develop new standards and advance the sharing of information within this community and for researchers worldwide. The handling of data is becoming increasingly important within life cycle management. Especially IFIP tries to contribute to the progress and further development of this field of research. As the research areas supported by IFIP are diverse, individual working groups have been set up within the organisation to deal with more specific research areas. In this organizational structure, Working Group 5.1 with the self-chosen name "Global product development for the whole life cycle" can be identified as the group responsible for research work on PLM [15].

The research areas of the group are directly related to the PLM and all its subcategories. The primary research areas are among others: Rapid product development; Global product development; Smart, intelligent and meta-products frameworks; Product life cycle management; product life cycle engineering; New organizational issues within product life cycle engineering. These areas can only be understood as a few sub-items, as the working group WG5.1 itself explicitly deals with product development and all its components including the entire product life cycle [15].

Since several years they have been holding a yearly conference about IT developments concerning Life cycle topics where they discuss new developments and the possibilities for new standards in this research field. With its conferences IFIP offers the possibility that new approaches from service and support areas of Life cycle Management spread globally and that these can be discussed and further developed.

Another group of researchers in this context deals with topics related to PLM from another point of view and with a slightly different focus. Increasingly, the aim of this group is to achieve closer networking between industrial manufacturing processes and the implementation of the data used in these processes within frameworks of PLM implementations. In most cases, the focus of these implementations is on the data communication between external engineering tools and the broad usability within PLM System, for example for the implementation of services or design processes [16].

"The International Academy for Product Engineering" abbreviated as CIRP was founded in 1951 by members from Belgium, France, England and Switzerland in the context of an international meeting with the common objective of advancing the development of production technologies. During the initial phase, the group still called itself: "The International Institution for Production Engineering Research", but later changed its name to CIRP.

In consideration of the connection to PLM the focus of the group is primarily on the implementation of computer-aided manufacturing processes. Likewise there can be seen a development within manufacturing towards intelligent machines. It is assumed that a big step in this direction has already been achieved by the help of autonomously acting computers with a sensible methodology, which have already been successfully implemented in industrial structures. However, only advances in technology and research are probably expected to provide the opportunity to establish intelligent machines [17].

The comparison of the IFIP and CIRP thematic fields from the Life cycle Management area show that the current research areas in the industrial field are all related to the use of data. Especially the topics of Industry 4.0, Internet of Things, as Industry of the Future and the digital transformation of industry are related to the industrial field. Due to this common basis of aspects from the industrial field there will be only a very abstract view possible to present relationships concerning Life cycle concepts, which could be generally described as the interdisciplinary use of data [18].

At the CIRP conference in 2016 a theoretical connection was already presented to realize the support of data in the form of combined solutions between product and service life cycle management. This was presented in the form of an approach for a PSS (Product Service System) concept. The difficulty was that PLM and SLM were not completely matchabel because of the PLM focus on products and of SLM on services. However, it was also obvious that the use of data in the context of life cycle management is nearly equal for both processes and that further development of IT concepts could possibly overcome this barrier [19, 20].

6.3 Options of Data Exchange

As a possible connection medium for exchanging data interdisciplinary between applications, the use of databases is a common tool. Therefore, it is important to focus on what a database is and afterwards to have a closer look at the option to use multiple databases or software tools cooperatively by exchanging data between them.

A database is used to organize and store data, while the structure and design of a database can be customized individually. With these features, databases are predestined to be considered a central component of data support, not exclusively to life- cycle processes but in general. As mentioned before, the main task of databases is to organize data. Organizing data in this context means storing data, access and rights management for users as well as conversion and interfacing options which could be provided by a database. This means that software tools, connected to a database, can be used in a variety of ways. The individual creation of a database structure or its interfaces can lead to limitations concerning the use by software tools. However, these limitations are negligible as long as the exchange of data from one database to another is possible. At this point, the use of exchange formats is essential. Exchange formats enable data to be transferred between databases via interfaces. The selection of exchange formats depends on the structure of the database and also on the connection to other software tools. As possible examples for exchange formats json and xml could be mentioned here.

On the one hand there will be a closer look on the format json. As an exchange format json provides a small file size with a variety of ways to transfer data from databases. It also allows the data to be used in several ways to be converted into other types of exchange formats such as XML.

On the other hand the XML format is not a pure exchange format, but is originally intended as a separate exchange language. Specified XML stands for 'Extensible Markup Language' which structures data in the format of a text file that is readable by both humans and machines.

To process data in this way provides the possibility to be used independently of implementations and platforms and additionally it allows to transfer data to all known computer systems. As a file format XML offers a very high interdisciplinary potential that could be particularly suitable for use in the life cycle environment.

6.4 Product Life Cycle Software Tools

From the preceding descriptions it can be seen that there exist many areas in which data support life cycle concepts. Furthermore, it is possible to assume that the processing of data is a core part of PLM processes. Since PLM concepts are always individually adapted to given requirements, it can be considered logically that almost all PLM concepts are to be understood as individual solutions. This is independent of PLM Software Tools as they are exclusively concerned with partial solutions of a PLM concept and can contribute to provide assistance to partial solutions. But these tools normally cannot support an entire PLM concept holistically. With individual life cycle concepts and a multitude of supporting tools, it is difficult to make a concrete statement about the data or data formats to be used. Since the software tools and offered partial solutions are constantly developing, it seems impossible at first to give concrete recommendations on how the handling of data can be supported holistically in the context of life cycle management. A possible point of reference at this juncture are the current research fields that are oriented towards the current changes in the industrial field, which is increasingly moving towards the digital trends of the future (to name them in the form of a few buzzwords: IT agile, scrum, Industry 4.0, digitalization …). By focusing on the database component and additionally orienting towards the analysis processes which helps to deliver adequate data, it is assumable that these two components will continue to exist even with future solutions within the framework of life cycle data support. Today exchange formats create platform independency and already lead to integrated cloud solutions by providing data that serve as support for life cycle concepts.

Consequently it can be deduced that one of the most efficient aspects of data within digital PLM concepts will be the exchangeability of data between different software solutions as well as a flexible use of the data, which will also be valid for future applications and conceptual new software solutions. This means that exchange formats (e.g. json, xml) across software boundaries are a good option for the holistic life cycle management area. Since these formats are usually compatible to software interfaces it could be possible that future applications or new combinations of database or software products can also provide these options. Most providers of PLM software specialize in certain parts of Product Life cycle Management. As it is very complex to combine all sections regarding data there are only a few providers on the market that cover a wider range of functions. These enterprises are for example EDS, Oracle, IBM/Dassault and SAP. “mySAP PLM” is SAP’s PLM solution and used by lots of enterprises of different sizes and branches. The software covers the following key areas, which are specified in the following [1].

- Data and document management
- Life Cycle Collaboration
- CAD Integration
- Project Management
- Quality Management
- Technical Assetmanagement
- Environment, Health and Safety.

6.4.1 Data and Document Management

A product life cycle covers many steps between the first idea of a product to end of production and recycling. In between this complex process a lot of information, data and documents accumulate. The following Fig. 6.5 described by Boos and Zancul shows the product life cycle of a product and the amount of data and documents emerging along. In producing enterprises, it’s mostly information about material, bills of material, drawings, CAD data, maintenance and recycling information [21].

SAP PLM is based on product data management which combines all product-, process-, and resource-data, i.e. specifications, bills of material, work schedules,

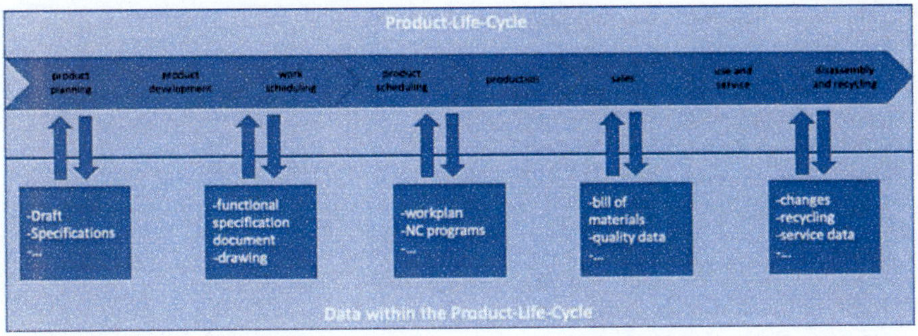

Fig. 6.5 Data within a product-life cycle [21]

resource data, recipes, CAD models and technical documents. SAP's essential elements are a structured browser and a document management system. Provided data and documents combined with different systems like CAD-, office- and graphic applications keep information transparent and available [1].

6.4.2 *CAD Integration*

Manufacturers are facing challenges like competitive pressure, shorter development times and faster Time-to-market. This means a necessary change for computer aided design and technical development as it used to be. As products get more complex and need to be individual sometimes, it's getting more and more important to merge data from all parts along the product engineering process. Combining technical and commercial departments with a direct CAD-ERP connection ensures fewer mistakes, availability of data and dynamic processes among the whole value chain. Having CAD data and corresponding documents available in the whole product life cycle efficiency increases in the following processes as well. The easiest way to keep data available is integrating CAD data in ERP processes. The example of SAP (CAx) illustrates clearly the advantages of integrating technical design and development data into the SAP system. The software links the provided technical data with logistical information and improves processes like triggering manufacturing orders and supplier enquires. Data access can be granted individually. SAP consolidates different data sources into one summarized software program, and combines CAD and ERP [22–24].

Integrating CAD files directly in SAP enables the company to manage the data and leverage the design information for following processes like costing, sourcing and manufacturing. Main benefits of CAD integration are listed on the SAP website [25]:

- Process integration
- Integrated into change and release process

- Integrated into SAP Business Suite
- Integrated Viewing Solution
- Simultaneous Collaboration
- "Single source of truth" for all participants
- Location-spanning access to information
- Real-time access to all relevant information
- Same Integration for different CAD-Systems
- Lower TCO
- No additional Data Management System (Island) necessary
- No additional system interfaces
- No Data Redundancy
- One central user management
- Sold and supported by SAP.

6.4.3 *Life Cycle Collaboration*

Life cycle collaboration enables employees to exchange, forward and edit information like project plans, documents, service sheets, technical drawings and product structures. It's an internet-based communication platform for development teams, business partners, customers and suppliers [1].

Shu and Wang describe the importance of collaboration and cooperation among departments and enterprises. An effective allocation of resources and harmonization of businesses is only possible if all information relevant to various stages of a product life cycle and all the information throughout a product life cycle can be accessed by everyone associated with its design, creation, sale, distribution and maintenance [10].

The Institute for Information Systems (IWi) points out that customer requirement in combination with technological improvements enables new or improved business processes. According to that fact new forms of cooperation like E-Collaborations arise. The institute describes an E-Collaboration architecture that shows how cross-enterprise, customer-driven processes can be planned, implemented and controlled. It's based on the differentiation of global and local knowledge in the widely used Architecture for Integrated Information Systems (ARIS). According to the institute's analysis it's important for enterprises to have software architecture that offers a set of integrated methods from the business concept level up to the implementation into ICT-systems. An appropriate graphic representation and user-friendly tools ensure a flawless connection between each level and support exchanging ideas and reconciliation of interests between the different recipients within the network [26].

6.4.4 Project Management

The project management plays an important role in product development and organization in as they facilitate the entire project. In this critical role, it is imperative to be integrated with the environment to where the product is implemented and created. Specifically, the project manager needs to formulate and implement guidelines to plan a project through its total life cycle. The use of PLM systems to support the project management activities in a meaningful and integrated way would help support efficiency of the organization. However, the PLM system needs to support the project management activity while concurrently integrating with the company's systems and environment [27]. To better understand the benefits of PLM with respect to Project Managers perceptive and work function groups it is good to understand the knowledge areas in the Project Management Body of Knowledge (PMBOK). The following Table 6.1 shows the knowledge areas of the PMBOK and shows the corresponding definitions for each area [28].

Table 6.1 Areas and definition of PMBOK

Areas	Definition
Integration	It comprises the processes and procedures required to identify, define, combine, unify and coordinate the various processes and project management procedures in the project management process groups [29]
Scope	Scope management includes the processes necessary to ensure that the project includes all the necessary work, but only this work, to complete it successfully. This primarily involves defining and controlling what is included in the project and what is not (See Project Management Institute, 2008, pp. 103–129)
Time	Time management is the process of developing the master timeline for the project and includes all of the necessary processes to develop and maintain a clear and concise timeline [29]
Cost	Cost management includes all the necessary activity to estimate and manage costs on a project [29]
Quality	Project Quality Management includes all the activities needed to determine and guarantee that the final product meets the customer's satisfaction throughout the planning process and into production of the product [29]
Human Resources	Project Human Resource Management entails the processes that plan for and lead the project team throughout the project life cycle [29]
Communication	Project Communication Management includes the processes required to ensure communication between the team members and stakeholders during the life of the project [29]
Risk	Risk Management includes all of the processes required to identify and assess risks on a project, including mitigation planning if required [29]
Procurement	Procurement Management includes all the necessary processes required to plan for the sourcing and purchasing of items required to implement a project into production [29]

Table 6.2 Quality management key functions [1]

Quality function	Description in refer to PLC
Quality engineering	Helps to define the right strategies for suitable quality planning in the first phases of the product life cycle
Quality control and assurance	Ensures the maintenance of a desired quality level by inspections during all phases of the cycle, as well as fast intervention in case of sudden problems
Quality improvement	Supports the continuous improvement of quality management processes
Audit management	Helps to carry out internal and external audits for processes, Systems and Products. The tests are carried out based on checklists. The software solution for the support of Quality audits help companies to operate in compliance with legal requirements and to comply with quality standards

6.4.5 Quality Management

To see the fundamental structure of a "Process-Oriented Quality Management System", according to DIN ISO 9000:2000, as a continuous improvement system which works in conflict management between customer requirements and the achieved customer satisfaction, an analogy with PLM cannot be overlooked. PLM focuses on product data and its integrity through all core and management processes. Here, synergies can be harnessed which, above all in the case of the already available process-oriented way of thinking of the personnel and of the documentation [28]. The following Table 6.2 describes the key functions of quality management.

6.4.6 Asset Management

Manufacturers in all different sectors are faced to increasing maintenance, repair and operations (MRO) costs at their plants. The largest cost increases are in energy, raw materials and from the negative impact of unscheduled production outages, often due to failure of assets. It is particularly important for companies to optimise the performance of individual production facilities and the availability of technical equipment, because the failure of a machine and its maintenance mean a downtime in production. If a company succeeds in minimizing downtimes with the help of suitable software, significant cost savings can be achieved.

Leading-edge companies discovered that they can improve production efficiencies and manage MRO expenditures by integrating their Product Life cycle Management (PLM) with Enterprise Asset Management (EAM) systems. Furthermore, PLM and EAM integration offers additional benefits when manufacturers enjoy in 'After Sales Services'. EAM technology can manage MRO tasks for the manufactured goods that customers buy or lease from Original Equipment Manufacturers (OEMs).

Normally, manufacturers have relied on distributors or parts suppliers to handle customer support. Service revenues to manufacturers were minimum and mostly from supplying parts and managing warranties and product upgrades. But nowadays, service after sales is a fast growing, high profit opportunity and many manufactures see advantages to using their PLM legacy data along with EAM solutions to offer direct customer service after sale. The profitable benefits include long-term customer relationships, additional revenue through service offerings and direct tracking of product performance and MRO in the field. Customers benefit too, through improved technical support, higher asset availability/performance levels and options for performance enhancement, field upgrades, etc. There is an emerging trend of OEMs offering their customers for-fee MRO services—that is maintaining their goods at their customer locations, extending further their services after sales as a new business mode [1, 30].

6.4.7 Environment, Health and Safety (EH&S)

The following Fig. 6.6 describes in which relation the Environment, Health and Safety department support with information, to ensure the fulfilment of requirements and regulations within the framework of the environmental protection [31].

An integrated PLM-system supports this department with helpful specific information. So is it for example necessary for the occupational safety and industrial safety management to know which hazardous materials does the product content. This information is also important for the waste management. With a Product Life Cycle Management System and an access for relevant information the Environment, Health and Safety issues can be handled in a much faster and more efficient way.

An efficient EH&S management significantly reduces costs, increases a company's image and thus its market opportunities [1].

6.4.8 Conclusion

Wannenwetsch reported that a company would gain four important benefits by an integrated PLM system [1].

- *Faster market maturity*
  - Sharing up-to-date information, decisions could be made faster and more accurate, and it has a better way for managing product development and production.
- *Cost reductions*
  - An efficient data management would lead to a faster implementation of changes.

Fig. 6.6 EH&S management support (*Source* Modified acc. to [31])

- *Improvement of the product quality*
  - The data share of the quality management system, which includes product development, production and maintenance, leads to an
  - improvement of product quality.
- *Improvement of the customer satisfaction*
  - Products that take the customer's wishes into account,
  - and an excellent maintenance service (the smooth and efficient
  - because it can access all relevant data, such as the bill of materials) lead to an improvement in customer satisfaction.

He refers his conclusion to a particle example of a company, Brose GmbH and Co. KG, which implemented mySAP PLM, a leading PLM-system by SAP. The company was able to reduce process times, i.e. the effective workload, by 25% [32]. The time savings are mainly achieved by tracking deadlines, checking incoming requests and compiling them in a quotation mirror. This leads to a high acceptance of the solution among the buyers. The benefit is offset by project expenditure of approx. 200 man-days, of which almost half for user training, after Wannenwetsch. So it would make an opinion, that PLM systems are a fast rentable solution for companies with near to anyone problems.

But data support in PLM systems are faced with amount of problems especially when it comes to "Big Data". The paper could give the reader the opinion, that more data would lead to a more effective company especially when they are saved in a PLM-system. When "Big Data" is applied in PLM, more specific challenges have emerged which seriously hold back the potential applications and problems [17, 33]:

- Data collection in PLM
- Data storage and transfer in PLM
- Data process based on manufacturing knowledge and experience
- System, service platform, and tool of "Big Data" in PLM
- Security of PLM data
- Data visualization.

Especially for small and medium-sized companies are these challenges hard to handle. In general, they do not have the knowledges and equipment to save, collect and visualize these high amounts of data. Therefor are specialist required, which these companies mostly cannot afford.

In general, it is possible to see a PLM-system as a communication tool, which provides information through different company's divisions. For big and maybe medium-sized enterprises with a high and straight hierarchy it could be a very useful tool and would raise the work-effectivity enormous, which leads to save money and gain a higher status in refer to customers and suppliers. For smaller companies is it not really required, because there already exist a hand in hand working philosophy. Information through different division can be easy to get because they know the responsible person and have the contact data about them. So, a PLM-system with the high data capacity and the massive training hours would be nonsense in refer to PLM-system's general benefit for small companies. Vice versa for bigger enterprises especially with a high and straight hierarchy PLM data support is a very useful tool to ensure a fast and efficient workload.

References

1. Wannenwetsch, H.: Vernetztes Supply Chain Management. SCM-Integration über die gesamte Wertschöpfungskette. Springer-Verlag, Berlin, Heidelberg (VDI-Buch) (2005). Online verfügbar unter https://dx.doi.org/10.1007/3-540-27507-X
2. Niemann, J., Tichkiewitch, S., Westkämper, E.: Design of Sustainable Product Life Cycles. Springer Verlag, Heidelberg Berlin (2009)
3. Niemann, J.: Life Cycle Management- as Paradigma der ganzheitlichen Produktlebenslaufbetrachtung. In: Spath, D. et al. (Hrsg.) Neue Entwicklungen in der Unternehmensorganisation. Springer-Vieweg, VDI Buch, Berlin (2017)
4. Niemann, J.: Ökonomische Bewertung von Produktlebensläufen-Life Cycle Controlling. In: Spath, D. et al. (Hrsg.) Neue Entwicklungen in der Unternehmensorganisation. Springer-Vieweg, VDI Buch, Berlin (2017)

5. Niemann, J., Tichkiewitch, S.: Westkämper Engelbert: Design of Sustainable Product Life Cycles. Springer Verlag, Heidelberg Berlin (2009)
6. Data Definition: Online: https://www.merriam-webster.com/dictionary/data; Checked 18 Dec 2018
7. Niemann, J., Pisla, A.: Sustainable potentials and risks assess in automation and robotization using the life cycle management index tool—LY-MIT. Sustainability **10**, 4638 (2018)
8. Kretschmar, D., Niemann, J., Deckert, C.: Digitalisierungsindex zur prozessnahen Analyse mittelständischer Unternehmen, In: ZWF Zeitschrift für wirtschaftlichen Fabrikbetrieb Apr 2019, Jg. 114, Ausgabe 4. S. 213–218
9. Reynolds, J., Glessner, M., Dury, J., Ward, N.: Product Life Cycle Intelligence: Turn PLM Data into Insights with Machine Learning. Online: https://viewpoints.io/entry/product-lifecycle-intelligence-turn-plm-data-into-insights-with-machine-lea; Checked 18 Dec 2018
10. Shu, Q., Wang, C.: Information Modeling for Product Life Cycle Management. In: Bernus, P., Fox, M. (eds.) Knowledge Sharing in the Integrated Enterprise. DIISM 2004, ICEIMT 2004. IFIP—The International Federation for Information Processing, vol 183. Springer, Boston, MA (2004)
11. Wang, R.: Research on data security technology based on cloud storage. Procedia Eng. **174**, 1340–1355 (2017). https://doi.org/10.1016/j.proeng.2017.01.286
12. Erkayhan, S.: PLM in der Cloud. Online: https://www.digital-engineering-magazin.de/plm-der-cloud; Checked 18 Dec 2018
13. Costello, H.: Global Cloud based PLM Market 2018 by Demand, Type, Size, Driver, Top Companies (Oracle, Siemens, Autodesk), Growth and forecast 2022. Online: https://www.reuters.com/brandfeatures/venture-capital/article?id=40751; Checked 18 Dec 2018
14. Adhikari, A., Adhikari, J.: Advances in Knowledge Discovery in Databases. Springer-Verlag, Berlin Heidelberg, Berlin, Heidelberg (2015)
15. International Federation for Information Processing: Online: https://www.ifip-wg51.org/. Checked 19 Dec 2018
16. Penciuc, D., Durupt, A., Belkadi, F., Eynard, B., Rowson, H.: Towards a PLM Interoperability for a collaborative design support system. Procedia CIRP **25**, 369–376 (2014). https://doi.org/10.1016/j.procir.2014.10.051
17. Lee, J., Kao, H. -A., Yang, S.: Service innovation and smart analytics for industry 4.0 and big data environment. In: Procedia CIRP **16**, 3–8 (2014). https://doi.org/10.1016/j.procir.2014.02.001
18. Meyes, R., Tercan, H., Thiele, T., Krämer, A., Heinisch, J., Liebenberg, M.: Interdisciplinary data driven production process analysis for the internet of production. Procedia Manuf. **26**, 1065–1076 (2018). https://doi.org/10.1016/j.promfg.2018.07.143
19. Marilungo, E., Coscia, E., Quaglia, A., Peruzzini, M., Germani, M.: Open innovation for ideating and designing new product service systems. Procedia CIRP **47**, 305–310 (2016). https://doi.org/10.1016/j.procir.2016.03.214
20. Niemann, J.: Die Services-Manufaktur, Industrielle Services planen –entwickeln – einführen. Shaker Verlag, Ein Praxishandbuch Schritt für Schritt mit Übungen und Lösungen. Aachen (2016)
21. Boos, W., Zancul, E.: PPS-Systeme als Bestandteil des Product Life cycle Management. In: Schuh, G. (eds.) Produktionsplanung und -steuerung. VDI-Buch. Springer, Berlin, Heidelberg (2006)
22. Lisse, R.: Effizientere Produktentwicklung durch Integration. Dig. Fact. J. Das Magazin für Industrie 4.0 & IoT, Feb 2018
23. Morar, L., Westkämper, E., Abrudan, I., Pisla, A., Niemann, J., Manole, I.: Planning and Operation of Production Systems. Fraunhofer IRB Verlag (2008)
24. Niemann, J.: Eine Methodik zum dynamischen Life Cycle Controlling von Produktionssystemen. Heimsheim: Jost-Jetter Verlag, 2007IPA-IAO Forschung und Praxis 459). Stuttgart, Univ., Fak. Maschinenbau, Inst. für Industrielle Fertigung und Fabrikbetrieb, Diss. (2007)
25. Hopf, C.: SAP Community Wiki. Online: https://wiki.scn.sap.com/wiki/display/PLM/CAD+Integration+in+PLM. Checked 18 Dec 2018

26. Adam O., Chikova P., Hofer A., Vanderhaeghen D., Zang S. : E-collaboration architecture for customer-driven business processes in inter-organizational scenarios. In: Funabashi M., Grzech A. (eds.) Challenges of Expanding Internet: E-Commerce, E-Business, and E-Government. IFIP International Federation for Information Processing, vol. 189. Springer, Boston, MA (2005)
27. Eastham, J., David T., Sumir V., Scott S.: PLM software selection model for project management using hierarchical decision. In: Proceedings of PICMET'13: Technology Management for Emerging Technologies, pp. 511–527. Springer, Portland (2013)
28. Feldhusen, J., Gebhardt, B.: Product Life cycle Management für die Praxis. Springer Verlag, Berlin, Heidelberg (2008)
29. Project Management Institute: A guide to the project management body of knowledge. (PMBOK® Guide), Project Management Institute, An American Standard ANSI/PMI 99-001-2008 (2008)
30. Popko, E., Luyer E.M.: IBM. ftp://public.dhe.ibm.com/software/plm/coe/asset_management_to_support_product_lifecycle_management.pdf, 18 Dec 2018
31. Bätz, C..: MYSAP PLM – VON PRODUKTDATEN UND PROJEKTEN. In Business Integration mit SAP Lösung, by Andreas Hufgard, Heiko Hecht, Wolfgang Walz, Frank Hennermann, Gerald Brosch and Sabine Mehlich, pp.184–205. Springer Verlag, New York (2004)
32. Österle, H., Senger, E.: Realtime Management – Fünf Fallstudien. Universität St. Gallen, Fallstudie, Gallen (2003)
33. Westkämper, E.; Niemann, J.: Digitale Produktion – Herausforderung und Nutzen. In: Spath, D. et al. (Hrsg.) Neue Entwicklungen in der Unternehmensorganisation. Berlin, Springer-Vieweg, VDI Buch (2017)

Chapter 7
Servitization and Modern Business Models

7.1 Introduction

In the past, many executives argued that they did not have the proper skills, it is too costly or anyway customers would never pay for extended services implemented within their products [1, 2]. Today, an increasing number of manufacturing firms are adding services to their traditional product offerings in response to economic and competitive pressures [3]. Moreover, new market trends push manufacturer towards not to sell the product by transferring the ownership, but rather to sell either the usage of the product. This phenomenon drives the evolution from a traditional business model, to a modern service-oriented business model [4]. The literature conceptualizes this shift from products to solutions through various concepts, such as servitization [5]. When servitization moves a manufacturer all the way to become a solution provider there are major changes on the business model and product utilization within the different life cycles, which have to be considered [6].

Therefore, the aim of this elaboration is to identify how servitization enables the shift to a modern service-oriented business model and the impact of the product life cycle. This elaboration starts with servitization as basic concept and moreover, as a roadmap to a modern service-oriented business model. This more comprehensive view of servitization, will enhance the understanding between traditional and modern business models. Moreover, this elaboration presents either an application as well as practical examples of servitization and modern business models.

7.2 Servitization

This chapter gives an overview about the definition, drivers and challenges of servitization. Additionally, it provides a basic concept on how to "servitize" among the

J. Niemann and A. Pisla, *Life-Cycle Management of Machines and Mechanisms*, Mechanisms and Machine Science 90,
https://doi.org/10.1007/978-3-030-56449-0_7

customer value life cycle. Moreover, it presents the capabilities of servitization, to move a manufacturer all the way to becoming a solution provider.

7.2.1 Definitions, Drivers and Challenges

Services are potentials or abilities of a service provider, which are created by people or machines [7]. The activity, undertaken by firms to add these potentials or abilities to their offered product is described as servitization. In the end, after the successful integration of service and product, a product-service system arises [5]. Figure 7.1 shows, that almost all manufacturing companies have started to consider servitization as an opportunity and more than 20% have already established servitization.

Key drivers for this shift are the intensity of competition, through associated pressure on prices, shorter product life cycles and increasing convergence of product offers. Moreover, the digitalisation as a megatrend leads to structural changes of markets, customers and companies. Firms believe that increasing services will deliver a possibility to react to changing market conditions, higher level of differentiation and increasing customer relationship [9]. Technology plays an important role in enabling servitization, the adoption of them allow the development of new, integrated product-service offers in manufacturing [10]. Table 7.1 summarizes the top technology drivers in servitization.

Fostering servitization also requires the involvement of customers and key partners in value co-creation [11]. Without alignment of business among these parties, a servitization initiative is likely to fail. Given the reliance of many product firms on different channel partners, cultural change may also have to encompass firms in the broader business network [12]. Moreover, servitization as business transformation of a company face increasing investment needs, which also have to be considered [13]. By adjusting their business structure and reassessing finances, manufacturers can incorporate servitization and obtain the benefits from it [14–16]. These benefits are outlined within the next subchapters.

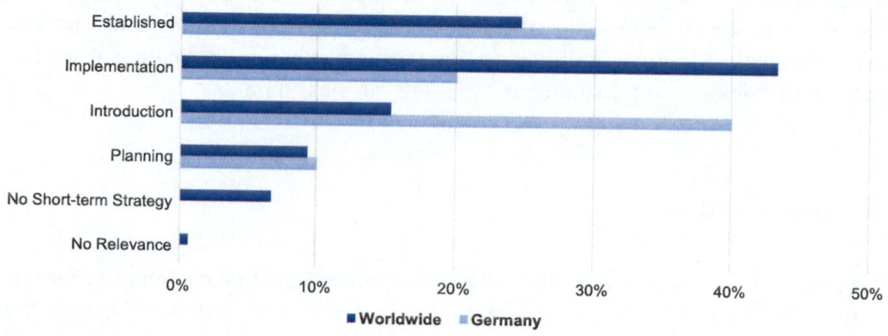

Fig. 7.1 Manufacturing companies on their way to servitization (*Source* Modified acc. to [8])

Table 7.1 Top technology drivers for servitization

Technologies	Operations
Predictive analytics	To predict specific failure modes
Remote communications	To remotely adjust, fix, or send software updates to machines/products
Mobile platforms	To receive internal data or customer information in real time
Consumption monitoring	To create consumption driven supply chains for consumer specific offerings
Dashboarding	To provide key performance Indicators to make services more tangible and visible
Sensor network integration	To facilitate big data analytics

Source Modified acc. to. [10]

7.2.2 *Basic Concept: Adding Services as Additional Offerings*

Servitization enables firms to add value to their product offerings, by adding service-based value propositions to their business model [17]. Most definitions of business models focus on the organisation, its value creation and offerings [18]. By analysing 17 different definitions, Al-Debei et al. consolidated the following:

The business model is an abstract representation of an organisation, be it conceptual, textual, and/or graphical of all core interrelated architectural, co-operational, and financial arrangements designed and developed by an organisation presently and in the future, as well as all the core products and/or services the organisation offers, or will offer, based on these arrangements that are needed to achieve its strategic goals and objectives [19]

Generally, a value proposition describes the benefits customers can expect from products and services. The value proposition map is an application to design new service-based value propositions. This map identifies the fit between specific value propositions from products and services with customer profiles. A good fit is indicated, when products or services produce gain creators and pain relievers that match one or more of the jobs, pains, and gains that are important for each customer [20]. Figure 7.2 shows the structure of a value proposition map.

The value proposition map identifies that commercial transactions take place when satisfactory value propositions brought to the right customer. Additionally, from the servitization perspective, value must be delivered in the appropriate moment to the customer. This is indicated through the customer value creation life cycle, which is built by the phases of requirements and acquisition, attributed to the "pre-sales cycle", and ownership and disposal, as part of the "post-sales cycle" (Fig. 7.3).

Each of the steps in every phase represents "a value creation moment" that will entail a certain degree of interaction between the manufacturer and the customer. This value life cycle offers a useful framework for analysing a manufacturer's different

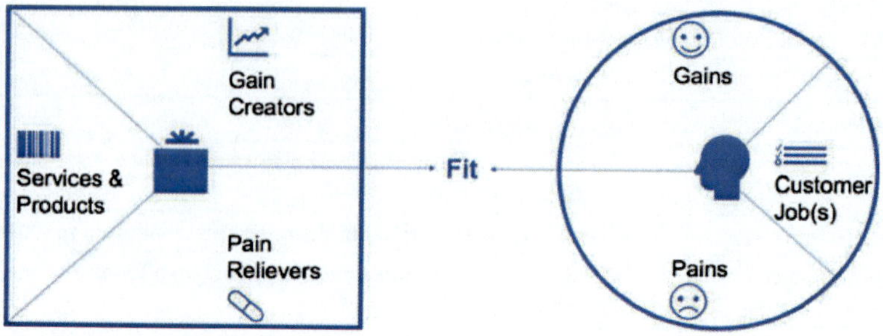

Fig. 7.2 Value proposition map [20]

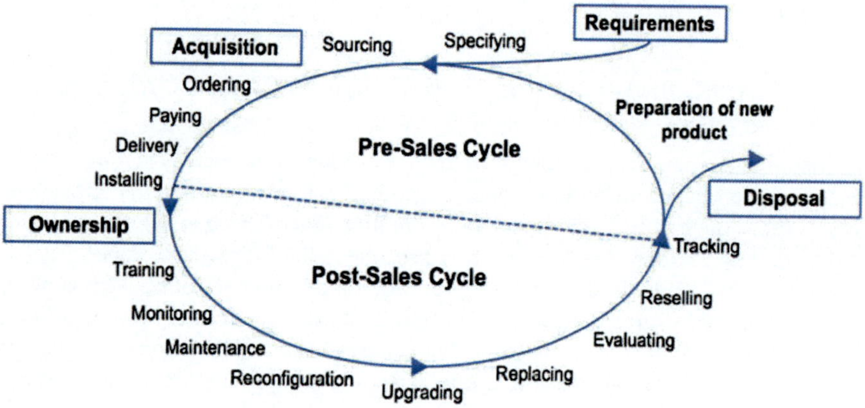

Fig. 7.3 Customer value creation life cycle (*Source* Modified acc. to [21])

alternatives for creating value and "servitizing" his product offering [21]. Furthermore, combining the phases with the value proposition map enables an integrated approach to manage different needs of customers with servitization.

7.2.3 Capabilities: From Manufacturer to Solution Provider

As outlined before, the basic concept of servitization is the objective to make a product more attractive by adding valuable services. There is an even more advanced view of servitization as business model innovation, where servitization is outlined as transition to deliver services instead of products [22]. The idea to take a complete transition to deliver services instead of products is based on the extended model of a product. This extended model is described by three layers, where the inner layer describes the standard core functions of a product. The middle layer describes the tangible assets of the product, like the shape and design. The outer layer summarizes services, as intangible assets, of the product (see Fig. 7.4).

By narrow sense, the product is considered as a tangible entity which is offered in a market whereas the broader sense gives an indication about the objective, which means satisfying a demand or solving a problem of the customer [23, 24]. Figure 7.5 shows that during a servitization process the tangible value-share of an extended product decreases while at the same time, the non-tangible value-share is increasing.

Customers are striving for benefits and solutions or even more they are requesting intangibles like success and leadership on the market. The consequence is decoupling of manufacturing of goods and selling of services. In this case the physical good can remain the property of the manufacturer, considered as an investment, while revenues come uniquely from the services. When servitization moves a manufacturer all the way to becoming a solution provider there are major changes on the business model that have to be considered [6]. This further development from a traditional to a service-oriented business model of manufacturing companies will be outlined in the next chapter.

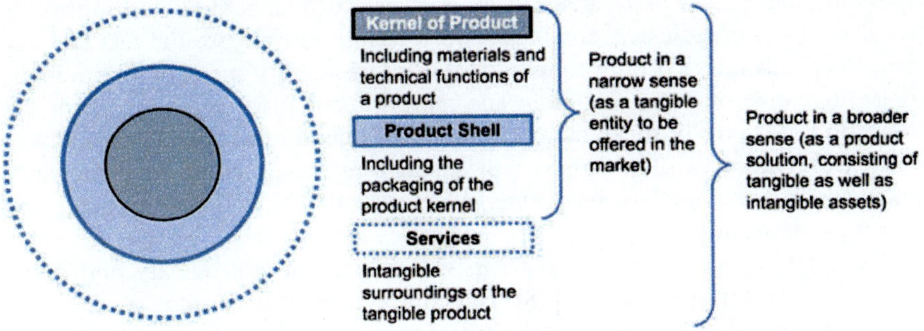

Fig. 7.4 Difference between products: narrow sense versus broader view (*Source* Modified acc. to [23])

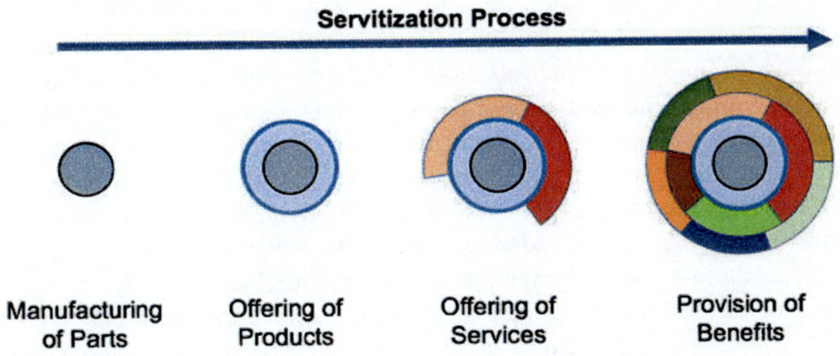

Fig. 7.5 From manufacturing of parts to provision of benefits (*Source* Modified acc. to [23])

7.3 Modern Life Cycle Business Models

In order to better understand the transition from a traditional to a modern service-oriented business model, this chapter starts with a basic concept of business modelling. After identifying different types and building a framework for modern business models, the focus will be set to the major impacts within the different life cycles.

7.3.1 Definition and Concept

In the age of information overload and ever faster growing technological advancements, companies must establish innovative and marketable business models in order to continue to assert themselves on the market [25]. Age, size, reputation or current sales figures are no guarantee that a company will be able to compete in the future market [26]. In the age of hyper competition, existing competitive advantages are repeatedly and increasingly attacked by measures of competitors and ultimately repealed. As a consequence, companies must adapt, regarding to the fact that any competitive advantage is only temporary. Instead of stability and equilibrium in the competitive environment, it is more and more important for companies in hyper competition to actively shake the prevailing status quo in the existing competitive landscape themselves [27]. This is why modern business models, also known as innovative business models, have emerged in recent years and have turned many industries upside down.

If you look at the literature on business models, you will quickly notice that there is no uniform description [28]. Nevertheless, there is a suitable method to describe the elements mentioned in the business model, the so called business model canvas (BMC). As shown in Fig. 7.6, the BMC divides the business model into nine basic building blocks [29]. This subdivision enables a company to analyse its current or future business model and collect new ideas in order to further develop the various segments. To create a definition for an innovative business model, it was first necessary to understand how a business model is structured. With this knowledge a definition for an innovative business model can be derived from the literature.

A business model innovation is a change to an existing business model or the creation of a new business model with the aim of meeting previously unsatisfied, new or hidden market needs [30]. When creating a new business model, care should be taken to ensure that the innovation is customer-oriented, future-proof and creates a fundamental change in the existing industry to ensure marketability [31]. The innovation can affect one or more elements of a business model [32]. If we consider the BMC for the definition, we can summarize that an innovative business model exists when at least one of the nine elements of the BMC is innovatively changed.

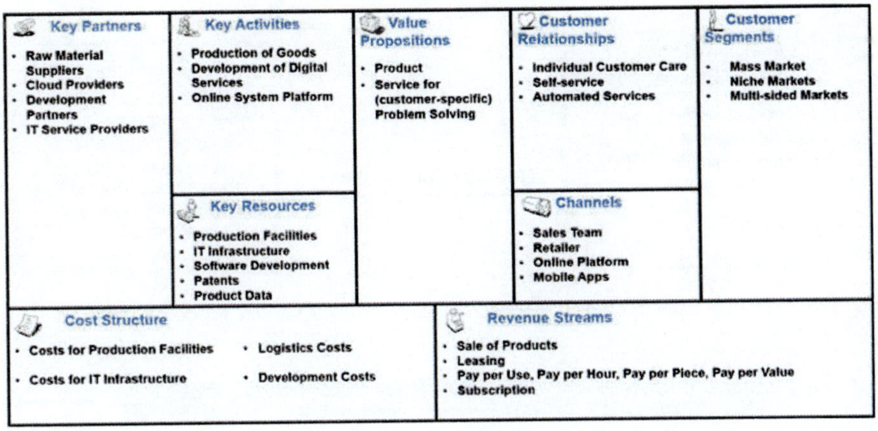

Fig. 7.6 Business model canvas (*Source* [33])

7.3.2 *Transition from Traditional to Modern Business Models*

The business model concept gained popularity during the dotcom bubble of the 1990s and was initially used to communicate complex business ideas to investors [34]. In the following years, the business model approach developed into a tool used for systemic analysis, planning and communication of the configuration and implementation of an organizational system [35]. Today, modern business models are increasingly seen as a source of outstanding organizational performance and competitive advantage [36] that either synergizes with the previous business model or completely replaces the previous strategy [37].

New business models such as pay-per-use (usage-based payment e.g.: Car2go), peer-to-peer (trade between private individuals e.g.: Airbnb) or performance-based contracting (payment for the final performance e.g.: Rolls Royce) have revolutionised entire industries [38]. Therefore, many companies have changed their model to move from pure product sales to the sale of problem solutions and services [39]. When servitization moves a manufacturer all the way to becoming a solution provider there are major changes on the business model [6]. Figure 7.7 illustrates the main aspects between traditional and modern business models.

For enabling this transition, several frameworks are described in the literature [40]. To keep this elaboration short, only the process model according to Bucherer is explained in more detail below. Figure 7.8 shows this model, which is applied for the development of a new business model on the basis of an existing model. It consists of several phases in which different activities are proposed. After each phase there is a gate which requires verification, if the planned solutions and the meaningfulness of the concepts are given. When this is fulfilled the next phase starts, otherwise there is a need to start from scratch with the previous phase. This model is to be understood as a cycle and should serve to question and optimize the business model during the entire life cycle.

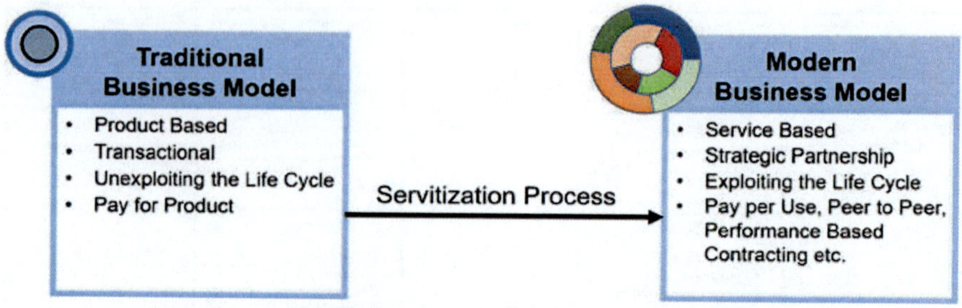

Fig. 7.7 Transition from traditional to modern business models (*Source* Modified acc. to [23])

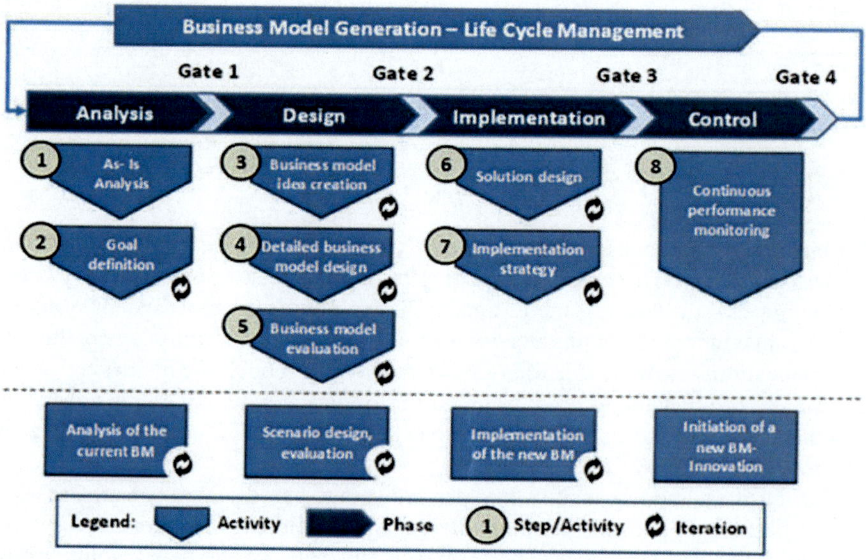

Fig. 7.8 Process model for business model innovation (*Source* Modified acc. to [41])

7.3.3 *Modern Business Models with Life Cycle Focus*

Nowadays, the simple sale of products does not create long-term relationships with customers. For this reason, the relationship between producer and customer has changed radically in recent years [42]. Manufacturers are trying to innovate their product range by offering beside products also services throughout the entire life cycle [43]. By offering services such as maintenance, upgrades and remanufacturing a company can extend the lifetime of their products and increase product sales accordingly [44]. For example, in the case of Electrolux A.B, which performed a life cycle analysis of a servitized floor-cleaning machine. The analysis shows that

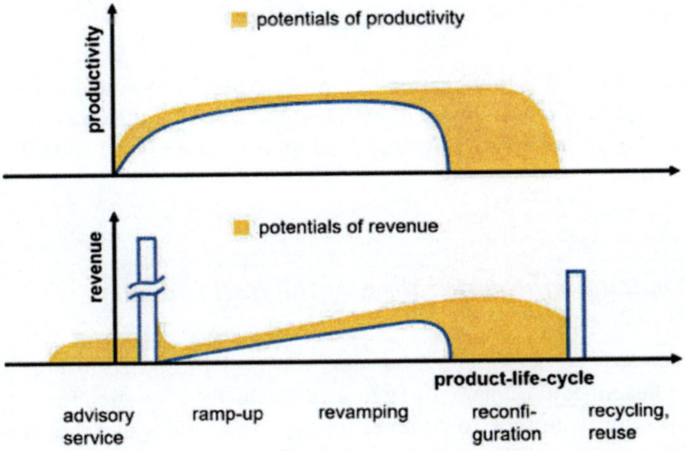

Fig. 7.9 Life cycle based economic potentials (*Source* Modified acc. to [47])

life cycle services reduced in-use (environmental) impacts as well as material and energy consumption through life extension, part re-use and recycling [45].

Additionally, services such as maintenance, repair, reconditioning and technological upgrading result in an extension of the product life cycle and therefore the overall ecological impact of product use [46].

The advantages of such modern business models are not only on the customer side. By offering services over the entire product life cycle, the distance between customer and supplier is not only reduced, but also the chance of getting to recognize and react to changes in customer requirements is increased. Furthermore, the life cycle of production means offers a range of opportunities for innovative services that not only increase productivity, availability and efficiency for the customer, but can also increase the potential revenues of the supplier [47]. Figure 7.9 visualizes the aforementioned advantages on the side of the customer and the supplier. The areas marked in yellow show the potential productivity increases on the part of the customer and the potential revenue increases on the part of the supplier in comparison to pure product sales (blue areas) without extended services. As you can observe, not only life cycle extensions are possible, but also revenues even before selling a product by offering pre-sale-services.

Because of the advantages described above, many companies in the capital goods market focus on creating various individual customer-based services that are tailored to the life cycle of the respective products [48].

7.4 Applications and Practical Examples

This chapter describes an application of a modern service-oriented business model within the rail industry. In addition, it provides practical examples of companies from different industries, which have already used servitization to innovate their business model.

7.4.1 *Application: Servitization of the Rail Industry*

A typical modern service-oriented business model is illustrated in Fig. 7.10. This application describes the interactions in a rail industry example. The customer is the train operator, who seek to contract for the use of trains and their availability, instead of the conventional product purchase. The manufacturer would contract to provide rolling stock, trains, maintenance and all support activities which are necessary to ensure the reliability and availability. The customer, the manufacturer and their financial and service partners have through-life interactions with each other and the product. These interactions are illustrated by the arrows in Fig. 7.10.

The customer is focused on the use of the capability to transport passengers. He has to provide some resources (1), such as train drivers and on-board staff to use the capability.

By delivering a schedule for the train journey (2), the train operator is able to receive the benefits from transportation of rail passengers (3). This benefit is translated to financial incomes, such as the sales of tickets and refreshment services

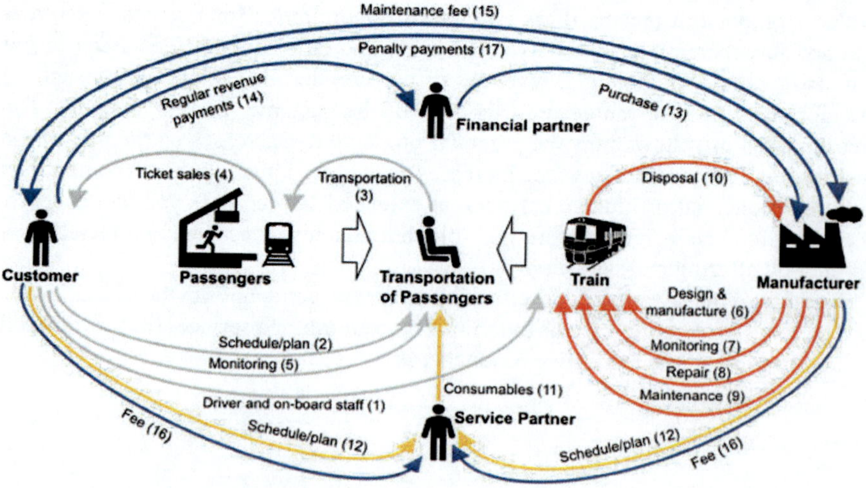

Fig. 7.10 Interaction in a modern service-oriented business model (*Source* [5])

during the transportation (4). Monitoring helps the customer to identify how well the capability to transport passengers is being provided (5), using measures that are directly related to his own business model. For example, the extent of train delays that passenger's experience. The manufacturer provides the design, manufacture and installation of the train (6). Besides that, he is responsible for the train availability, the monitoring of the train performance (7), repair (8), maintenance (9), and disposal (10) of the equipment, during the life-time of the contract. Both the manufacturer and the customer engage partners to provide support services like consumables such as catering or fuel (11). These are initiated by plans or schedules (12) from the customer or manufacturer, depending on their contractual agreements. The funds that enable the product purchase (13) can be provided through a financial partner, which returns therefore regular revenue payments, dependent to the usage of the capability by the customer (14). A similar fee is paid from the customer side for the train maintenance (15) to the manufacturer. Both the customer and the manufacturer have to make payments, for any additional services, that have been consumed, to the service partners. Should the train fail to perform, that the capability is not available to the train operator, then the manufacturer will be liable for the financial penalties (17) [5].

This application outlines that there are more revenue streams for the manufacturer, besides selling the product, because of increased through-life interaction within a modern service-oriented business model. Additionally, value co-production requires that the customer plays an active role in developing the service offering to co-create value and drawing upon different resources to attain desired outcomes [49].

7.4.2 *Practical Examples: Service-Oriented Business Models*

Table 7.2 shows various trends in modern service-oriented business models and the associated practical implementations by various companies.

The practical examples above, point out that many companies from different industries have already innovated their business models in recent years. The examples represent only a fraction of the companies that have already servitized their business. This shows the high relevance and topicality of this subject [52].

7.5 Conclusion

The importance of the service sector and the corresponding modern business models has steadily increased in recent years. Digitalisation and changed customer behaviour contribute to this. These trends mean that companies must question their current business model and offer innovative products, services and modern business models in order to be able to compete in the market. In doing so, companies have to consider the life cycle and offer individual services tailored to the customer for each phase.

This not only increases the benefits for the customer and extends the life cycles of the products, but also establishes the relationship with the customer on a strategic level. With the increasing relevance of services and modern business models, the latest literature already contains methods for developing innovative business models and innovating existing business models. These methods are mostly built up cyclically and should be continued by companies even after the introduction of innovations.

Table 7.2 Practical examples for service-oriented business models

Industry	Trend	Example
Aerospace and Defence	On the way to performance-oriented service and logistics concepts with customers, including guaranteed availability and reliability of devices, modules and entire platforms over long periods of time	Rolls-Royce has long been focused on selling "Power by The Hour." Customers can pay a fixed warranty and operational fee for the hours that their engines are running, so Rolls-Royce must focus on the entire package, from products, installation, after-sales maintenance, repair and overhaul, to overall service and parts management to ensure profitable long-term growth
Automotive, Commercial Vehicles	Excellence in service and parts management is the key to the brand image and overall business models of some companies	Hyundai and Kia vehicles are sold with warranties of up to 10 years/100,000 miles, and the service and parts operation must function at the highest level of efficiency to avoid customer service problems and excessive warranty costs, and to sustain profitable growth
Diversified Manufacturing and Industrial Products	Customers view provider service offerings as an integral part of the business	ABB provides its industrial automation customers "performance service" tailored to their needs. The offerings range from simple, product-focused maintenance and field services to "Automation Performance Management" where ABB guarantees the performance levels and assumes the risks of the customer's equipment over its life cycle whether or not it was originally made by ABB

(continued)

Table 7.2 (continued)

Industry	Trend	Example
High Tech and Telecom Equipment	Downtime can cost $100,000 or more per hour. As the industry continues to consolidate, many customers today have a cumbersome mix of installed hardware and software that often need to be serviced for decades	Printer makers, such as Hewlett-Packard and Xerox, have derived more revenue and profit in selling ink and after-sales services than in initial printer sales for quite some time. ABB provides a remote monitoring system for the gas industry for real-time optimization of control systems, raising the overall efficiency and productivity of processes

Source [50, 51]

The trend is moving more and more in the direction of offering complete solution bundles instead of a single physical product. With business models such as performance-based contacting, it even goes so far that a company no longer owns the machines or other operating equipment, but only pays for the availability of a system. This raises the question of whether anyone will be able to buy products as such in the future, or whether they will only pay for the services as solutions of the actual problems, without transferring the ownership.

References

1. Vandermerwe, S., Rada, J.: Servitization of business: adding value by adding services. Euro. Manage. J. (1998)
2. Niemann, J.: Die Services-Manufaktur, Industrielle Services planen –entwickeln – einführen. Shaker Verlag, Ein Praxishandbuch Schritt für Schritt mit Übungen und Lösungen. Aachen (2016)
3. Vladimirova, D.: Made to Serve. How manufacturers can compete through servitization and product-service systems. In: Production Planning and Control (2015)
4. Adrodegari, F., Alghisi, A., Ardolino, M., Saccani, N.: From Ownership to Service-oriented Business Models: A Survey in Capital Goods Companies and a PSS Typology. Procedia CIRP (2015)
5. Al-Debei, M., El-Haddadeh, R., Avison, D.: Defining the business model in the new world of digital business. In: Proceedings of the Americas Conference on Information Systems (AMCIS) (2008)
6. Karlsson, C.: Servitization. Extended Business Model for more Revenue and Profit. Online: https://www.effektivitet.dk/magasin/nr-2-2018-servitization/servitization-extended-business-model-for-more-revenue-and-profit.aspx. Checked 19 Dec 2018
7. Hentschel, B.: Dienstleistungsqualität aus Kundensicht: vom merkmals- zum ereignisorientierten Ansatz. DUV, Wiesbaden (1992)

8. IFS Deutschland GmbH & Co. KG.: Viele Fertigungsunternehmen nutzen die Chancen service-orientierter Geschäftsmodelle noch nicht. Online: https://www.ifsworld.com/de/news/presse/2017/11/30/studie-servitization/. Checked 19 Dec 2018
9. Bruhn, M., Hadwich, K.: Servicetransformation – Eine Einführung in die theoretischen und praktischenProblemstellungen. In: Bruhn, M., Hadwich, K. (Eds.) Servicetransformation. Springer Fachmedien Wiesbaden, Wiesbaden (2016)
10. Dinges, V., Urmetzer, F., Martinez, V., Zaki, M., Neely, A.: The Future of Servitization: Technologies that will Make a Difference. University of Cambridge, Cambridge (2015)
11. Aarikka-Stenroos, L., Jaakkola, E.: Value co-creation in knowledge intensive business services: a dyadic perspective on the joint problem solving process. In: Industrial Marketing Management (2012)
12. Kowalkowski, C., Gebauer, H., Kamp, B., Parry, G.: Servitization and deservitization: overview, concepts, and definitions. In: Industrial Marketing Management (2017)
13. Zhang, W., Banerji, S.: Challenges of Servitization: A Systematic Literature Review. Industrial Marketing Management (2017)
14. The Manufacturer: The Importance of Servitization in Manufacturing. Online: https://www.themanufacturer.com/articles/the-importance-of-servitization-in-manufacturing/. Checked 19 Dec 2018
15. Niemann, J., Tichkiewitch, S.: Westkämper Engelbert: Design of Sustainable Product Life Cycles. Springer Verlag, Heidelberg Berlin (2009)
16. Morar, L.; Westkämper, E., Abrudan, I., Pisla, A., Niemann, J., Manole, I.: Planning and Operation of Production Systems, Fraunhofer IRB Verlag (2008)
17. Vendrell-Herrero, F., Bustinza, O. F., Parry, G., Georgantzis, N.: Servitization, Digitization and Supply Chain Interdependency. Industrial Marketing Management (2007)
18. Lambert, S. C., Davidson, R. A.: Applications of the business model in studies of enterprise success, innovation and classification: an analysis of empirical research from 1996 to 2010. Euro. Manage. J. (2013)
19. Osterwalder, A., Pigneur, Y., Bernarda, G., Smith, A., Papadakos, P.: Value Proposition Design: How to Create Products and Services Customers Want. Get Started With. Wiley, Hoboken, NJ (2014)
20. Verstrepen, S., Deschoolmeester, D., Berg, R. J.: Servitization in the automotive sector: creating value and competitive advantage through service after sales. In: Mertins, K., Krause, O., Schallock, B. (Eds.) Global Production Management. Springer US, Boston, MA (1999)
21. Oliva, R., Kallenberg, R.: Managing the transition from products to services. Int. J. Serv. Ind. Manage. (2003)
22. Thoben, K., Jagdev, H., Eschenbächer, J.: Extended Products: Evolving traditionals product concepts. In: Thoben, K. D., Weber, F.; Pawar, K. (Eds.) Proceedings of the 7th International Conference on Concurrent Enterprising: Engineering the Knowledge Economy Through Co-operation. Bremen, Germany, June 2001
23. Ducq, Y., Chen, D., Alix, T: Principles of servitization and definition of an architecture for model driven service system engineering. In: van Sinderen, M., Johnson, P., Xu, X., Doumeingts, G. (Eds.) Enterprise Interoperability. Springer, Berlin, Heidelberg (2012)
24. Nagl, A., Bozem, K.: Geschäftsmodelle 4.0. Springer Gabler, Wiesbaden (2018)
25. Ismail, S., Malone, M. S., Geest, Y. van, Diamandis, P. H.: Exponentielle Organisationen: das Konstruktionsprinzip für die Transformation von Unternehmen im Informationszeitalter. (M. Kauschke, Trans.). Verlag Franz Vahlen, München (2017)
26. Eckert, R.: Business Innovation Management: Geschäftsmodellinnovationen und multidimensionale Innovationen im digitalen Hyperwettbewerb. Springer Gabler, Wiesbaden (2017)
27. Zollenkop, M.: Geschäftsmodellinnovation: Initiierung eines systematischen Innovationsmanagements für Geschäftsmodelle auf Basis lebenszyklusorientierter Frühaufklärung, 1st edn. Deutscher Universitäts-Verlag, Wiesbaden (2006)
28. Übelhör, J.: Industrieunternehmen und die Transformation von Geschäftsmodellen im Kontext der Digitalisierung – Eine empirische Studie über die Auswirkungen anhand des Business Model Canvas. HMD Praxis der Wirtschaftsinformatik (2018)

29. Mast, C.: Neuerfindung einer Industrie: Evolution von Organisationen und Märkten durch die Innovation des Geschäftsmodells. Springer Gabler, Wiesbaden (2017)
30. Schallmo, D. R. A., Brecht, L.: An innovative business model: the sustainability provider. In: XXII ISPIM Conference: Sustainability in Innovation: Innovation Management Challenges (2011)
31. Casadesus-Masanell, R., Ricart, J. E.: From Strategy to Business Models and onto Tactics. Long Range Planning (2010)
32. Mitchell, A., Bruckner, C., Coles, T.: Establishing a continuing business model innovation process. J. Bus. Strat. (2004)
33. Osterwalder, A., Pigneur, Y., Clark, T.: Business Model Generation: A Handbook for Visionaries, Game Changers, and Challengers. Wiley, Hoboken, NJ (2010)
34. Zott, C., Amit, R., Massa, L.: The business model: recent developments and future research. J. Manage. (2011)
35. Doleski, O.D.: Integrated Business Model: Applying the St. Gallen Management Concept to Business Models. Springer Gabler, Wiesbaden (2015)
36. Chesbrough, H.: Business Model Innovation: It's not Just About Technology Anymore. Strategy and Leadership (2007)
37. Baines, T., Lightfoot, H.: Made to Serve: How Manufacturers Can Compete Through Servitization and Product Service Systems. Wiley, Chichester, West Sussex, United Kingdom (2013)
38. Lorenz, H.: Innovative Geschäftsmodelle von Pay per Use bis Performance-Based Contracting, 2015. Online: https://das-unternehmerhandbuch.de/innovative-geschaeftsmodelle-von-pay-per-use-bis-performance-based-contracting/#Beispiele_bekannter_Geschaeftsmodelle. Checked 17 Dec 2018
39. Stott, R. N., Stone, M., Fae, J.: Business models in the business-to-business and business-to-consumer worlds—what can each world learn from the other? J. Bus. Ind. Market. (2016)
40. Geissdoerfer, M., Savaget, P., Evans, S.: The Cambridge business model innovation process. Procedia Manuf. (2017)
41. Bucherer, E.: Business Model Innovation—Guidelines for a Structured Approach. Shaker-Verlag, Aachen (2010)
42. Cohen, M. A., Whang, S.: Competing in product and service: a product life cycle model. Manage. Sci. (1997)
43. Kastalli, I. V., Van Looy, B.: Servitization: disentangling the impact of service business model innovation on manufacturing firm performance. J. Oper. Manage. (2013)
44. Mont, O.: Product-Service Systems: Panacea or Myth? IIIEE, Lund (2004)
45. White, A. L., Stoughton, M., Feng, L.: The Quiet Transition to Extended Product Responsibility. U.S. Environmental Protection Agency Office of Solid Waste (1999)
46. Brouillat, E.: Recycling and extending product-life: an evolutionary modelling. J. Evol. Econ. (2009)
47. Meier, H., Massberg, W.: Life cycle-based service design for innovative business models. CIRP Annal (2004)
48. Bhandiwad, S., Sundarajulu, V., Hariharan, S., Shelar, A.: Servitization in Manufacturing—The Final Frontier (White Paper). Tata Consultancy Services Ltd., Mumbai (2016)
49. Green, M. H., Davies, P., Ng, I. C. L.: Two strands of servitization: a thematic analysis of traditional and customer co-created servitization and future research directions. Int. J. Prod. Econ. (2017)
50. Glueck, J.J., Koudal, P., Vaessen, W.: The Service Revolution: Manufacturing's missing Crown Jewel (Deloitte Review). Deloitte Development LLC, Zürich (2007)
51. Berry, B.: An eye on the field: remote natural gas measurement and automation systems. In: ABB Review Special Report Instrumentation and Analytics (2006)
52. Niemann, J.: Life Cycle Management-das Paradigma der ganzheitlichen Produktlebenslaufbetrachtung. In: Spath, D. et al. (Hrsg.) Neue Entwicklungen in der Unternehmensorganisation. Springer-Vieweg, VDI Buch, Berlin (2017)

Chapter 8
Big Data

Current estimations expect an annual volume of 40 exabytes of electronic data worldwide in 2020 [1]. This is mainly due to the Internet (social networks, IoT, etc.), which increases besides the data volume also the data complexity [2]. The various sources (RFID data, sensors, etc.) will also include different data formats, [3] which must be taken into account from the beginning when implementing Big Data projects. In the context of the complexity of the production environment, large amounts of data are collected from different sources (in particular from sensors) [4]. The data is contemplated from a horizontal, a vertical and an end-to-end integrated digital engineering perspective. In general, two main aspects can be associated with Big Data in the framework of Industry 4.0 and production environments:

(1) Existence of data with a high volume and complexity
(2) New technologies which efficiently proceed/analyze a high volume of data [5]

Given these two aspects, there is ample space for a definition of Big Data. In conformity with the elaborated aspects and the scope of Industry 4.0, the following definition of Big Data is determined. This self-stated definition will be subsequently the reference point for further investigations regarding Big Data and its surrounding areas:

Big Data describes the economical asset of electronical information, which is present in a high volume and a high complexity and needs new technologies to be efficiently proceeded and analyzed in order to optimally adjust internal and external business activities.

From a company's perspective, data is especially an economic asset [6]. Hence, the pro-ceeding and analyzing of Big Data is a competitive factor which delivers access to new business opportunities [7]. One example is the technology data center of the company 'Trumpf', where machines are supplied with data through access to internal and external technology know-how, thus ensuring that orders can be produced in time and in high quality [8]. That aspect is also consistent with the holistic approach of Industry 4.0 as it includes the market, which reflects the far end of the value chain.

J. Niemann and A. Pisla, *Life-Cycle Management of Machines and Mechanisms*, Mechanisms and Machine Science 90,
https://doi.org/10.1007/978-3-030-56449-0_8

8.1 Big Data Analytics

The term "analytics" was already introduced in the 1940s in connection with the decoding of German messages during the Second World War [9]. Over time, Analytics encompassed more and more approaches such as Enterprise Resource Planning (1980s), Dashboards (1990s) and Cloud Computing (2000s). Big Data Analytics [10, 11] emerged in the 2010s and is one of the latest approaches [12]. It is called this way because it focuses on the challenge of processing and analyzing Big Data using new and advanced technologies. Big data analytics can therefore be seen as the next level of Business Intelligence [5]. Analytics not only includes the act of analysis, it also includes additional complementing tasks and processes [12]. Although the terms Business Intelligence and Data mining are often used synonymously for Analytics it is necessary to distinguish between them [12, 13]. Accordingly, this is performed in the following chapters, with respect to the available data basis, the knowledge that can be gained from it and the associated use of technologies.

8.2 Business Intelligence

Business Intelligence focuses primarily on structured information such as ERP-data. Whereas, Big Data Analytics focuses on handling a high volume of semi unstructured data, sourced internally and externally along the value chain [14]. Typical business intelligence activities include the creation of reports, the definition of key performance indicators (KPIs) and the compilation of dashboards. They are carried out as necessary or on request. However, these activities primarily provide information about historical events [14]. In contrast, Big Data Analytics provides insights to anticipate problems or to solve them immediately [13]. Following the understanding of Big Data Analytics, it is necessary to make optimal decisions, while at the same time the requirements for the procedure and analysis of data are becoming more and more difficult to fulfill. Thus, this is no longer possible by means of conventional approaches [5]. Therefore, powerful tools, new methodologies and human expertise are required [13]. These include, for instance, Data mining techniques [10].

8.3 Data Mining

The processing of Big Data as part of Analytics is usually performed by means of data mining. The relationships between the different data sets are analyzed with the help of certain methods [15]. The methods applied depend on the overall problem, which must be clarified beforehand [5]. Mathematical operations, such as statistics, can then be made use of. For instance, a regression model can be used to estimate a certain development [11]. This can then lead to the identification of patterns, dependencies, trends and outliers [4, 5]. Data mining is therefore the key enabler to gain insights from large data sets and to ensure the achievement of objectives [12, 16]. In addition, this trend is even steadily increasing due to the aforementioned effects of the Internet. Standardized processes, such as Cross-Industry Standard Process for Data Mining (CRISP-DM), have been developed for that purpose. The different processes have different scopes and therefore different pros and cons [12]. However, the descriptions of the many processes are not given at this point as it is beyond the scope of this work.

8.4 Data Characteristics

Big data analytics not only facilitates many issues, it comes also with many challenges, such as the capturing and processing of data. As a matter of fact, that exceeds the obvious circumstances of data volume and complexity. Figure 8.1 illustrates the general characteristics of big data.

Since big data analysis is a major challenge for existing IT-infrastructures, it is important to understand the potential data sets [11]. There are three main terms used in literature to distinguish between conventional data and big data. Volume, velocity and variety are mainly mentioned and are known as the 3V's [18]. Additionally, two more terms are frequently mentioned. These are the characteristics veracity and value, which then establish the 5V's [16, 19].

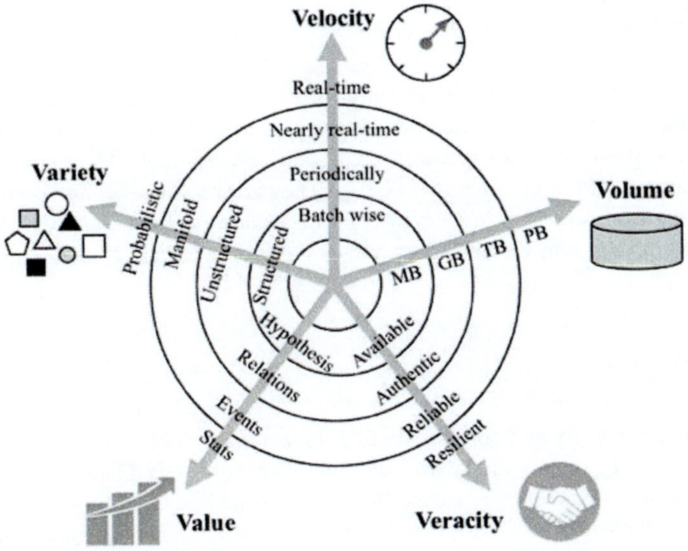

Fig. 8.1 Big data characteristics [5, 17]

8.4.1 Volume

The volume describes the quantity of data that needs to be processed and analyzed. The reason for the constantly growing volume is mainly the Internet and the associted networking of objects (Chap. 2). Industrial objects, such as workpieces, tools, machines and processes, supported by sensors generate an immense amount of data [20]. Smart factories will intensify this effect in the future as more and more data is needed to make processes more efficient [16].

8.4.2 Velocity

Since more data is required in order to optimally design the processes in Smart Fatories, more insights are required in a shorter time. As described in chapter 2.7, the basis for CPPSs is a real-time knowledge of the mutual conditions of all objects. This leads to the difficulty of collecting, processing and analyzing data much faster than in conventional production environments. In particular, the irritative and powerful ability to analyze data with different models and algorithms is a key feature of Big Data. All this makes it possible to gain ubiquitous insights in short cycles, which corresponds to the CPPS demand of real-time decision-making in the broader sense [21].

8.4.3 Variety

The characteristic variety refers not only to the different data sources, but also to different data formats. This includes structured, semistructured and unstructured data [11]. In particular, unstructured data such as user data, documents, e-mails, images, etc. are difficult to process [11, 18]. They are very diverse and do not have a uniform structure, which requires new approaches to process and analyze them. The combination of these three data types makes it possible to gain holistic insights [5].

8.4.4 Veracity

As the volume, variety and complexity of data increases, it becomes more and more difficult to ensure their validity and quality. Therefore, Big Data Analytics must evaluate the data before it can be used for decision-making. Credibility is ensured by reliable data cleansing processes that detect errors, inaccuracies and anomalies [22].

8.4.5 *Value*

How much knowledge can be gained from Big Data depends above all on the data itself. Even if the data can be accessed and the necessary IT-infrastructure is available to further process and analyze the data, it is still a waste of time if the data has no value for the company. Concrete application scenarios enable the evaluation of cost–benefit relations, which are essential for Big Data projects [10].

8.5 Requirements for Data Processing in Industrie 4.0

By understanding the requirements for handling data-driven production environments (i.e. Smart factories), the foundation will be established to realize Big Data Analytics projects. Based on the following two questions, this chapter discusses the generic requirements to process data within the production-based framework of Industry 4.0:

(1) What requirements entail data for Industry 4.0?
(2) What requirements need to be fulfilled to process data industry 4.0? [5]

Resulting from this, three categories for data requirements and four categories for data processing, each with several sub-categories, can be determined. However, it must be stated that there are more categories in literature, especially dealing with the aspects of decision-making requirements related to autonomy, decentralization and overall optimum within smart factories.

8.5.1 *Data Model*

8.5.1.1 Unified Knowlede Management

One of the objectives of knowledge management is to give the several entities (i.e. systems, CPPSs and Augmented Operators) within the production environment a unified access to the available data [23]. Semantics for information technology in Smart factories focuses thereby on the unified description of data in order to interpret its meaning and purpose in a standardized manner [24]. In order to get an overview of the available (and unified through semantics) data, it is necessary to portray it. That can be reached through ontology, which establishes terms and describes relationships between information (e.g. the Entity Relationship Model).[25–28]. An additional requirement is the transformation of data. Thereby, it must be converted into standardized units [29–33]. When collecting sensor data of a punching machine, for instance, it is needed to process the recorded data later on (e.g. for analytical purposes) in a standardized and consistent format, such as the SI-unit 'kN'.

8.5.1.2 Unified Interfaces

Unified interfaces are important for vertical integration within Smart factories and horizontal alignment between different smart factories [24, 34, 35]. It is the prerequisite for smooth communication between the elements of smart factories such as Smart Objects and IT-systems. Thus, it requires universal and independent standards and protocols [24, 29, 36]. In order to enable communication between different objects and systems, it is also necessary to enable access to uniform interfaces via communication channels [37–39].

8.5.2 Data Content

8.5.2.1 Autorisations

The authorization of objects is required to identify smart objects and assign them roles with specific permissions and access rights within the smart factories [40]. This facilitates data security and prevents unauthorized access and manipulation. Procedures such as encryption are used for this purpose [41]. Based on the assigned roles and rights, data can be provided accordingly [34].

8.5.2.2 Specification of Objects

The specification of structures, attributes and functionalities of smart objects in CPPSs is necessary because they play an active role within Smart factories, Smart Products in particular act as information carriers (e.g. using RFID-tags). However, this requires information about the structure of Smart Products (e.g. bill of materials) [35] and attributes such as geometry data (e.g. CAD model), quality data (e.g. tolerances), material data (e.g. material strength) and functional data (e.g. performance data) [29, 42]. The reason for this is that the data is required for communication with other smart objects [39, 43].

There also other smart objects like machines, tools, and means of transport which have to be described. The scope includes structures, attributes and functionalities such as process flow, geometric constraints, manufacturing technology [5]. Essentially everything that is necessary to create a uniform database for big data analytics.

8.5.2.3 Production and Sensor Data

The different sensors in smart factories collect huge amounts of data. Among other things, this includes the processing of data from smart objects (e.g. the impinging force of a stamping machine on a component). The sensor data can then be used, for instance, to monitor the quality parameters of a process and, if necessary, to adjust them using actuators [44]. In addition, production data such as planned process steps, assigned equipment, cycle times, parts lists and measurement criteria are required to create a holistic data foundation [42, 45].

8.5.2.4 Condition Data

Condition data is the result of processing sensor data [46]. Historical sensor data is used to understand processes and environmental influences. Based on this, correlations can be derived and rules for decisions can be established based on models [47, 48]. The value of information is based on the selected analytical approach. Condition data contains current information about orders, entities, processes and much more. For instance, ideal smart products must know their processing status (e.g. fulfillment of quality parameters) in order to decide on the next process step. Status data from smart machines, for instance, enables autonomous monitoring, diagnosis and reaction [49].

8.5.2.5 Knowledge

Knowledge is especially important because it is the key to understanding contexts. This includes dependencies between the data categories described above, such as specifications and condition data. This data can be based on historical and real-time sources. To be of benefit, the data must be prepared and made accessible [48]. To gain full insight, it also requires the knowledge of experts who interpret the data in an advanced and contextualized way.

8.5.3 Data Integration

8.5.3.1 Vertical Integration

In order to react to changes in the various hierarchy levels of the classical automation pyramid, a vertical integration of the data is required. The data on the levels are usually not static and therefore are subject to changes. In order to react to these changes at a certain level, it is necessary to process the new data at the next level. Accordingly, the data must be consistent and granular [34, 36, 49]. This ensures accuracy when processing data at the next level. An example of this is a quality problem at the shop floor level that slows down production and thus influences the delivery quantity (i.e. delivery residues). To adjust the communicated delivery quantity automatically and adequately, it is necessary that all data is available (i.e. consistency) in a suitable way (i.e. granularity) at the ERP-level. The same applies vice versa if the information flow progresses from a high level to a lower level. The information model describes the integration and synchronization process of the data of the different levels and shows how data can be accessed [25, 45, 50].

8.5.3.2 End to End Digital Integration of Engineering

The end-to-end digital integration of engineering concerns the requirements for data integration during the life cycle of cyber physical production systems CPPSs. In particular, the information from machines and other entities is important for development, production and service processes. This requires the reproduction of the physical world into the cyber world. Ideally, this includes information such as mechanics, electronics, software and much more. The integration of vertical and horizontal data must also be taken into account [46, 51]. Smart objects used in CPPSs must be described throughout their life cycle in order to obtain consistent data [25]. This results in three major life cycle phases: Development, manufacturing and use [52]. For instance, in the development phase of a product, all structures, attributes and functions must be described. This allows the information to be reproduced in the cyber world and machines and other processes (e.g. knowledge-based decision making processes) can access the data as needed. This is just one of many examples and can also be used in the manufacturing and utilization phase of an object.

8.5.3.3 Horizontal Integration

The requirements for the horizontal integration of data describe the integration along the value chain. These include internal processes (internal logistics, production, quality management, etc.) and cross-company processes (supplier production, external logistics, after-sales services, etc.) [24]. Therefore, a comprehensive coordination of the activities of all participants along the supply chain is necessary [51, 53]. Considering the above-mentioned problem with regard to delivery quantities, customers could be informed quickly and even autonomously in order to reduce influences on the supply chain. Another important point worth mentioning is the access to relevant data by all parties involved in the supply chain.

8.5.4 Decision Making Process

8.5.4.1 Condition Monitoring

As previously described, condition data results from the intelligent processing of sensor data. CPPSs require Smart Objects to perceive their relevant environment. This allows state data to be used to monitor Smart factories in real-time. Based on this, Smart Factories processes can be assessed, planned and controlled. Examples are the approval of production orders and the planning of maintenance work [36, 54].

8.5.4.2 Real Time Reaction

Due to the complexity of Smart factories and their rapidly changing surrounding conditions, CPPSs need to know about the conditions of their physical components and environment in order to adapt. This is made possible by monitoring conditions, which immediate reactions to current situations [55]. An example of this is the ad-hoc reconfiguration of a machine to ensure a sufficient quality level of products.

8.5.4.3 Model Utilisation

The use of analytical models is necessary to evaluate opportunities. The models contain real information about the individual processes and originate from historical and real-time data as well as expert knowledge [38, 47]. The models must be constantly adapted and evaluated as they are subject to constant changes in the production environment [29]. Such changes can be an increase in production volume or the installation of new equipment. Some approaches in literature therefore describe the deployment of automated models [53].

8.5.5 Knowlede Processing

8.5.5.1 Model Creation and Adaption

In order to use models, they must first be developed. Models need information from process data as well as expert knowledge. By means of Big Data Analytics, insights can be gained. The resulting patterns can be used to establish rules that identify process deviations, dependencies between parameters, and the root causes of problems [25, 34, 56]. Predictive models, for instance, can be used to anticipate incidents [56]. This allows machine failures to be predicted [57] and maintenance work to be proactively planned [49]. In addition, negative trends in process and product parameters can be identified, such as the steadily decreasing durability of a material or product. The rule is, the more data and the higher its value, the more beneficial it is. When developing models, it is important to ensure that they can adapt to the specific conditions of the real world, such as ongoing changes in the production environment [58]. For this reason, the more topical the data, the more accurate the model. But this also means that it becomes more difficult to adapt the model automatically, which is due to the necessary and immediate processing of large amounts of data.

8.5.5.2 Expert Knowlede Processing

As already mentioned, efficient models also require the use of expert knowledge. This is necessary to interpret trends, situations and conditions of processes, machines and products [52, 59]. This is also necessary to define and select situational measures [26]. Due to the many different but related disciplines within Smart factories such as quality management and maintenance, it is all the more important to combine different sources of expert knowledge [5]. In order to benefit from this knowledge, formalization known as knowledge engineering is required [60].

8.5.6 Real Time Processing

8.5.6.1 Data Access

Smooth access to data is essential for the efficient use of resources within Smart factories. Process parameters and conditions can be analyzed immediately. This applies in particular to the horizontal integration of Big Data Analytics between several Smart factories [61]. In order to identify all relevant objects within Smart factories, the objects must be clearly identifiable and able to communicate at any time [46, 52]. It also requires that coherent information from different sources, such as the process flow and position of a product, is transmitted simultaneously [62].

8.5.6.2 Connectivity and Communication

The real-time processing of data for communication is determined by the response time required by Smart Objects and CPPSs [5]. The Association of German Engineers (VDI) states that a response within milliseconds is required to fulfill real-time communication [52]. The real-time communication between Smart Objects and CPPSs of Smart factories is especially important to make fast and target-oriented decisions based on holistic data foundations.

8.5.7 Security

8.5.7.1 Network Security

In particular, the horizontal integration of Big Data Analytics across company boundaries requires solid network security. Therefore, protective measures are required that also guarantee data security. These focus on the identification, authentication and authorization of objects [29, 40].

8.5.7.2 Data Security

Data is a form of intellectual property. It is therefore a valuable asset and competitive advantage for a company and must be secured. The manipulation of data could be a catastrophe and therefore two security aspects are important. First, the mechanisms for encrypting and encapsulating the data [40, 41]. Secondly, the assignment of roles for users and Smart Objects with different permissions and consequential rights. When data is accessed, these mechanisms are constantly applied, resulting in greater data transfer and processing complexity [5].

8.6 Classification of Big Data Analytics Maturity

Big data analytics can be used in a wide range of areas. Hence, there are many methodologies to structure the procedure. Nevertheless, there are many similarities regarding the classification of value and complexity [12, 13]. The various attempts to classify the maturity of big data analytics focus on the value to the business created by big data analytics and the complexity of the analytics procedure. This is then reconciled with the time perspective of the data. The following principle applies: the more one focuses on topicality, the more difficult it gets. The most frequently mentioned classifications are descriptive analytics, predictive analytics and prescriptive analytics [12, 13]. Additionally, some authors mention diagnostic analytics [63]. The different methodologies will be explained in the following chapters.

8.6.1 Descriptive Analytics

Descriptive analytics is the initial method of big data analytics and shall answer the question: *What happened or what is happening?* [12, 13]· It is a simplified approach by which operational processes are represented in reports, dashboards, or other data viewing tools. The features are similar to business intelligence and therefore both terms are used synonymously [12].The overall goal is to present structurally underlying and mostly historical information without profound analyses.

8.6.2 Diagnostic Analytics

Coherences between influence factors and occurrences are analyzed by diagnostic analytics. It shall answer the question: *Why did it happen?* [63]. Methods such as visualization and data mining techniques are used to identify the effects of the influence factors [63]. The derived knowledge can be used to learn from previous incidents and to anticipate issues through the given coherent insights. However, predictions are not given through the analytics procedure itself, because diagnostic analytics merely shows the correlations and gives no particular prediction.

8.6.3 Predictive Analytics

Predictive Analytics is used to anticipate future events based on data. Predictions can be made through answering the question: *What will happen?* [12]. Thereby, entire system conditions, KPIs and much more can be predicted. The difference towards diagnostic analytics is the higher complexity through future related models based on consistent assumptions. These probability forecasts of events are achieved through data mining [12, 13].

8.6.4 Prescriptive Analytics

Prescriptive analytics is the most advanced approach of analytics. The main question is: *How can it be made?* [63]. Therefore, it focuses on how processes can be designed and steered in order to generate the highest possible benefit. That will be achieved through complex Data mining models, supported by methods such as artificial intelligence, simulations, statistical operations and many more. These models take structured and unstructured data as well as internal and external data into account. The resulting outcomes show the different effects of actions and hence also the best solutions for the processes. Based on predictive analytics, prescriptive analytics provides also recommendations for actions [13].

References

1. Statista (no date) Wertschöpfungsanteil der Automobilzulieferer am weltweiten Automobilbau in den Jahren 1985 bis 2015. Available at: https://de.statista.com/statistik/daten/studie/162996/umfrage/wertschoepfungsanteil-der-automobilzulieferer-am-automobilbau-weltweit/. Accessed 31 Jan 2019
2. BITKOM Big Data im Praxiseinsatz – Szenarien, Beispiele, Effekte. Berlin. Available at: https://www.bitkom.org/sites/default/files/pdf/noindex/Publikationen/2012/Leitfaden/Leitfaden-Big-Data-im-Praxiseinsatz-Szenarien-Beispiele-Effekte/BITKOM-LF-big-data-2012-online1.pdf. Accessed 24 Jan 2019
3. Woo, J., et al.: Developing a big data analytics platform for manufacturing systems: architecture, method, and implementation. Int J Adv Manuf Technol **99**(9), 2193–2217 (2018). https://doi.org/10.1007/s00170-018-2416-9
4. Schöning, H., Dorchain, M.: Data Mining und Analyse. In: Industrie 4.0 in Produktion, Automatisierung und Logistik. Springer Fachmedien Wiesbaden, Wiesbaden (2014), pp. 543–554. https://doi.org/10.1007/978-3-658-04682-8_27
5. Gölzer, P.: Big Data in Industrie 4.0 - Eine strukturierte Aufarbeitung von Anforderungen, Anwendungsfällen und deren Umsetzung. University of Erlangen, Nuremberg (2017)

6. Helmke, S., Uebel, M., Dangelmaier, W.: Effektives Customer Relationship Management, Effektives Customer Relationship Management (2017). https://doi.org/10.1007/978-3-8349-4176-3
7. Dapp, T., Heine, V:. Big Data–Die ungezähmte Macht, Deutsche Bank Research, 2014. Available at: hhttps://www.dbresearch.de/PROD/RPS_DE-PROD/PROD0000000000444436/Big_Data_%E2%80%93_die_ungez%C3%A4hmte_Macht.PDF. Accessed: 13 Jan 2019
8. Forschungsunion: Umsetzungsempfehlungen für das Zukunftsprojekt Industrie 4.0, BMBF (2013)
9. Alexanderson, G.L.: Dark hero of the information age: in search of Norbert Wiener, the father of cybernetics. Math Intell **28**(2), 70–73 (2006). https://doi.org/10.1007/BF02987160
10. Chen, M., Mao, S., Liu, Y.: Big data: a survey. Mobile Netw Appl **19**(2), 171–209 (2014). https://doi.org/10.1007/s11036-013-0489-0
11. Schmarzo, B.: Big Data: Understanding How Data Powers Big Business. John Wiley & Sons, Indianapolis, USA (2013.) https://doi.org/10.1017/CBO9781107415324.004
12. Dursun, D.: Real-World Data Mining: Applied Business Analytics and Decision Making. FT Press, New Jersey (2014)
13. Mortenson, M. J., Doherty, N. F., Robinson, S.: Operational research from taylorism to terabytes: a research agenda for the analytics age. Euro. J. Oper. Res. (2015). https://doi.org/10.1016/j.ejor.2014.08.029
14. Holsapple, C., Lee-Post, A., Pakath, R.: A unified foundation for business analytics. Decis. Support Syst. **64**. https://doi.org/10.1016/j.dss.2014.05.013
15. Gersch, M.: Customer Relationship Management. In: Kleinaltenkamp, M. et al. (eds.) Geschäftsbeziehungsmanagement: Konzepte, Methoden, Instrumente. Gabler Verlag, Wiesbaden, pp. 309–357 (2011). https://doi.org/10.1007/978-3-8349-6928-6_8
16. Emmrich, V. et al.: Geschäftsmodell-Innovation durch Industrie 4.0, 2015. Available at: https://www.wieselhuber.de/migrate/attachments/Geschaeftsmodell_Industrie40-Studie_Wieselhuber.pdf
17. Klein, D., Tran-Gia, P., Hartmann, M.: Big data. Informatik-Spektrum **36**(3), 319–323 (2013). https://doi.org/10.1007/s00287-013-0702-3
18. Chen, H., Chiang, R. H. L., Storey, V. C.: Business intelligence and analytics: from big data to big impact. MIS Q. **36**(4), 1165–1188 (2012). https://doi.org/10.2307/41703503
19. Gadatsch, A.: Die Möglichkeiten von Big Data voll ausschöpfen. Controlling & Management Review: Big Data - Zeitenwende für Controllerm 2016. https://doi.org/10.1007/978-3-658-13444-0_8
20. Gupta, V., Ulrich, R.: How the Internet of Things will reshape future production systems, 2017. Available at https://www.mckinsey.com/business-functions/operations/our-insights/how-the-internet-of-things-will-reshape-future-production-systems. Accessed 13 Jan 2019
21. Dinter, B. et al.: Big Data und Geschäftsmodell-Innovationen in der Praxis: 40+ Beispiele. Berlin:BITKOM, 2015. Available at: https://www.bitkom.org/sites/default/files/file/import/151229-Big-Data-und-GM-Innovationen.pdf. Accessed 13 Jan 2019
22. Hand, D.J.: Principles of Data Mining. Drug Saf. **30**(7), 621–622 (2007). https://doi.org/10.2165/00002018-200730070-00010
23. Davenport, T.H., De Long, D.W., Beers, M.C.: Successful knowledge management projects. Sloan Manage. Rev. **39**(2), 43–57 (1998)
24. Kagermann, H., Wahlster, W., Helbig, J.: Umsetzungsempfehlung für das Zukunftsprojekt Industrie 4.0. Forschungsunion Wirtschaft und Wissenschaft (2013)
25. Reinhart, G., et al.: Cyber-Physische Produktionssysteme. Produktivitäts-und Flexibilitätssteigerung durch die Vernetzung intelligenter Systeme in der Fabrik. Werkstattstechnik (WT) **103**(2), 84–89 (2013)
26. Lentes, J., Eckstein, H., Zimmermann, N.: amePLM – eine Plattform zur Informationsbereitstellung in der Produkt- und Produktionsentstehung. Werkstattstechnik (WT) **104**(3), 146–150 (2014)
27. Pal, K., Karakostas, B.: A multi agent-based service framework for supply chain management. In: International Conference on Ambient Systems, Networks and Technologies. Elsevier, pp. 53–60 (2014)

28. Niemann, J., Tichkiewitch, S.: Westkämper Engelbert: Design of Sustainable Product Life Cycles. Springer Verlag, Heidelberg Berlin (2009)
29. Klocke, F., Veselovac, D., Keitzel, G.: Cloudbasierte Informationssysteme. Werkstattstechnik (WT) **103**(2), 90–95 (2013)
30. Niemann, J., Dehmer, J.: Value chain management through cloud-based platforms. Procedia Soc. Behav. Sci. **238**, 177–181 (2018). https://doi.org/10.1016/j.sbspro.2018.03.021,April2018
31. Dehmer, J., Bolte, S., Niemann, J.: Digital Leadership 4.0. In: 2018 International Conference on Production Research—Africa, Europe, Middle East 5th International Conference on Quality and Innovation in Engineering and Management, 25–26 July 2018, Cluj-Napoca, Romania
32. Kretschmar, D., Niemann, J., Deckert, C.: Digitalisierungsindex zur prozessnahen Analyse mittelständischer Unternehmen. In: ZWF Zeitschrift für wirtschaftlichen Fabrikbetrieb April 2019, Jg. 114, Ausgabe 4. S. 213–218
33. Niemann, J., Fussenecker, C., Schlösser, M.: Measuring the impact of Life cycle management and service performance. In: International Conference on Competitive Manufacturing (COMA´16), Stellenbosch, South Africa, 27–29 Jan 2016
34. Sauer, O.: Informationstechnik für die Fabrik der Zukunft, Industrie Management **29**(1), 11–14 (2013). Available at: https://www.iosb.fraunhofer.de/servlet/is/21752/Informationstechnik_fuer_die_FabrikderZukunft_IM_2013_1.pdf?command=downloadContent&filename=Informationstechnik_fuer_die_FabrikderZukunft_IM_2013_1.pdf. Accessed 2 Feb 2019
35. Aruväli, T., Maass, W. and Otto, T.: Digital object memory based monitoring solutions in manufacturing processes. In: DAAAM International Symposium on Intelligent Manufacturing and Automation. Elsevier, pp. 449–458 (2014)
36. Franke, J., et al.: Intelligente Steuerungskonzepte für wandlungsfähige Produktionssysteme. Industrie Manage. **26**(2), 61–64 (2010)
37. Jentsch, D. et al.: Fabrikaudit Industrie 4.0. ZWF Zeitschrift für wirtschaftlichen Fabrikbetrieb. Carl Hanser Verlag **108**(9), 678–681 (2013)
38. Krückhans, B., Meier, H.: Industrie 4.0–Handlungsfelder der Digitalen Fabrik zur Optimierung der Ressourceneffizienz in der Produktion. HNI-Verlagsschriftenreihe, pp. 31–40 (2013)
39. Höme, S. et al.: Semantic Industry: Herausforderungen auf dem Weg zur rechnergestützten Informationsverarbeitung der industrie 4.0. Automatisierungstechnik (AT). De Gruyter Oldenbourg **63**(2), 74–86 (2015)
40. Holtewert, P. et al.: Virtual fort knox federative, secure and cloud-based platform for manufacturing. In: CIRP Conference on Manufacturing Systems. Elsevier, pp. 527–532 (2013)
41. Landherr, M., et al.: Virtual Fort Knox. Werkstattstechnik (WT) **103**(2), 146–151 (2013)
42. Anderl, R. et al.: Integriertes Bauteildatenmodell für Industrie 4.0. ZWF Zeitschrift für wirtschaftlichen Fabrikbetrieb. Carl Hanser Verlag **109**(1–2), 64–69 (2014)
43. Scheifele, S. et al.: Flexible, self-configuring control system for a modular production system. In: 2nd International Conference on System-Integrated Intelligence. Elsevier, pp. 398–405 (2014)
44. Gronau, N.: Bionic Manufacturing-Steuerung der Fabrik mit Mitteln der Natur im Zeitalter von Industrie 4.0. Industrie Manage. **29**(6), 12–16 (2013)
45. Abele, E., Liebeck, T., Wörn, A.: Measuring flexibility in investment decisions for manufacturing systems. Manuf. Technol. https://doi.org/10.1016/S0007-8506(07)60452-1
46. Herkommer, O., Hieble, K.: Ist Industrie 4.0 die nächste Revolution in der Fertigung? Industrie Manage. **30**(1), 42–46 (2014)
47. Auerbach, T., et al.: 'Digitales Technologiewissen durch intelligente Fertigungs- systeme', ZWF Zeitschrift für wirtschaftlichen Fabrikbetrieb. Carl Hanser Verlag **108**(7–8), 561–565 (2013)
48. Stark, R. et al.: Notwendige Voraussetzungen für die Realisierung von Industrie 4.0: Ein Beitrag aus der Sicht der Industriellen Informationstechnik. ZWF Zeitschrift für wirtschaftlichen Fabrikbetrieb. Carl Hanser Verlag **110**(3), 134–141 (2015)
49. Posada, J., et al.: Visual computing as a key enabling technology for industrie 4.0 and industrial internet. IEEE Comput. Soc. IEEE **35**(2), 26–40 (2015)

50. Morlock, F., Dorka, T.: Automatisierungspyramide für die integrierte Sach- und Dienstleistungserbringung. Werkstattstechnik (WT) **104**(7/8), 448–452 (2014)
51. Brettel, M., et al.: How virtualization, decentralization and network building change the manufacturing landscape: an industry 4.0 perspective. Int. J. Mech. Ind. Sci. Eng. **8**(1), 37–44 (2014)
52. VDI/VDE Statusreport Referenzarchitekturmodell Industrie 4.0 (RAMI4.0). Düsseldorf, 2015. Available at: https://www.vdi.de/fileadmin/user_upload/VDI-GMA_Statusreport_Referenzarchitekturmodell-Industrie40.pdf. Accessed 26 Jan 2019
53. Schuh, G., Potente, T., Thomas, C., et al.: Steigerung der Kollaborationsproduktivität durch cyber-physische Systeme. In: Industrie 4.0 in Produktion, Automatisierung und Logistik. Springer Fachmedien Wiesbaden, Wiesbaden, pp. 277–295 (2014). https://doi.org/10.1007/978-3-658-04682-8_14.
54. Engelhardt, P., et al.: Situationsbasierte Steuerung modularisierter Produktionsabläufe. ZWF Zeitschrift für wirtschaftlichen Fabrikbetrieb **108**(12), 987–991 (2013)
55. Overmeyer, L., et al.: Kognitive Transportsysteme durch optische Technologien. Werkstattstechnik (WT) **103**(2), 139–145 (2013)
56. Schmiedgen, P., Noennig, J.R., Nyhuis, P.: Resilienzsteigerung durch Wissenstransfer in CPPS. Werkstattstechnik (WT) **104**(3), 164–168 (2014)
57. Scholz-Reiter, B., et al.: Selbststeuerung in der betrieblichen Praxis. Industrie Manage. Citeseer **23**(3), 7–10 (2007)
58. Lee, J., Kao, H. -A., Yang, S.: Service innovation and smart analytics for industry 4.0 and big data environment. In: CIRP Conference on Industrial Product Service Systems. Elsevier, pp. 3–8 (2014)
59. Elsweier, M., Nyhuis, P., Nickel, R.: Assistenzsystem zur Diagnose in der Produktionslogistik. ZWF Zeitschrift für wirtschaftlichen Fabrikbetrieb **105**(6), 562–568 (2010)
60. Studer, R., Benjamins, V.R., Fensel, D.: Knowledge engineering: principles and methods. Data Knowl. Eng. **25**(1), 161–198 (1998). https://doi.org/10.1016/S0169-023X(97)00056-6
61. Plorin, D., et al.: Ambient Assisted Production; Konzepte für die Produktion 4.0 unter Berücksichtigung demographischer Entwicklungen. Werkstattstechnik (WT) **103**(2), 135–138 (2013)
62. Abele, E. et al.: Effiziente Fabrik 4.0. ZWF Zeitschrift für wirtschaftlichen Fabrikbetrieb. **110**(3), 150–153 (2015)
63. Banerjee, A., Bandyopadhyay, T., Acharya, P.: Data Analytics: Hyped up Aspirations or True Potential? Vikalpa. SAGE Publications, **38**(4), 1–12 (2013). Available at: https://journals.sagepub.com/doi/pdf/10.1177/0256090920130401

Chapter 9
Smart Life Cycle Services

> Uber, the world's largest taxi company, owns no vehicles. Facebook, the world's most popular media owner, creates no content. Alibaba, the most valuable retailer, has no inventory. And Airbnb, the world's largest accomodation provider, owns no real estate. Something interesting is happening. [1]

Tom Goodwin is right about that, something interesting really is happening. Services in the way known from before are changing; they are becoming "smart". Smart Service Life cycles are a very recent topic, as services are also affected by the new industrial revolution and the digitalization. In the following paper, the topic of "smart life cycle services" will be explained.

The first step will be to provide background information on the topic of the Internet of Things and Industry 4.0, as "smart" service life cycles are emerging because of the given technological changes and the increased digitalization. Understanding those terms is therefore necessary to understand why smart services exist. After that, the smart service life cycles will be explained in detail, i.e. what is a smart service, what is the difference between a "normal" and a "smart" service and where smart services are applied. Then, the concept of smart service life cycles will be portrayed further through the example of Amazon, a company which works very much in the field of offering smart services. Last but not least, the conclusion will hold a critical discussion of the results.

9.1 The Industry 4.0 and Internet of Things

The terms Industry 4.0 and Internet of Things (IoT) are always mentioned together. That is because the IoT is the driving technology behind the fourth industrial revolution, known as Industry 4.0. In the following chapter, both terms will be explained shortly to provide a basic understanding of the terms.

J. Niemann and A. Pisla, *Life-Cycle Management of Machines and Mechanisms*, Mechanisms and Machine Science 90,
https://doi.org/10.1007/978-3-030-56449-0_9

9.1.1 The Internet of Things

The term Internet of Things (IoT) was first used by Kevin Ashton in 1999, a US Scientist who was back then an employee at Procter and Gamble. In an article which he published ten years later in the RFID Journal, he writes the following: "If we had computers that knew everything there was to know about things—sing data they gathered without any help from us—we would be able to track and count everything, and greatly reduce waste, loss and cost. We would know when things needed replacing, repairing or recalling, and whether they were fresh or past their best." [2]. Ashton's explanation of the IoT already hints to the changes in the industry: tracking and counting everything through the collection of data which the customer himself provides is what is shaping the way services are being offered.

Before 1999, the idea of the IoT already existed. In 1991, Mark Weiser had the idea of "ubiquitous computing", an environment which has computers and sensors in form of tablets that communicate with each other [3]. This fits another definition which explains the IoT as a "pervasive presence around us of a variety of things or objects (e.g., RFID tags, sensors, etc.) that cooperate with their neighbors to reach common goals" [4].

There are many more definitions that ultimately contain a similar content, which leads to the conclusion that the IoT is a concept where physical objects called things are supplied with embedded systems and then communicate to each other through the internet. When machines communicate to each other, then this is called Machine to Machine communication (M2M) [5]. M2M is the central idea of the IoT. The aim is to connect many things over the internet for data exchange, which provides new opportunities in the industry, including the way services are offered. Whenever things are connected, they are called "smart". After getting a basic understanding of what is meant by the IoT, the next step is to look at the industry 4.0.

9.1.2 Industry 4.0

Life cycles are becoming "smart" and this is because the industry has entered what is called the fourth industrial revolution, rather known as "Industry 4.0". It originates from the german word "Industrie 4.0" which is a marketing term that was used to describe a "future project" from the German government. The main characteristic is the connection between a product and its service as well as the integration of the customer into the value chain. This will function through the integration of embedded systems and autonomous machines that function and take decisions on their own [6].

The term industry 4.0 was first used in 2011 at the Hannover Messe in Germany [7]. It is short for "fourth industrial revolution" and describes the digitalization of the entire value chain, i.e. from the genration of the idea until the recycling of the product, therefore the entire life cycle [8].

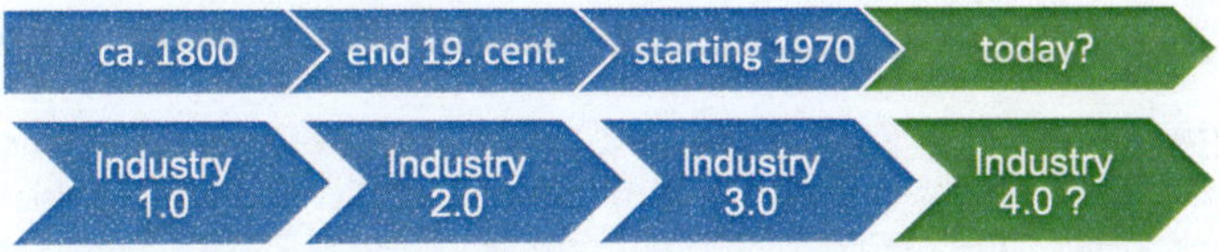

Fig. 9.1 Timeline of the industrial revolutions [9]

In contrast to the past industrial revolutions, the industry 4.0 is the first to be announced beforehand. Since this is the Industry 4.0, it is obvious that there must have been three former industrial revolutions. The prior industrial revolutions were related to the introduction of mechanical production sites powered by hydro- and steam power, mass production through the availability of electricity and the use of electronics and informationtechnology as a driver for automatization, respectively [9]. Figure 9.1 shows a timeline of the industrial revolutions.

Today, a transition to the industry 4.0 is happening, or has happened, there is no clear decision on that. The most important thing at this point is to understand that the industry 4.0 is here, and the industry is adapting to the changes. Life Cycle Services are also adapting to the changes now. Digitalized services are called "Smart" services. In the next chapter, the smart life cycle services will be explained.

9.2 Smart Life Cycle Services

The following chapter explains the theoretical background of smart life cycle services. The first section gives a short introduction to smart services. The differences between classic and smart services will be explained. In the following, the basic strategy of smart services will be presented. Subsequently, the life cycle of a smart service will be described. For this purpose, a process model is used to illustrate "the path to smart services". In the last chapter, some examples of currently implemented smart life cycle solutions will be presented.

9.2.1 Smart Services

According to Freitag's definition, smart services can be defined *"as data-based, individually configurable offers of services, digital services and products organized via platforms"* [10]. The term "smart service" describes the demand-oriented provision by a combination of Internet-based and physical services. *"The main difference between standard and smart services is the potential of smart services to enable service providers to establish closer ties with customers by identifying customers' needs intelligently by using IT"* [11]. While traditional standard services use service efforts by adding them to the (physical) product or core services, smart servives are embedded in the core product itself.

In the literature Standard Services are called "must-be-service". This includes services such as technical consulting of the product and is expected by the customer to be offered [11]. A detailed description of the standard services can be found in the appendix. However, smart services offer customers a wide range of services.

They can be seen as connected components that offer at least one of the following add-ons:

- "Diagnostic: The application enables a device to self-optimise and allows troubleshooting, monitoring, and repair.
- Replenishment and commerce: The application monitors the consumption of a device and consumer behaviour.
- Status: The application reports on performance and usage of the product or services.
- Upgrades: The application optimises the performance of a given device.
- Location mapping and logistics: The service support system can be optimised with this application" [11].

From the perspective of the service provider the goal should be to charge for the smart service. Therefore, it is important to make the case for customer value and to visualise the benefits.

9.2.2 *The Strategy of Smart Services*

The use of Smart Services leads to a variety of changes in the corporate structure. This structure leads to a repositioning or adaptation of the corporate strategy. Figure 9.2 illustrates four areas that are affected seriously by Smart Services.

9.2.2.1 New Business Models

The increased service orientation of companies requires a rethink and leads to create new business models. The traditional buyer–seller-relationship has already changed. Customers do not want to invest in products, machines and plants anymore. Instead, there is an increasing demand for new business models that focus on smart services. Rather than selling a physical product, the aim is to provide customers an adequate range of product service packages at any time and place ("everything-as-a-service"; i. e. Sale of flying hours instead of aircraft engines at Rolls-Royce) [1]. This can increase customer satisfaction. In addition, it is important to verify the economic feasibility. Conditions are assessed in order to offer cost-covering service. Various legal restrictions as well as cost and revenue models are taken into account. An important tool is the business model canvas, which can be used to get an overview of all.

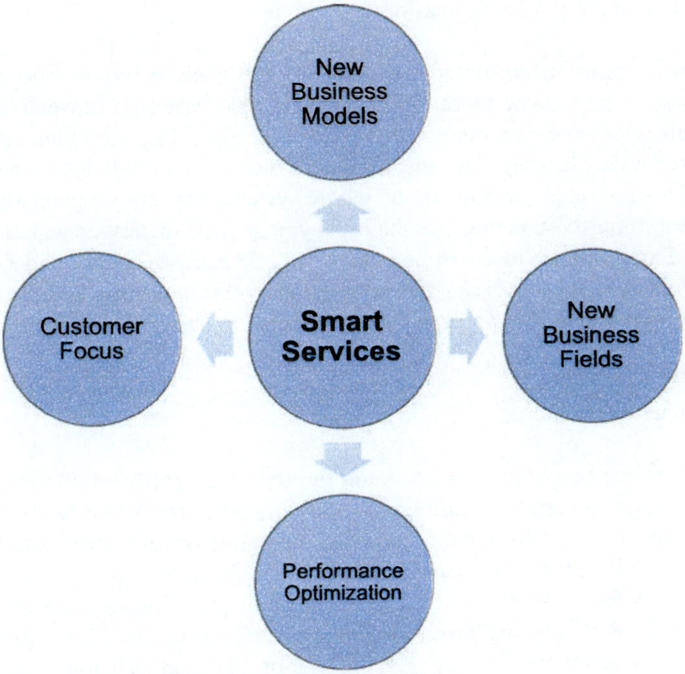

Fig. 9.2 Influenced fields by smart services (Own representation based on [1])

9.2.2.2 New Business Fields

The utilization of smart services creates new opportunities. Companies are able to offer completely new products and services that did not previously exist in the business portfolio [1]. This means, that new business models will be built around services. By offering smart services, customers receive additional functionality besides the physical products. For example, the Amazon Kindle allows its users easy access to Amazon services such as music streaming, books, videos and apps. Jeff Bezos, the founder and currently the CEO of Amazon, is quoted for stating that "The Kindle is not a device, it is a service" [12]. While the Kindle is a tablet for reading electronic books, the offering doesn't stop with the product. The syncing through the Kindle app allows customers to use the Kindle service across all their devices with complete continuity. This is the decisive aspect, why Mr. Bezos views the Kindle as an interface to provice customers with Amazon's services on demand.

9.2.2.3 Performance Optimization

Performance optimization includes digital services such as remote and predictive maintenance. These allow more efficient operations, based on preventive maintenance of machines and systems, sometimes even from a distance. Digital networking enables predictive planning. This means that problems can be anticipated before they occur and lead to major damage in the whole system. This allows preventive maintenance, enormous cost savings, as the deployment costs of service technicians can be reduced and time savings can be realized [1]. In addtion, unplanned failures of plants, products or machines can also be avoided by this approach. Service offerings are highly individualizable and can thus be adapted to customer needs in real time.

9.2.2.4 Customer Focus

The value proposition of companies is not limited to the acceleration, optimization and quality improvement of existing processes and products. Smart Services should create a "transition" to completely new and adaptable or individual offerings that generate new value propositions and ensure sustainable profitability for the customer. Smart services focus on customers more and more [1]. This situation gives them an active role in structuring products and services. Companies learn more about preferences and requirements by analyzing user profiles and have the opportunity to tailor offerings to target groups and customers. A stronger personalized service leads to a closer customer relationship and leads to higher customer loyalty.

The four areas emphasize that companies can benefit from new products and services and stronger customer orientation while using Smart Services. Modifying the business model is a key aspect for the success of Smart Services.

9.2.3 The Life Cycle of "Smart Services"

A life cycle is defined as various successive periods of time that describe the process of a product or service [13]. The provider of a smart service is usually responsible for the entire life cycle of the underlying service system. Therefore, the process model of a smart service life cycle management will be analysed in more detail, because it's a planning and management tool that visualizes the workflow and shows a sequence of events that result in a final result [10]. The process model of smart service life cycle management consists of seven phases which is presented below (Fig. 9.3).

The first phase in the service life cycle is the generation of ideas. The "idea finding" phase consists of merging all ideas for new services. Ideas for new services can be developed internally as well as by external partners, e.g. network partners. "*To achieve the maximum advantage of an inter-enterprise environment, it is essential to focus on core competencies while integrated in a network of partners, each of which brings its core expertise to the table.*" [14]. These are independent partners with

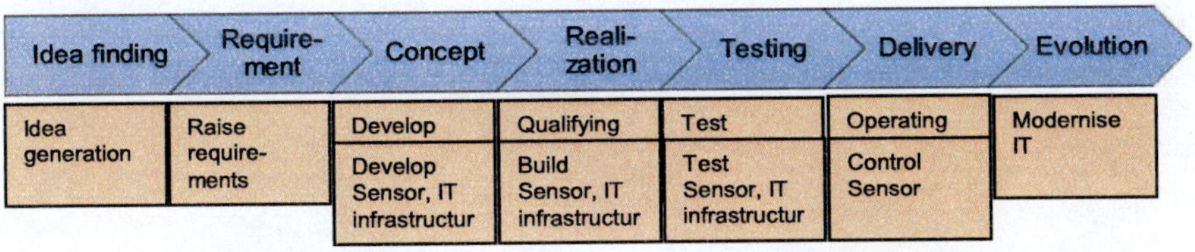

Fig. 9.3 Process model of smart service life cycle management (Own graph acc. to [10])

similar interests and strengths who support each other. The relationship between the partners consists of the exchange of products (goods, services, etc.), information and mutual trust. So, in order to increase flexibility of production capacities, it's useful to arrange network partners. Afterwards, these ideas are evaluated and selected with suitable criteria, which are predefined by taking the area of application of the smart service into account. Subsequently, the idea is evaluated in terms of implementability and expected profits.

The second phase aims to define specific IT-requirements (system, hardware and software) for a smart service. Furthermore, existing standards and legal requirements have to be considered. A detailed documentation of all requirements identifies the result of the requirements analysis [15].

The third phase deals with the concept of smart services. The main focus is the detailed documentation of the new service. First, the underlying business model should be worked out and described in detail. Afterwards, the smart service should be developed and the necessary sensor and IT infrastructure should be designed [16].

After the conception, the smart service can be implemented. For this purpose, the employees, who are involved in this process, will be trained. Well-qualified employees who are well acquainted with the software can also satisfy customer needs. Furthermore, missing resources will be procured and required sensor and IT infrastructure will be built [17–19].

In the test phase, the smart services will be tested to determine its most significant weak points. These can be at the business model level, at the performance optimization level and at the economic level. During this phase, an interaction between the smart service and the customer can also be tested. Customer opinions and reviews can change the current state of development and make services more customer-optimized [20].

The first step in the delivery phase is the acquisition of customers. This requires target group-specific marketing. It is important that the Smart Service works at this stage. Furthermore, from an economic point of view, it is important to control the revenue and costs of the smart service [10].

In the last phase, the smart service is completed. Subsequent improvements are decided in this phase.

Smart life cycle services are an actual approach with digitalization emerging in every field of the industry, which is not only limited in production. In the following, the application areas of Smart Services will be presented.

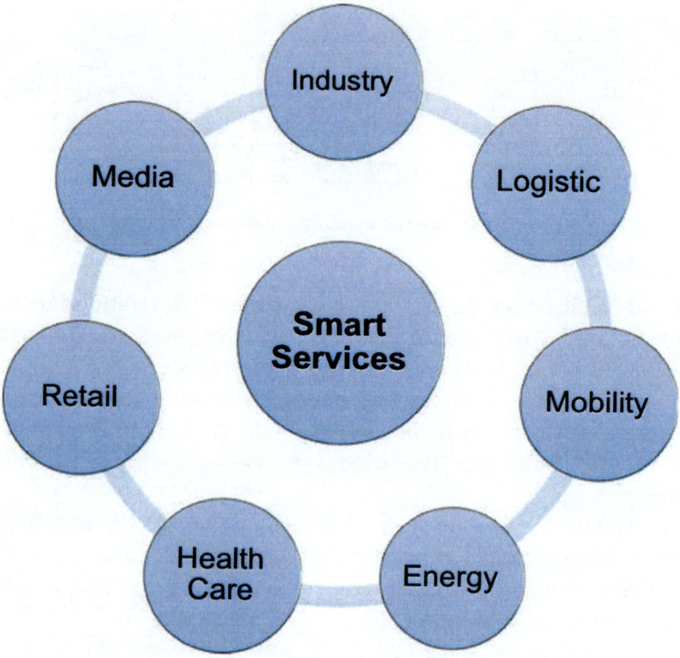

Fig. 9.4 Applications fields of smart services (Own representation based on [1])

9.2.4 Fileds of Application

Smart services can be used in a variety of industries and branches of business and daily life. The next Fig. 9.4 shows these branches.

Industry: In the industrial sector, especially in the production, smart services are used to control machines and plants. In order to minimize failures and downtimes, preventive services are used to control the status of cross-linked objects. By that, companies can optimize the runtimes of these objects [1, 17].

Logistics: Smart services in logistics are designed to optimize the flow of goods and transport chains. Through the analysis of real-time data, routes can be planned. For instance, congestions can be avoided by that [1].

Mobility: The smart service also continues its influence in the segment of mobility. Also in the automotive industry, the basic idea of smart service has become popular. The "connected car", the intelligently networked automobile, is informed about traffic jams and accidents by real-time data and can communicate with other vehicles [1].

Energy: By integrating electricity suppliers, end customers, grid operators, electricity producers and other actors an efficient energy supply can be enabled that can keep supply and demand synchronised. An intelligent power grid can predict consumptions and optimize the electricity supply [1].

Health Care: The healthcare sector has seen many changes which are driven by Smart Services. For example, rooms can be equipped with sensors to send an emergency call signal in the unfortunate case of a fall. In addition, the construction of medical platforms supports diagnostics and therapy and uses information collected worldwide [1].

Retail: Smart services play a major role in retail. It allows providers, such as Amazon or Alibaba, to offer their customers individualized proposals for suitable products based on their user profiles in order to create a personalized shopping experience for them. Another benefit is the usage of GPS functions. The use of location-based services, which locate the user via the GPS function of the smartphone, enables the precise transmission of information, e.g. weather forecast, sights, restaurant recommendations and retail discount campaigns [21].

Media: In the age of digitalization, (social-)media has become a major part of daily and business life and is one of the main areas that benefits from smart service. Music or video streaming providers such as Spotify, Netflix, Amazon Prime or Google Play offer customers the access to many music titles or films, for instance [22]. Customers have the opprtunity ti creat their individual library on these digital platforms. This allows the provider to make proposals for additional personalized titles based on customers' preferred search criteria.

The application areas show the different application possibilities of smarter services. At this point, it is already clear that various factors and actors must play together in order to create smart services.

In order to give a detailed example, the next chapter will focus on the the usage of Smart Services in the field of media, especially Amazon which is a multinational technology company focusing in cloud computing, e-commerce and artificial intelligence.

9.3 Smart Services

The term "smart services" is closely linked to the terms "digital transformation", "Internet of Things (IoT)" and "Industry 4.0". While the "digital transformation" describes the development of new technologies, after the introduction of the Internet, which has changed both the corporate world and the society, the "Internet of Things" represents the connection of a physical thing or object to the internet. Thereby, data gets generated, that can be used afterwards. By 2020, more than 50 billion devices are to be connected to the internet. The term "Industry 4.0" characterizes the fourth industrial revolution, driven by information and communication technologies and the "Internet of Things" [23, 24].

Building on these developments, a new term has emerged, called "smart services". Smart services are services that are tailored to specific use cases with the help of data and intelligent processing. The basis for smart services is an integrated platform. This consists of three levels. The sensor infrastructure, the data infrastructure and the service platform, see Fig. 9.5.

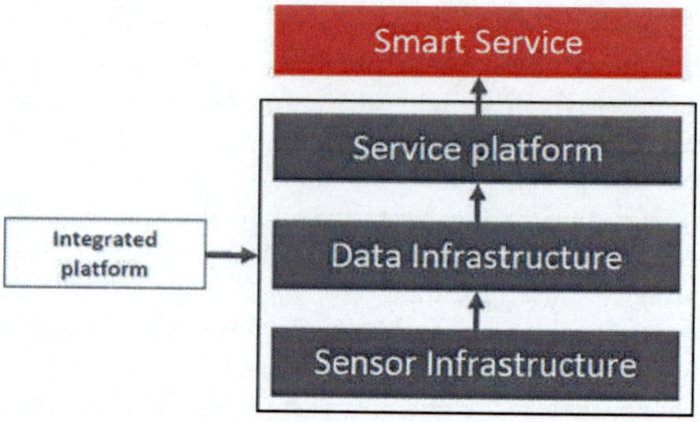

Fig. 9.5 The basis for "smart services", the integrated platform [25]

In the lowest level, the sensor infrastructure, data is continuously collected via different sensors. These data are sent encrypted and anonymized to the data platform where they are stored. At the top level, the service platform, the results are made available and visualized as needed in applications [23].

The use of these applications that are based on the collected data makes it possible to upgrade existing services or to offer new intelligent services. These services are then called smart services. The successful design of smart services requires further consideration of various aspects, such as designing and testing business models, qualifying the staff and integrating potential customers. That's why the implementation of smart services is quite complex and time-consuming [25].

Smart services have different areas of application that can be mainly classified into:

- Production (e.g. services to offer maintain information)
- Mobility (e.g. services to provide up-to-date traffic information)
- Housing and Living (e.g. service to support home cleaning)
- Medicine (e.g. service to offer an insurance, based on generated health data).

The more these "smart services" interact, the more attractive and user friendly they get. It can therefore be assumed that Smart Services will communicate even more with each other in the future and will enable a completely new dimension of services offered by companies [26].

9.4 Influence of Smart Services on Business Models

Smart Services not only influence the economy, they also cause big changes inside of the companies itself, which have to change their strategic orientation and partially their whole business model with the application of Smart Services.

By using these "smart services" to offer a higher service potential to their customers, many companies transform from a product manufacturer to a service provider. Figure shows this transformation which exemplary represents the service

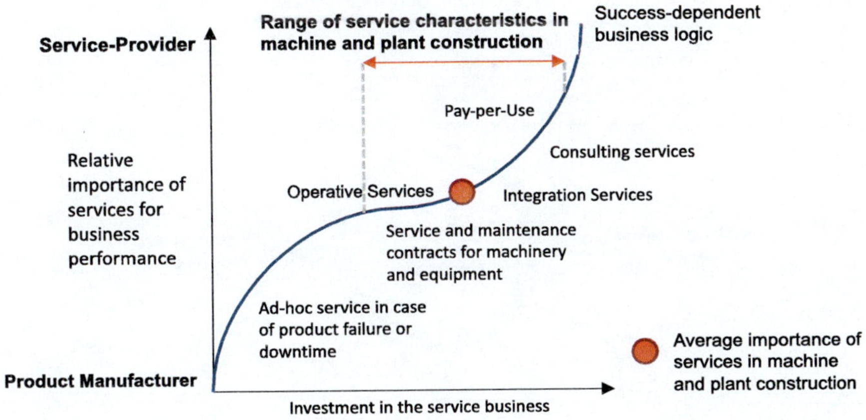

Fig. 9.6 Smart services and the change of business models [10]

development inside of companies for machine and plant construction. For the service to play a more and more important role, high initial and linear investments are necessary, see Fig. 9.6 [23].

After reaching the turning point, the meaning of services becomes more and more important for the company and the importance of service for the business performance changes exponentially. The company is shifting to a service provider [27].

Previous and traditional services, such as "ad-hoc services", in which for example a machine service only gets applied after the machine has failed, are replaced by "Smart Services" in which sensors inside the machine generate data, that shows the customer or service provider in advance whether a machine service is necessary or not. Thereby the product manufacturer that has evolved into a service provider can handle and offer its service in a much early stage than before [28].

In this context the five forces by Porters can be included. They say that companies have to compete with five forces within the market they are located, see Fig. 9.7. With the digitization and introduction of Smart Services, the companies even have to face new challenges. Traditional and so far, conservatively managed companies reach new markets and locate their company e.g. as a service provider, data analyst or IT service provider [27, 28].

Regarding the five forces, smart services especially influence the competition, the unique market position and the power of a company. As the company participates more and more in the value creation process, they receive important market knowledge, which they can use for their advantage. This makes their customers dependent and makes it almost impossible for them to change the supplier or service provider.

On this background, smart services offer a lot of opportunities for companies, which also explains, why more and more companies are shifting from a product manufacturer to a service provider.

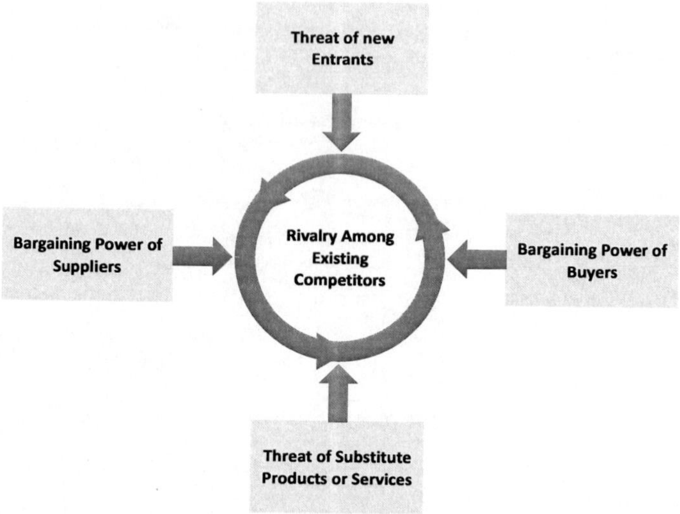

Fig. 9.7 The five forces that companies have to compete with [28]

9.5 Smart Life Cycle Service Management

The development of smart life cycle services can be divided into different levels. At first the business model needs to be created. The first step of this level is to generate and evaluate ideas. Therefore, new services are identified on the basis of the existing data or executing questionnaires, interviews and workshops with possible customers. All in all new knowledge about further or future needs of the customers are gained. The next step is to collect market requirements which include the market situation, feasibility studies and develop scenarios with the goal of getting a broad view about the environment and minimize the entrepreneurial risk. As the last step, a business model is developed where the realization is checked by using a real application case. This includes an analysis under which conditions the service will cover its costs by looking at the legally restrictions, costs and revenues [10].

After the management of the business model, the smart services need to be developed. In the beginning, the IT requirements, that include the system, service, hardware and software, need to be collected to define a specific business case with its smart services. Examples for those requirements are the connection to the internet, required hardware as for example sensors, design of the service processes, and training of the service employees. Those requirements will be the basis of later validation and verification tests. Now the smart services are developed by describing the services in detail. Therefore, the process, resources, infrastructure and equipment must to be defined. After this definition, the services have to be tested. The tests are containing tests of the infrastructure and processes and also acceptance tests that are checking the acceptance of the focused target group. The next step is to implement the smart

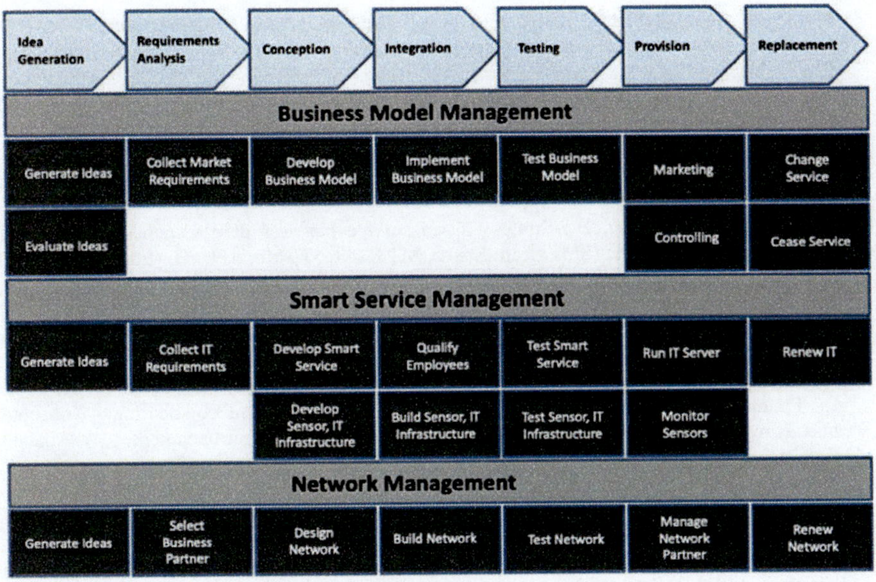

Fig. 9.8 Process of smart life cycle management [10, 29]

services in the company. This implementation contains the realization of the IT concept, the implementation of organizational measures, the personnel measures and the marketing measures [10, 29].

After the implementation in the company, the smart services can be introduced in the market by the roll out of the services. In this phase it is important to monitor the startup and success of the services (Fig. 9.8) [10, 17–19, 29, 30].

References

1. Paluch, S.: Smart Services—Analyse von strategischen und operativen Auswirkungen. In: Bruhn, M., Hadwich, K. (Eds.): Dienstleistungen 4.0. Geschäftsmodelle – Wertschöpfung – Transformation. Springer-Gabler Verlag, Wiesbaden (2017)
2. Ashton, K.: That 'Internet of Things' Thing. RFID J. **1**(1) (2009)
3. Adelfinger, V. P., Hänisch,T.: Industrie 4.0: Wie cyber-physische Systeme die Arbeitswelt verändern. Springer Fachmedien, Wiesbaden (2017)
4. Batalla, J.M., Mastorakis, G., Mavromoustakis, C.: Beyond the Internet of Things Everything Interconnected. Springer International Publishing, Switzerland (2017)
5. Främling, K., Kubler, S., Buda, A.: Universal messaging standards for the IoT from a life cycle management perspective. IEEE Internet Things J. **1**(4) (2014)
6. Bendel, O.: Gabler Wirtschaftslexikon, Stichwort: Industrie 4.0. Springer Gabler Verlag (Eds.) (2017). Online: https://wirtschaftslexikon.gabler.de/Archiv/-2080945382/industrie-4-0-v2.html. Checked 19 Dec 2018
7. Drath, R.: Industrie 4.0 - eine Einführung, Bericht, o. O. (2014)

8. Krings, B. -J.: Digitalisiert, effizient and global? Die fortlaufende Technisierung Der Erwerbsarbeit, in: Hausstein, A., Zheng, C. (Hrsg.) Industrie 4.0/Made in China 2025, KIT Scientific Publishing, Karlsruhe (2018)
9. Frick, T.W.: Von Industrie 1.0 bis 4.0 – Industrie im Wandel der Zeit, 2017. Online: https://industrie-wegweiser.de/von-industrie-1-0-bis-4-0-industrie-im-wandel-der-zeit/. Checked 17 Dec 2018
10. Freitag, M., Hämmerle, O.: Smart Service Life Cycle Management. WT Werkstattstechnik (2016)
11. Kreuzer, E.; Aschbacher, H.: Strategy-based service business development for small and medium sized enterprises (SMEs). In: Snene, M., Ralyté, J., Morin, J. -H. (Eds.) IESS 1.1—Seconde International Conference on Exploring Services Sciences. Genf 211. Springer-Verlag, Heidelberg (2011)
12. Shiftbase Research Ltd., 2015: Online: https://guides.shiftbase.net/amazon/. Checked 02 Dec 2018
13. Anke, J.; Thoben, K. -D., Wellsandt, S.: Modellierung der Lebenszyklen von Smart Services. In: Thomas, O., Nüttgens, M., Fellmann, M. (Eds.) Smart Service Engineering – Konzepte und Anwendungsszenarien für die digitale Transformation. Springer Fachmedien, Wiesbaden (2017)
14. Niemann, J.: Life Cycle Management-das Paradigma der ganzheitlichen Produktlebenslaufbetrachtung. In: Spath, D. et al. (Hrsg.) Neue Entwicklungen in der Unternehmensorganisation. Springer-Vieweg, VDI Buch, Berlin (2017)
15. Niemann, J.: Die Services-Manufaktur, Industrielle Services planen –entwickeln – einführen Ein Praxishandbuch Schritt für Schritt mit Übungen und Lösungen. Shaker Verlag, Aachen (2016)
16. Freitag, M., et al.: Smart service life cycle management in der Luftfahrtindustrie. In: Borgmeier, A., Grohmann, A., Gross, S.F. (eds.) Smart Services und Internet der Dinge: Geschäftsmodelle, Umsetzung und Best Practices, pp. 73–89. Carl Hanser Verlag GmbH & Co. KG, Munich (2017)
17. Niemann, J., Pisla, A.: Sustainable potentials and risks assess in automation and robotization using the life cycle management index tool—LY-MIT. Sustainability **10**, 4638 (2018)
18. Niemann, J., Schemann, T., Erkens, J.: Servitization—pathway of transformation from product manufacturer towards a service provider. In: 2018 International Conference on Production Research—Africa, Europe, Middle East 5th International Conference on Quality and Innovation in Engineering and Management, July 25–26 2018, Cluj-Napoca, Romania
19. Stöhr, Carsten; Janssen, Monika; Niemann, Jörg: Smart Services. In: 14th International Symposium in Management, Challenges and Innovation in Management and Entrepreneurship, 27–28 Oct 2017, Timisoara, Romania
20. Jin, J., Gubbi, J., Marusic, S.: An information framework for creating a smart city through internet of things. IEEE Internet Things J. **1**(2) (2017)
21. Li, F., Tao, Y., Cheng, L.: Big data in product life cycle management. Int J. Adv. Manuf. Technol. **81**(1) (2015)
22. Isaksson, O., Larsson, T. C., Öhrwall Rönnback, A.: Development of product-service systems: challenges and opportunities for the manufacturing firm. J. Eng. Des. Spec. Issue Product-Serv. Syst. **20**(4) (2008)
23. Borgmeier, A., Grohmann, A., Gross, S.F.: Smart Services und Internet der Dinge: Geschäftsmodelle, Umsetzung und Best Practices. Hanser Verlag, München (2017)
24. Jalali, S.: IEEE Recent Advances in Intelligent Computational Systems. M2M Solutions—Design Challenges and Considerations. Trivandrum, India (2013)
25. Fraunhofer-Institut für Arbeitswirtschaft und Organisation IAO: Fraunhofer IAO. Fraunhofer IAO. Online: https://wiki.iao.fraunhofer.de/index.php/Smart_Services. Checked 15 Dec 2018
26. Bundesministerium für Wirtschaft und Energie: Smart Service Welten. Online: https://www.bmwi.de/Redaktion/DE/Artikel/Digitale-Welt/smart-service-welt.html. Checked 19 Dec 2018
27. Porter, M.E.: Competitive Strategy: Techniques for Analyzing Industries and Competitors. Free Press, New York (2004)

28. Porter, M. E., Heppelmann, J. E.: How Smart, Connected Products Are Transforming Competition. Harvard Business Review (2014)
29. Bullinger, H. -J., Meiren, T.: ICPR, International Conference on Production Research, vol. 23. Smart Services in Manufacturing Companies. Manila, Philippines (2015)
30. Niemann, J.: Eine Methodik zum dynamischen Life Cycle Controlling von Produktionssystemen. Heimsheim: Jost-Jetter Verlag, 2007IPA-IAO Forschung und Praxis 459). Stuttgart, Univ., Fak. Maschinenbau, Inst. für Industrielle Fertigung und Fabrikbetrieb, Diss. (2007)

Chapter 10
System Operators

In an economy where the possibility of short-term access to far-reaching resources forms the basis of commercial success, the entire potential of a product's life cycle moves into the centre of strategic focus. It will no longer be a question of selling a single product to as many customers as possible, but rather of looking after a single customer and supplying him/her with as many products as possible.

This new paradigm focuses primarily on the maximum exploitation of a single product instead of the maximisation of total sales. Businesses will concentrate more on building up long-term relationships with individual customers. Success will be measured by the amount of value-adding realised to the customer or to the product(s) sold over the total duration of the relationship [1]. These alliances and networked partnerships will also be given a push as a result of a dramatic acceleration in technical innovation cycles. Also, due to the fact that technical equipment, manufacturing processes, production sequences, goods and services all age faster in a highly-electronic environment, the long-term ownership of a production facility becomes less and less attractive. In order to be able to constantly maximise performance and precision, short-term access will become an ever-increasing option. Leasing, rental contracts or performance contracts will become more attractive than buying and owning. Accelerated innovation cycles and speeded-up product turnover will dictate conditions for the new network economy. Whereas the traditional market is characterised by the exchange of goods, access to holistic concepts will include material aspects in a networked economy.

As a result, companies will only be able to exist if they are explicitly capable of increasing product profit by using other additional products. The paradigm of product lifetime value, i.e. the evaluation of commercial success over the entire life span of a product, therefore demands an specific focus to be placed on the requirements of individual customers (Fig. 10.1). The linking-up of products to modern information and communication technology instruments offers an excellent pre-requisite for researching and recording specific customer needs. Among other things, these technologies will also enable value-added services to be offered to the customer

J. Niemann and A. Pisla, *Life-Cycle Management of Machines and Mechanisms*, Mechanisms and Machine Science 90,
https://doi.org/10.1007/978-3-030-56449-0_10

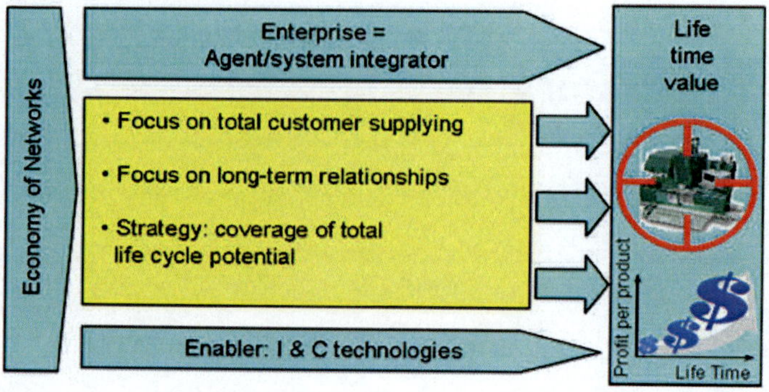

Fig. 10.1 The paradigm of lifetime value [1]

directly or machine optimisations to be carried out over great distances. The change in paradigm no longer places the focus of attention on the maximum use of resources implemented by companies, but rather on the maximum technical and economical exploitation of products during their life cycle. This will also be forcefully demanded due to an alteration in society's conception of value with regard to environmental compatibility and to the closed circulation of materials. The technical conditions required for this already exist and will bring massive structural changes with them. Under the pressure of international competition, it is no longer possible for many companies to survive just by manufacturing and selling goods. More and more often, enterprises are transferring added-value activities towards the areas of product design, assembly and service.

10.1 Cooperation for Life Cycle Benefit

The customer-orientated service concept means that manufacturers are closely involved in every phase of a product's life cycle, starting with product specification. Life cycle cooperation provides the manufacturing industry with an extensive range of services which speed up and boost product development and production stages.

The following observations can be made from Fig. 10.2:

−1–0 The customer has a new product idea or a completed product which requires special know-how from outside the company in order to finish it or complete its final exterior.

0–1 In the product specification stage, the parties agree upon details regarding different tasks, sub-areas and responsibilities which need to be taken care of as the project advances.

1–2 In the product specification stage, the customer's product is developed in cooperation with experts. At this point the product is already quite detailed and

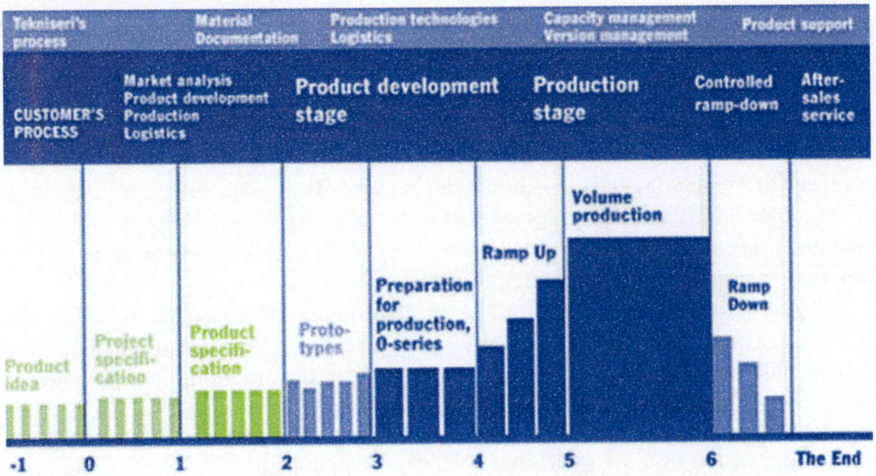

Fig. 10.2 Life cycle cooperation [1]

the suitability of different materials and production methods as well as different kinds of technical solutions are considered, taking cost effectiveness into account. It is here that the choices which have the most impact on a product's manufacturing costs are often made.

2–3 With the product development stage being carried out in cooperation, it is possible to reduce unnecessary testing and prototype series to a minimum, saving time and money. In order to achieve the greatest certainty of duplication and cost-effectiveness, details are polished with production.

3–4 The product development stage is the preparation for production and 0-series. During this time the product is brought towards final implementation and the design of the exterior is completed. Also, production provides additional feedback regarding processability, risk of error and lead-time. These factors can still be taken into consideration before starting serial production. Carrying out this critical stage requires close and confidential cooperation.

4–6 In volume production, production amounts are set as precisely as possible to match the customer's forecasts. Buffer stocks and storage services can be used if needed. The customer's production runs without interruption even during production peaks by using real-time and smooth logistics solutions. It is also extremely important to manage changes in product generations and the related documentation. A company's ability to invest and increase capacity is becoming more and more valuable.

6 Product support together with documentation is important for customer after-sales services. At the end of the product's life cycle, a controlled ramp down is essential. During the life cycle cooperation, valuable information is collected continuously. These data are stored and documented for future projects. This permits a successful cooperation model for future projects to be constructed

as well. The data also paves the way for additional product and production optimisation services.

Surveys estimate that over the next five years, up to 70% of the production planning and control (PPC) systems used in today's companies will be replaced by integrated order management systems. In this field alone, huge potentials exist for reducing costs by implementing new methods and techniques. In the future, a high level of customer-orientated dynamics will have to be achieved in complex networked systems. Long-lasting improvement will only become possible if methods as well as structures are revised and integrated into networks.

Lengthy pathways and intensified work distribution are obstacles to agility and dynamics. For this reason, the objective of market- and customer-orientated businesses must be to reduce restricting factors such as lengthy administrative procedures, non-aligned interfaces or great distances in material flows. As a result, when organising production networks, preference must be given to transformable structures with fast decision-making pathways. The virtual network process as shown in Fig. 10.3 actually should be created through configuration rather than through ad hoc procedures and rules or exceedingly complex and time-consuming software design and implementation. This is because the final objective is to create and re-create efficient networks within a very short period of time. Preliminary ideas for creating such dynamics using virtual capacities already exist. However, these only permit the temporary inclusion and utilisation of resources for a short time when required.

Methods for carrying out improvements exist due to the availability of faster, cheaper and, in principle, more open information and communications systems. These enable the entire flow of information from customer to supplier to be completely integrated and production is only commenced on receipt of a customer order. Additionally, formal and informal information can be made available almost

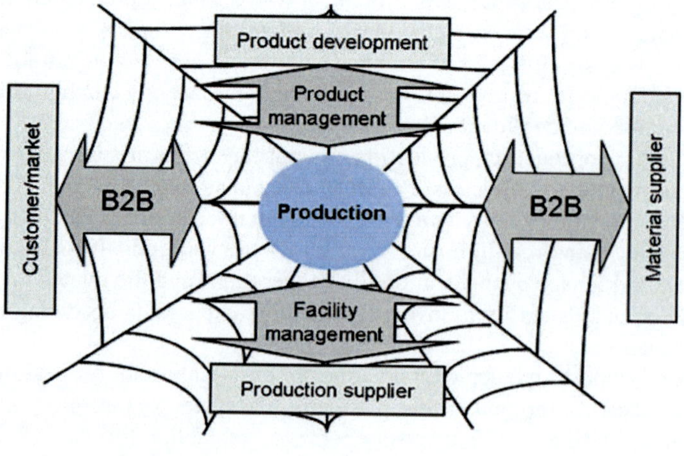

Fig. 10.3 Network partners for production excellence [2]

anywhere in the world. The process chains run through in an order sequence could be significantly accelerated using modern means.

By networking productions, there is a huge opportunity to uphold competitiveness, as this is targeted at achieving high synergic effects and simultaneously at attaining a high level of dynamics. History has shown that successful organisations are those which keep their networks transformable and adaptable and which are able to master these networks totally. Modern information and communication technologies provide the opportunity to use these possibilities in the interest of manufacturing technology. Consequently, we should take this chance and develop it further. In the future, it will be possible to extend this system and to integrate it into a web-based platform for holistic product support concerned with all aspects of performance optimisation.

A web-based platform will provide the backbone for constantly-optimised machine operations (Fig. 10.4). The platform will integrate the manufacturer, machine operator and various other engineering service providers right up to and including additional research institutions. These strategic alliances will accompany a product for the duration of its entire life cycle. Through this, the complete optimisation of a single product becomes the focus of attention in all business activities. The decisive criterion for future market success will be the ability to establish, organise and promote such networks for product support. These networks will be created in order to monitor a product during its entire life-time. They may be extended or reduced while any desired service can be performed on demand. The ability to control and monitor machines digitally constitutes the foundation for mastering these constraints in the digital age. On the other hand, the machine-user will benefit from such holistic networks which help to optimise his machine or machine park. The additional benefit will generate surplus profits which the user will share among the network partners.

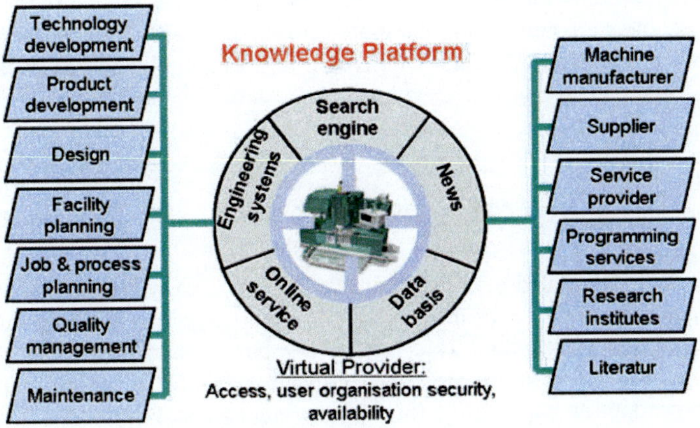

Fig. 10.4 Networks and knowledge platforms [2]

10.2 Integrated Product-Service Systems

Many manufacturing industries are shifting their production process from conventional mass production towards knowledge-based and service-orientated manufacturing which enables production on demand, mass customised solutions, rapid reactions to market changes and quick time-to-market for innovative solutions. It is now becoming increasingly crucial to base product development on a quick and reliable innovation process in order to overcome both the strategic and production aspects of product life cycle management. These changes on the suppliers' side are being forced due to increasingly complex and multifaceted requirements expressed by a wider group of customers (both business and consumers). The improvement demanded on the suppliers' side affects almost all levels and functions of the company, from the definition of an innovative vision/mission and the identification of a proper strategy right up the determination of consistent operative goals [3].

The development of integrated Product-Service Systems (hereinafter: PSS) can be seen as a new paradigm, an innovative approach as a solution to the mutations monitored in the market. A PSS can be defined as being the result of an innovation strategy, shifting the business focus to designing and selling tangible products and intangible services combined, enabling them to jointly fulfil customer needs. According to this vision, physical products (cars, white goods, mobile phones …) are combined with complementary services which integrate the functions and performances of the physical products (e.g.: financial plans for buying the solution, maintenance, embedded services …). PSSs are one of the most promising opportunities which innovative companies can rely on in order to improve their interaction with the market. They would even permit improvements in the production of standard products, which are no longer able to autonomously fulfil complex and variable customer requirements. Moreover, companies need to gain or maintain position in the market by improving not only the product itself but also the services related to the product. Finally, customers are demanding integrated solutions where products are sustained in all phases of their life cycle and where their performance standards are always at the top [4–6].

10.2.1 Developing Product Service Systems

Especially SMEs (but also many big companies) are usually focused on a specific area, while the development of integrated PSS requires a broad spectrum of competencies, skills and knowledge. This is why companies interested in the development of this kind of solution are asked either to improve their internal competences and production capabilities (hiring new employees, buying new production machinery, …) or to improve contacts with external entities (research centres, universities, consultants or other companies) which provide complementary experiences and develop corresponding services or products. For these reasons, improvements in communication between different functions within the same company or between actors from different bodies are one of the goals to achieve. Companies are also asked to adapt their approach and internal organisation to the development of solutions in order to simultaneously manage both the tangible and the intangible side of their offer. In such a context, companies need to receive support both to add value to conventional products and also to implement completely new solutions. Therefore, the capability of already having a complete view of the product life cycle in the design phase is fundamental for the producer to enable the use of multiple Designs for X strategies. Design for assembly, design for maintenance and design for environment emerged to encompass a wide range of approaches to product design. Business planning and aligning the development of new solutions with the business focus is the first stage towards successful innovation. Process planning to establish control methods and the allocation of resources is also a preliminary step towards real innovation [7, 8].

The main changes which conventional companies are expected to manage in order to take up the shift towards integrated PSS include [9, 10]:

- Redefinition/adaptation of the innovation process in order to strengthen interactions between the different departments and actors of a company
- Redefinition of the innovation process in order to involve external entities right from the early phases in the development process
- Identification of a consistent and worthwhile strategy for facing the market
- Improvements in the organisation of the company functions
- Enhancement of activities devoted to the identification of new market opportunities and to the redefinition of company business models.

Figure 10.5 shows a practical example from ABB Automation. The example clearly demonstrates how the performance of a system develops if no appropriate measures are taken to maintain efficiency over the life cycle. More and more malfunctions occur due to ageing processes, resulting in increased downtimes and greater performance losses. In order to counteract this, the company has developed a wide range of services for its customers—so-called *life cycle services*—which improve the creation of value over the entire life cycle.

The first group of traditional services reduces interruptions in production in the event of a malfunction. On the next level, preventive services help to avoid production downtimes through maintenance and regular plant overhauls keep the creation of value constant. On the top level, pro-active services increase the value of the plant through continuous modernization (e.g. upgrading and replacement).

For the customer, performance losses mean a loss in turnover which is reflected in so-called *opportunity costs*. Opportunity costs represent a measure of the amount of lost profit and describe the difference between the actual performance level of a manufacturing system and the performance level which is theoretically available on the market. Although these costs are only of a theoretical nature, they represent a significant threat if one considers the fact that potential competitors could invest in such technologies themselves. For this reason, efficient life cycle management not

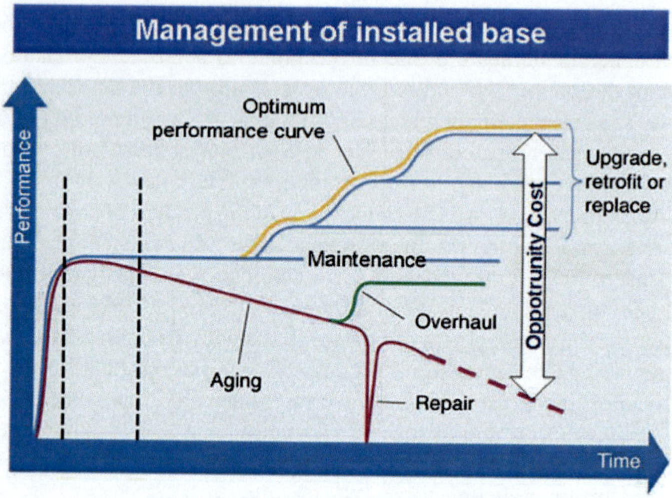

Fig. 10.5 Strategies to leverage operational excellence for our customers [2]

only provides financial advantages but also protects a company from being overtaken by competitors from a technological point of view.

If the example is considered abstractly, three fundamental strategies can be derived for developing and providing successful services or even complete service packages. The first strategy includes services which are orientated towards the entire life cycle of a product and aim at extending product lifetime (e.g. maintenance contracts). A second strategy comprises services which improve the efficiency of a product and thus increase the production yield per unit of time (performance services). The third strategy demands both dimensions to be taken into account as well as continuous services to be provided with the aim of improving performance. Only in this way is it possible to maximise product efficiency and product benefit and achieve operational manufacturing excellence.

10.2.2 Supporting Activities and Modules

In order to support companies in managing the improvements mentioned above, a platform offering tools and methods to innovative firms has been developed within the EC-funded RTD project ProSecCo (Product and Service Co-design, G1RD-CT-2002-00716). This platform is based on a high-level frame integrating both methodological and commercial viewpoints where ICT and non-ICT tools and methods are conceived to support SMEs in innovation management. Three main modules have been created in order to provide full assistance to companies interested in developing integrated solutions:

A *diagnosis* module is meant to support the whole PSCD (Product and Service Co-Design) process by assessing needs in innovation fields and evaluating the PSS ideas. The module is partly integrated into the other modules and major parts of the assessment questionnaires are implemented in and accessible via the ICT platform. With the diagnosis module, the platform provides both companies and consultants with a logical, structured framework for analysing the different functional areas, identifying real problems and establishing priorities for solving the problems detected. Four main areas are addressed through this set of tools:

- Identification of present status of the company and major improvements required
- Definition of the most tailored approach for improving performance
- Definition of an intervention plan, with a fixed list of results and a set of indicators and dimensions to be monitored and measured during the improvement process
- Current and retrospective evaluation of achieved objectives and analysis of possible derivations from planned times, costs and/or quality.

An opportunity recognition module, to assist firms in finding innovative PSS ideas. This is particularly important from a strategic and development point of view and is dedicated to extended market analysis, future scenario definition and idea generation. It is specifically targeted for PSS according to the results of the diagnosis and applies defined methodologies for the creation of PSS. Two main sub-modules form the opportunity recognition module:

- The *mind setting* sub-module explains a certain way of thinking and other conditions needed to set the scene for the workflow and the process of defining opportunities. In order to be able to go through the process, a different way of thinking is required which enables a different perspective to be viewed.
- The *methodology and process* sub-module explains the proposed steps to be taken for the process of opportunity detection. The steps provide a path to organise the thinking process in order to achieve innovative insights for product-service solutions.
- A process implementation module supports the user from an organisational point of view during the implementation procedure in order to re-define the approach of the company to the innovation process and organisation according to the requirements of the new PSS. The main activities performed by this module include:
- Acquisition and analysis of current practices
- Formalisation of present innovation process
- Proposal of improvements through the involvement of new actors
- Definition of a new innovation process
- Testing and simulation.

The results achieved in the project enable new PSSs to be generated where companies intend to innovate beyond the state of art. It also targets existing systems if improvements both of the system and the organisation are required. The experiences gained with SMEs and large companies confirm that PSSs can be successfully integrated if tangible and intangible assets are properly structured and managed by the company. Their co-design patterns are strictly linked to organisational processes. The need to reformulate already-existing working schemes through successful re-engineering often reflects new organisational and technological structures, the management of change and motivation, the need for training and the commitment of people to an open vision regarding the entire life cycle of the product. Companies are rarely interested in adopting the whole range of solutions simultaneously: step-by-step changes and improvements are usually favoured. Companies with a valid product and solution portfolio generally prefer to improve their innovation process in order to better exploit their technology and solutions, especially through the involvement of customers and other partners in their innovation process. SMEs are the main customers: they usually lack management and innovation management skills and this tool supports them in finding the right complementary partners. Large companies on the other hand tend to use only some of the tools, coupling them with approaches, methodologies and systems they have already adopted.

10.3 Selling the Benefit Instead of the Equipment

The fields involved with the initial steps of processing basic materials or manufacturing parts, components and equipment are being dislocated. By using information and communications systems, suppliers become involved in product development. Howcver, the so-called *system management* will remain in the hands of the OEMs operating directly with the market. The field of after-sales where long-term customer relationships can be established is gaining strategic importance. This development can be extended right up to the level of so-called performance contracts. Here, the manufacturers of technical products will also take over operation of their products and will only sell their usage (Fig. 10.6).

This not only leads to increased manufacturer responsibilities but also to a stronger and more durable relationship with the OEM. The OEM, for example, purchases the service (instead of the machine) or pays only for inspected parts. This implies a conversion of fixed costs into variable costs for the OEM along with the extension of the value-adding chain and an opportunity for the manufacturer to increase profits. The manufacturer becomes a system operator who offers his services worldwide. All machines are connected and operated by modern information and communication technologies and controlled in a central surveillance centre where all incoming data are acquired, analysed and evaluated.

The availability of these tools will enable system operators to benefit from economies of scale. Once a database has been set up, statistical performance evaluation becomes possible and the *best practice* can be determined. Through worldwide learning, system operators will be able to rapidly ascend the learning curve. By applying information and communication technologies, new knowledge will become immediately available which can be implemented on all machines worldwide.

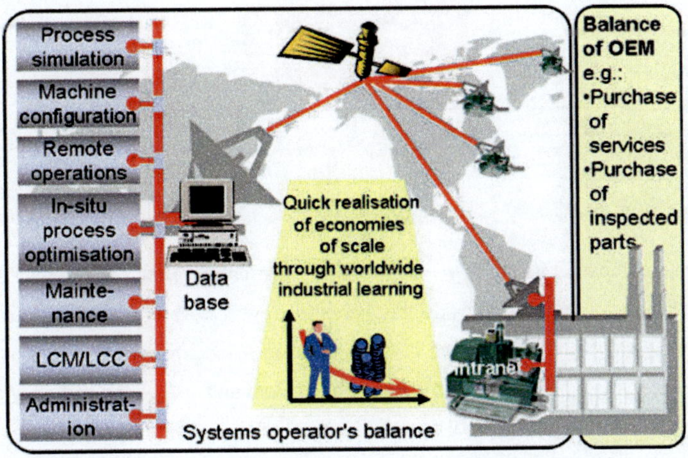

Fig. 10.6 Systems operators and performance contracts (Own graph according to [1])

10.4 Full Service Concepts

10.4.1 Business Model

A business model describes in simplified form how a company creates economic value and is an abstract representation of how a "business" works (business logic) [11]. The design of a business model is primarily dependent on the type of industry, the target group and the product/service. In general, each business model has 3 main components that theoretically represent the main idea and goal behind the business of a company. Figure 10.7 provides more detailed information about those 3 main components.

10.4.2 Service Management Terms

The terms defined and explained in this chapter are essential for the further understanding of the full-service concepts elaborated in the following chapters.

10.4.2.1 Full-Service Contractor

> …is the one, who can fulfill all requirements (architecture, engineering, inspection, management, procurement, waste disposal, etc.), associated with a contract. [12]

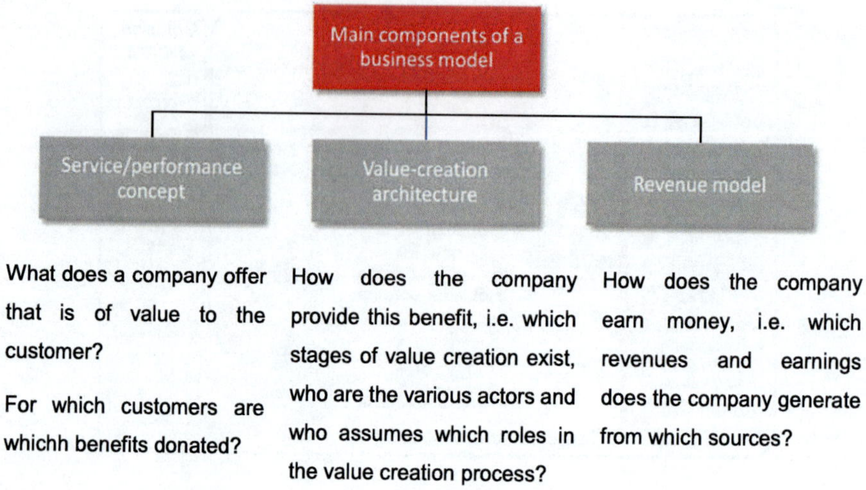

Fig. 10.7 Main components of a business model

10.4.2.2 Full-Service

> Providing customers with a complete range of services. [13]

10.4.2.3 Product as a Service

> Product as a service is the concept of selling the services and outcomes a product can provide rather than the product itself… In its purest form, the manufacturer continues to own and maintain the product, and the customer leases it for use or subscribes to a menu of services. In other scenarios, the customer owns the product but is not responsible for maintenance, or such responsibilities are divided according to the license agreement or warranty. In all cases, the manufacturer uses the product as a platform for delivering additional services to the customer (Fig. 10.8). [14]

Service Life Cycle Management (SLM) is *"…a strategic way to look at service planning and delivery as an integral part of the overall equipment life cycle management. SLM enables the service organization to manage all the service aspects of a product from design phases until it is no longer in service."* [15, 16].

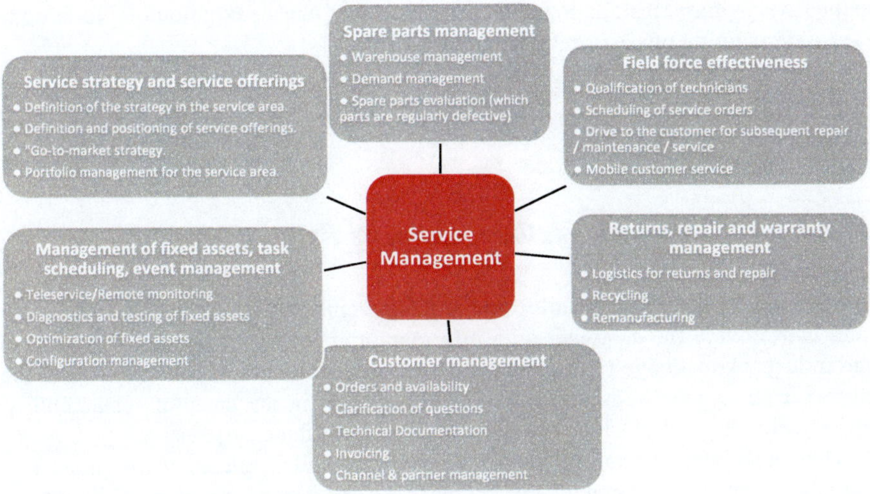

Fig. 10.8 Six main components of service management

10.5 Holistic Facility Life Cycle Management

10.5.1 Holistic Life Cycle Management

> …Holistic PLM is characterized by an integrated view on processes and business activities addressing not only product data, but also people and organizational structure, strategies and performance, working methods, information systems, etc., as well as their alignment to one another in order to monitor and control complexity along the product life cycle. [17]

After all the terms have been explained, it can be said that an implementation of a full-service concept for a holistic life cycle management means an adoption of service life cycle management.

10.6 Why Service Life Cycle Management?

This chapter sheds light on the main idea and the objectives of the service life cycle management (the integration of a full-service concept in a business model) and reveals the advantages and the disadvantages due to this implementation from a customer's and a producer's/provider's point of view.

Service life cycle management (SLM) means that the customer receives service, support and advice from the producer right from the start, in some cases even before the start of the asset's life cycle—during the so-called orientation phase. The service continues through the planning and development phase, procurement and delivery phase, installation and commissioning phase, exploitation phase up to the modernization phase or recycling phase [18].

10.7 Advantages

10.7.1 Advantages from the customer's Point of View

The customer benefits above all from the manufacturer's knowledge. The manufacturer can provide the customer with information about trends, future technologies and industry knowledge right from the start. The manufacturer can also inform the customer about legal regulations and standards, e.g. in the areas of sustainability, safety technology, explosion protection and energy efficiency [18].

The manufacturer could also undertake all activities related to the asset (for example: concept development, project planning and design, maintenance and operating, delivery service and commissioning, application programming, repair and conversion, etc.), which could save the customer time, money and organizational effort as well as the need for training of employees or additional staff. Experts, specialized in a certain field/product, achieve faster the desired result than non-experts. While internal employees would still have to familiarize themselves with the subject matter, the external expert could already deliver high-quality results. In addition, using of external forces means savings for companies in terms of social security expenditure. The use of an expert also means an increase in the quality of a job and risk minimization [19].

> When considered early enough in the life cycle, design engineering and service planners can consider modifying the physical architecture, design special tools and devise appropriate techniques to ensure fast as well as safe service. These techniques can be proven by employing the same simulation tools that are used in manufacturability assessments [15, 16].

A further advantage is the saving of money for maintenance tools and machines as well as for their maintenance, exchange and repairs, which leads to a reduction of the ongoing costs and of the commitment of capital, also to opening-up liquid scope for other (maybe profitable) investments [20],

Today, in the era of the Internet of things and the digitization, most of the big manufacturers use online sensor technology to track, control and monitor their facilities in real time. There is a continuous flow of data between the facility and the manufacturer. The collected data could be used for various analyses and this is another advantage for the customer. The service life cycle management contributes to increased transparency through standardized access to all relevant data regarding the controlling and the operation of a facility. The through the service life cycle management collected data can also support the quality management and the compliance with standards and guidelines. In a long term, the collected data (e.g. through risk early warning systems, maintenance forecasts, etc.) could be used for strategic planning purposes, for example, to identify improvement potentials, weak points and for more efficient capacity planning [21].

10.7.2 Advantages from the Producer's/Provider's Point of View

The main goal of every company is to be successful, competitive and better than its rivals. One of the main indicators that can provide valuable insight for the overall business success and the ROI (Return of Investment) of a company is the sales volume [22]. Every company strives to achieve better sales results and sales growth, because this usually means better business results.

We live in a time of globalization and digitalization and for today's customers there are almost no limits. The modern customer has fast access to almost all markets in the world and they are only 1 phone call or a few mouse clicks away. This is game changing and hence the companies try to adapt to this "new world" by offering new services and changing their business models.

> As profit margins in manufactured goods continue to be challenged due to increased global competition, price deflation and rising commodity costs, product companies must seek alternative sources of profitable revenue growth. One area of growth opportunity and profit protection is product service, and many companies are examining ways to increase service revenue or offer additional add-on services to supplement product revenues. [15, 16]

Jordan Belfort, the author of the memoir *"The Wolf of Wall Street" says in an interview for Success Resources that there is no chance for a company to be successful if it can't sale as well as "...At the highest level, sales is the transference of emotion. And the primary emotion you're transferring is certainty. The certainty that the product will fulfil my needs as a customer".*

One way for the companies to give their customers the sense of certainty and security is by offering full-service for their products during their whole life cycle, or in other words by adopting the strategy of the service life cycle management. To be able to transfer this feeling to its customers is a huge benefit for each company, because in this way the satisfaction of the customers is positively influenced. Many studies and analysis prove that it is very profitable and very beneficial for the company to retain its current customers and to make them long-term customers.

> As the old verse goes, "Make new friends, but keep the old. One is silver, the other gold." Similarly, a long-term customer is of more value than a single-deal customer, and it's a lot less expensive to keep a current customer than to acquire a new one. This is not to say that we shouldn't go out and get new customers, but if we can keep a larger percentage of those customers for a longer life cycle, we build on a revenue foundation that is more profitable and predictable; two factors that have created tremendous wealth for entrepreneurs. [23]

Another advantage for the producer is that, due to the, in best case, constant (depending on the region and the legal requirements) data flow between his facilities (located with his customers) and his own system, useful data can be collected for the goals of product optimization, weak point identification, defect detection, efficiency increase, quality improvement and development.

10.8 Disadvantages

10.8.1 Disadvantages from the Customer's Point of View

Perhaps one of the biggest disadvantages from the customer's point of view is the dependence on the manufacturer of the corresponding facility. Everything that concerns the facility and its work, down to the smallest detail (no matter whether process change, new process numbers, relocation etc.), must be discussed with the manufacturer. This causes additional communication and organizational effort [24].

According to Mr. Michael Schulz, leader of the Siemens' Gas and Power division in Mülheim an der Ruhr, another big issue that worries the customer is related with the security of the data transfer and with the accuracy of the data. Many companies (especially those in the automotive, chemical, energy, space and defense industries) and even some countries have very strict data security requirements. Therefore, there are often difficulties to ensure the provision of all the agreed services when not all the data relating to a relevant facility may be collected/transmitted. This is a big disadvantage for the customer, because he cannot receive 100% service and cannot benefit from all the services a certain facility or manufacturer could provide [3].

10.8.2 Disadvantages from the Producer's/Provider's Point of View

The disadvantages from producer's/provider's point of view are primary associated with the additional work and the additional personal that he needs to maintain the offered services.

> …it is critical to provide service technician training, and it is even more important to make available top-notch service information, such as theory of practice, diagnostics and troubleshooting, work instructions and illustrations, safety and compliance guidelines and part catalogs. But since the configuration of each deployed piece of equipment can be different enough to impact the accuracy of the troubleshooting and repair activity, effective service information should be configurable, comprising only information pertinent to the specific configuration of the equipment under service [15, 16].

The SLM requires a lot of communication and organizational effort in order to provide professional, adequate and fast service. In order to achieve a high level of communication quality the producer has to invest in modern and complex IT- and communication systems, which will allow him to keep a big data base with all of his customers and their facilities as well as with the already done, planned, in process and future services regarding each facility.

10.9 Full-Service Concepts in the Business Models

The aim of this chapter is to present an example from the real business life and to reveal the disadvantages and the advantages for both, customer and producer, as well as the changes in the producer's business model in this particular case. After a careful and extensive research and after interview with Mr. Jochen Gaertner, team leader of the Facility Management Research and Development department at Daimler AG, a very well-known company, has been selected as an example for a producer who offers a wide full-service pallet to its customers-Siemens. Siemens is a German international company known for various range of products and services in different industries. The company has a pioneering role in the fields of digitalization, automation, internet of things and Industry 4.0.

The Siemens product, chosen for an object of this chapter, is a wind turbine. Siemens designs, manufactures, assembles and installs its wind turbines by itself and in addition, the company monitors and maintains the active wind turbines during their whole life cycle, so this is a very good example for Service Life Cycle Management approach.

In order to explain the Siemens' full-service concept in depth, first the underlying business model is analyzed by the means of the so-called canvas model. The canvas model is a tool to visualize and test a business model [25, 26]. The model consists of the following nine elements (see Table 10.1) [27].

Table 10.1 The nine elements of the canvas model [27]

Elements of the canvas model	Explanation
Customer segments	In this segment, the potential target group of the product or innovation is defined
Value proposition	The second segment should show the values that can be offered to the target group
Channels	How is the value handed over to the future customer?
Customer relationships	How should the relationship to customers be built up?
Revenue streams	How should customers pay? What are the pricing strategies?
Key resources	What resources are needed to produce the product?
Key activities	What activities are required to produce the product?
Key partners	Which strategic partners are needed to produce and market the product?
Cost structure	This segment includes financial planning

The represented nine elements of the canvas model help for a clear identification of all the players involved in the wind turbine business. The opportunities offered by the full-service concept and the resulting necessities also become apparent.

The first section deals with the customer segments. In the case of Siemens wind turbines, these are operators of wind farms who need reliable solutions over the entire product life cycle [15, 16]. The changes affecting the business model brought about by the full-service concept are very significant in this segment. By offering a full-service solution for its customers, Siemens addresses on the one hand customers without knowledge in the area of wind turbines. Siemens delivers this missing know-how. On the other hand, it is possible to offer existing customers or the existing customer segment new services through the service of all assets (planning, production, control, repairs, etc.) and to strengthen customer loyalty to Siemens.

The segment of value proposition contains major innovations when comparing full-service concepts with a normal purchase of a wind turbine. Instead of just selling the product and transferring the whole responsibility to the customer, the wind turbine is accompanied by Siemens through all phases of the product life cycle together with the customer. *"From planning and initial commissioning through to productive operation, the optimization of the available resources plays an important role"* [15, 16]. Above all the automation of the turbines, turbine protection and cloud-based condition monitoring are important innovations when considering the full-service concept. On its website, Siemens asks the question: "Do you know the health state of your plant?" Cloud-based monitoring ensures the availability of wind turbines at the highest level and enables the customer to keep the cost of energy low [15, 16].

The adoption of Service Life Cycle Management strategies also changes the section "Channels". Apart from the usual channels, further ways are being developed to pass on values to the customers. On the one hand, supportive activities are already necessary during the process of finding ideas. To this purpose, capable employees will support the customer. Building on this, the planning and implementation of the wind turbines will also be carried out by Siemens employees. The maintenance and control of the wind turbines will also be carried out via new channels by the full-service provider Siemens. This requires cloud-based channels, which must also be provided by the infrastructure of the full-service provider.

The fourth segment of the canvas model looks at the customer relationships. This raises the question of which model is suitable for full-service concepts. If one considers the purchase of a wind turbine as a single transaction, the full-service concept is not fulfilled. It is important to establish a strategic partnership with the customer that will last till the end of the products life cycle and that will share the strategic influence on the business as well as the risk of the business [3].

The next segment that is considered is the revenue streams. The question here is how customers should pay and what strategies should be used in pricing. Primarily, the full-service concept is a model in which the customer's relationship with the manufacturer creates a certain dependency. In the special case of Siemens, this dependency is created by the knowledge of the Siemens employees and the cloud-based monitoring and control of the plants and wind turbines. Furthermore, the full responsibility for repairs and guarantee of availability by Siemens employees creates a possibility to generate permanent and long-term revenue opportunities. This raises the question of whether it makes sense to have the customer pay for the availability or to set the purchase price in the form of a transaction at a level that ensures a profitable business over the entire life cycle of the product despite the full-service concept. If the customer is to be charged for the availability of the product, a regular staggered payment method is the appropriate option.

If we now look at the key resources, it becomes clear that considerable changes are necessary for a full-service concept. First there are the employees. Siemens needs additional personnel to carry out the consulting activities before and during the planning phase. Additional personnel are needed for installation, remote maintenance via cloud-based monitoring and for repairs and dismantling after the end of the life cycle. These additional employees must be trained to comply with availability guarantees. Furthermore, the company's infrastructure must be adapted both inside and outside the company. The availability of the product must be the top priority. KPIs, data and facts must be permanently available in order to comply with service agreements. Industrial communication must also be standardized to meet the latest requirements. Siemens also offers solutions for this [16].

The so-called key activities, which must be carried out for the product, are broadly diversified. On the one hand, a lot of planning effort is required to support the customer already in the process of idea generation. This is followed by engineering tailored to the customer, production, assembly on site and then all maintenance and repair work.

The last two segments are Key Partners and Cost Structure, which have to be defined by the Siemens specialists. For the full-service wind turbine product, Siemens has included the most important components in its product portfolio. Financial planning must be carried out in order to ensure that the full-service concept as such is profitable and self-supporting.

After examining the nine segments of the Canvas model, it becomes clear that the business model of Siemens AG's full-service concept for wind turbines has significant effort to follow to realize the full-service concept. Many changes must inevitably be made in order for the company to be able to guarantee all services to the customer. This includes above all trained, capable personnel and an outstanding digital infrastructure as well as well working intra-logistical and extra-logistical infrastructure in the company.

10.10 Advantages and Disadvantages Over the Business Life Cycle

The life cycle of an investment, in this case a wind turbine is examined in this chapter and advantages and disadvantages from the customer's point of view are explained.

The service life cycle management by Siemens with its advantages shown in Fig. 10.9 enables Siemens to relieve customers of many difficult decisions. In addi-

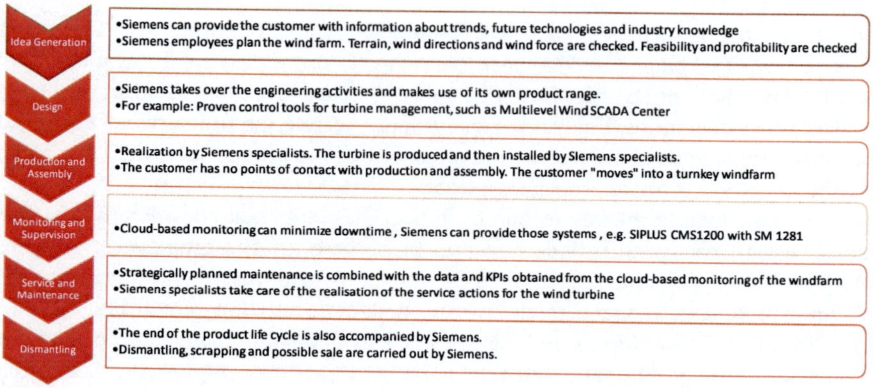

Fig. 10.9 Advantages of the full-service over the investment's life cycle [28]

tion, many activities with a great deal of know-how are taken over so that the customer receives the turnkey plant. The big goal is to supply the customer with a highly available product. The cost of energy is minimized through performance, reliability and high efficiency. In addition, Siemens enables the customer to keep the cost of operating the turbines to a minimum through turbine protection, turbine automation, turbine and wind farm management and cloud-based monitoring. Risks are minimized and downtimes are best avoided or planned [28].

While many advantages result for the customer of a full-service product, the biggest obvious problem or disadvantage shown in Fig. 10.10 is the dependency relationship with Siemens. In every stage of the wind turbine life cycle, it becomes clear that the customer must rely on expertise and know-how of Siemens. In addition,

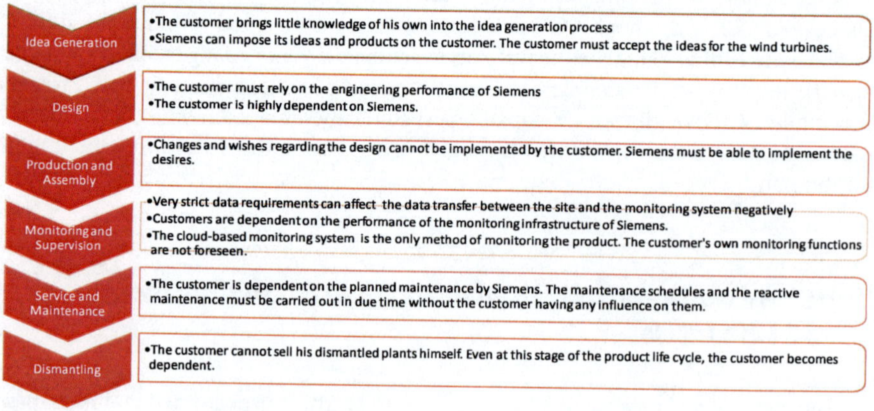

Fig. 10.10 Disadvantages the full-service concept over the investment's life cycle [28]

significant efforts must be made to ensure data security and digital infrastructure to enable the successful transfer of the operation, control and maintenance of the plant.

10.11 Summary and Conclusion

Many advantages could be worked out for the customer. One of the many advantages is the fact that the customer can focus on his core business. He does not have to waste his capacities on information procurement, engineering services or maintenance and repairs. This in turn is usually accompanied by an increase in effectiveness and productivity. Only a few aspects are negative for the customer. One of the few is the fact that the customer enters into a strong dependency relationship with the full-service provider. For the producer, however, this relationship of dependence is a positive aspect. Through high-quality products and the fulfilment of the full-service model, it is possible to achieve a good position on the market. In addition, it is possible to identify and improve weak points of the product at an early stage through a constant data flow between the product and the control and maintenance departments. However, this is only possible with considerable expenses in the personnel segment and in the company's infrastructure. Employees must be trained, data highways set up and technical know-how constantly improved [29–32].

The service life cycle management strategies and approaches are the next step in offering a good service and in creating and adding value to a product. The SLM not only supports the customer by doing his core business tasks, but also facilitate the efficiency improvement, the sustainable development and the optimization of the customer's business through a continuous flow of information and regular subject-specific analyses. A SLM strategy can be applied across many industries, from selling water dispenser on a leasing basis to selling high-tech customized solutions in the fields of production, telecommunication and power generation, and helps a producer to build close long-term relationships with its customers, which has a crucial importance in today's world of digitalization and globalization. Furthermore, an implementation of a SLM strategy changes the entire business model of a manufacturer by adding new responsibilities and jobs (even departments) and associated with them interfaces, required technologies and processes (primarily in terms of communication and logistics).

References

1. Morar, L., Westkämper, E., Abrudan, I., Pisla, A., Niemann, J., Manole, I.: Planning and Operation of Production Systems. Fraunhofer IRB Verlag (2008)
2. Niemann, J., Tichkiewitch, S.: Westkämper Engelbert: Design of Sustainable Product Life Cycles. Springer Verlag, Heidelberg Berlin (2009)

3. Niemann, J.: Die Services-Manufaktur, Industrielle Services planen –entwickeln – einführen. Ein Praxishandbuch Schritt für Schritt mit Übungen und Lösungen. Aachen, Shaker Verlag (2016)
4. Fraunhofer-Institut für Arbeitswirtschaft und Organisation IAO: Fraunhofer IAO. Retrieved from Fraunhofer IAO. https://wiki.iao.fraunhofer.de/index.php/Smart_Services, 15 Dec 2018
5. Feldhusen, J., Gebhardt, B.: Product Life Cycle Management für die Praxis. Springer Verlag, Berlin, Heidelberg (2008)
6. Freitag, M., Hämmerle, O.: Smart Service Life Cycle Management. WT Werkstattstechnik, 477–482 (2016)
7. Berkovich, M., Leimeister, J. M, Krcmar, H.: Requirements engineering for product service systems. A state of the art analysis. Bus Inf Syst Eng. https://doi.org/10.1007/s12599-011-0192-2 (2011)
8. Mont, O.: Clarifying the concept of product–service system. J. Clean. Prod. **10**(3), 237–245 (2002)
9. Meier, H., Roy, R., Seliger, G.: Industrial product-service systems—IPS2. CIRP Ann. **59**(2), 607–627 (2010)
10. Meyer, K., Thieme, M.: Theory and practice for system services providers in complex value and service systems. In: Proceedings of the International Symposium on Service Science ISSS, Leipzigiger Beiträge zur Informatik, Band 41, University of Leipzig (2013)
11. Schwens, C.: Entrepreneurship. Cologne, Germany (2019)
12. Business Dictionary: Retrieved from https://www.businessdictionary.com/definition/full-service-contractor.html, May 2015.
13. Cambridge Dictionary: Retrieved from https://dictionary.cambridge.org/dictionary/english/full-service, May 2015
14. Rouse, M.: TechTarget. Retrieved from https://searcherp.techtarget.com/definition/product-as-a-service, 2015
15. Siemens: Service Life Cycle Management. Plano, TX 75024 USA. Siemens PLM Software (2017)
16. Siemens. The Key to the Digital Enterprise. Retrieved 23 May 2019, from Siemens. https://new.siemens.com/global/en/products/automation/industrial-communication.html (2017).
17. Corallo, A., Di Biccari, C., Lazoi, M., Marra, M.: A methodology for product life cycle management assessment. Retrieved from https://www.researchgate.net/publication/320886933_A_Gap_Analysis_Methodology_for_Product_Lifecycle_Management_Assessment (2017)
18. Sew-Eurodrive: Sew-Eurodrive. Retrieved from https://www.sew-eurodrive.de/unternehmen/ihr_erfolg/life_cycle_services/life_cycle_services.html#panel-e32a2360-e2bd-4dc1-a749-0a573910da3e-5, Retrieved Oct 2019
19. Schulz, M.: Full-service concepts for a holistic life cycle management. In: Kozovski, I. (Interviewer) Mülheim an der Ruhr, Germany (2019)
20. Orga-Pannhausen.: Orga-pannhausen.de. Retrieved from https://www.orga-pannhausen.de/faq-outsourcing/. Retrieved Oct 2019
21. Daimler, A.G.: Computer Aided Facility Management Strategie DE. Daimler AG, Stuttgart (2019)
22. Verizon: Verizon Connect. Retrieved from https://www.verizonconnect.com/resources/article/how-to-measure-business-success/. Retrieved Oct 2019
23. Myler, L.: Acquiring New Customers Is Important, But Retaining Them Accelerates Profitable Growth. Forbes (2016)
24. Kiel, C.: digitaler-mittelstand.de. Retrieved from https://digitaler-mittelstand.de/trends/ratgeber/outsourcing-vorteile-und-nachteile-von-auslagerung-3198. Retrieved Oct 2019
25. Startplatz: Retrieved from Business Model Canvas: https://www.startplatz.de/startup-wiki/business-model-canvas/. Retrieved Oct 2019
26. Osterwalder, A.: Alex Osterwalder. From https://alexosterwalder.com/. Retrieved Oct 2019
27. Sammer, W.: Up to Eleven, from Das Business Model Canvas: Dein Geschäftsmodell kompakt. https://ut11.net/de/blog/dein-geschaftsmodell-kompakt-der-business-model-canvas/. Retrieved Oct 2019

28. Siemens: From Das Maximum aus Wind herausholen. https://new.siemens.com/global/de/branchen/windenergie/equipment.html. Retrieved Oct 2019
29. Niemann, J., Pisla, A.: Sustainable potentials and risks assess in automation and robotization using the life cycle management index tool—LY-MIT. Sustainability **10**, 4638 (2018)
30. Niemann, J., Schemann, T., Erkens, J.: Servitization—pathway of transformation from product manufacturer towards a service provider. In: 2018 International Conference on Production Research—Africa, Europe, Middle East 5th International Conference on Quality and Innovation in Engineering and Management, 25–26 July 2018, Cluj-Napoca, Romania
31. Stöhr, C., Janssen, M., Niemann, J.: Smart services. In: 14th International Symposium in Management, Challenges and Innovation in Management and Entrepreneurship, 27–28 Oct 2017, Timisoara, Romania
32. Niemann, J.: Eine Methodik zum dynamischen Life Cycle Controlling von Produktionssystemen. Heimsheim: Jost-Jetter Verlag, 2007IPA-IAO Forschung und Praxis 459). Stuttgart, Univ., Fak. Maschinenbau, Inst. für Industrielle Fertigung und Fabrikbetrieb, Diss. (2007)

Chapter 11
Tools for the Digital Business Transformation

11.1 Business Model Dimensions and Trends

The primary economical goal of a company is to create value for the stakeholder, in order to generate profit. Value is created by combining the knowledge and competences of the employees with resources (tools, assets, technology, and information) and satisfying a market need [1]. The technological progress and the changing market bring new potentials to provide value, as new customer and market demands arise. However, it also brings high risks for existing industries and markets, as their business models can become redundant [2].

11.1.1 Dimensions and Strategies

Creating value means to provide a solution to a current or future issue, satisfying a customer need or to provide a public good that the customer is willing to pay for [3]. A business model describes the basic logic of a company on how to create that value for customers or partners and describes how the value generates revenue flowing back to the company. The provided value enables a differentiation toward the competitors resulting in customer loyalty and competitive advantages [4].

There are certain dimensions of a business model:

- The **customer** dimension contains the various customer segments, customer channels and customer relationships.
- The **value** dimension describes the purpose of the company and its activities, resources, competences and processes needed to create value.
- The **partner** dimension contains business partner, partner channels and partner relationships.

J. Niemann and A. Pisla, *Life-Cycle Management of Machines and Mechanisms*, Mechanisms and Machine Science 90,
https://doi.org/10.1007/978-3-030-56449-0_11

- The **financial** dimension deals with revenues and costs and describes how the value generates profit.
- The **distribution channel** dimension describes the channels through which the value is offered and distributed to the customer.
- The **internal organisation** dimension describes the organisational system that exists to form the frame of the business model [4].

When developing a new business model, these elements are combined to achieve a new way of providing value for the customers and partners and creating a differentiation to the competitors [5].

The literature has shown that business models can be developed into different directions. According to Ansoff, whose product-market-matrix is the basis for many strategy concepts, there are 4 different growth directions concerning market factors and product factors, however, his matrix neglects factors associated with service. The first direction is to further develop the current business model within the current market and with the current product portfolio (market penetration). This means that a strategy is set on how to become a market leader or on how to improve the product, or access more customers. The second direction is market development. This means that the current product portfolio is brought to a new market. This could include geographical expansion or new market segments. For example, a smartphone developer could introduce a smartphone specialized on the needs of elderly people. The third direction is to develop new products that satisfy a market need. The last direction is said to be the most difficult, as it includes developing a new product or service for a totally new market. This is also referred to as the "blue ocean" strategy [6, 7].

Figure 11.1 shows Ansoff's product-market-matrix. However, the original model was developed many years ago, and did not include the service perspective, which has been included in Fig. 11.1.

Furthermore, these development directions can underline different perspectives:

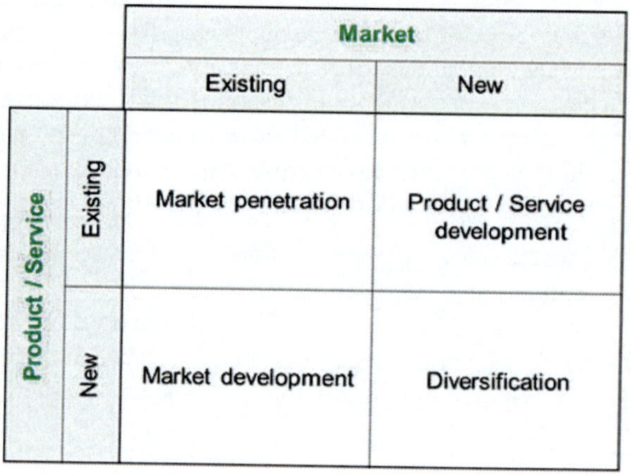

Fig. 11.1 Adaption of Ansoff's product-market matrix [6]

- Market driven strategies (A business model is developed to satisfy a need given by the market or enabled through new technologies)
- Supply driven strategies (Change of the value chain, usage of digitalization or new work and organization form e.g. crowd sourcing)
- Finance driven strategies (Cost models are changed, low-price strategies are put into place) [6].

Red Ocean Versus Blue Ocean Strategy

In the last 20 years, European companies have primarily concentrated on improving quality, costs and process time. This has led to a supply surplus and a highly competitive environment, known as a "red ocean", which is defined by pricing pressure and competition. The "blue ocean" stands for the opposite, meaning new markets with no competition [8].

If a new product or service is to be developed, a thorough market analysis must be made to ensure that there is no similar product on the market. Furthermore, it should also be examined in which market segments there is a customer demand. If so, an evaluation should be made to see if it is worthwhile to continue in this field.

Entering a competitive market can be difficult, so solutions should be developed on how to stand out from the existing ones. The blue ocean strategy says that it is easier to enter a new market to ensure sustainable business success by providing an innovation. However, this new market has to be created first and a solution for the market needs has to be found. Following the red ocean strategy means that there are many competitors that may have been in the business for a very long time and have more expertise. Here we speak of a crowding out market [7]. When following the red ocean strategy, the competition must be thoroughly analysed, as well as the customer demands. It is the goal to outperform the competition through quality, price, design or value adding service [8].

11.1.2 *Trend to Digitalisation as New Business Model Enabler*

Todays trends such as lean supply chain, smart manufacturing, cloud platforms, big data management, artificial intelligence, augmented and virtual reality, mobility, smart e2e transparency, additive manufacturing, customization, service-orientated business models and outsourcing etc. are all based on the changes of customer mentality and technological advancements. Through the faster means of communication caused by the introduction of worldwide accessible internet, trends form and can spread much faster than ever before in history. Figure 11.2 shows the different driving factors on business model trends.

Shorter product life cycles, continuous changes of business processes and higher customer expectations lead to a new kind of relationship between the customer and

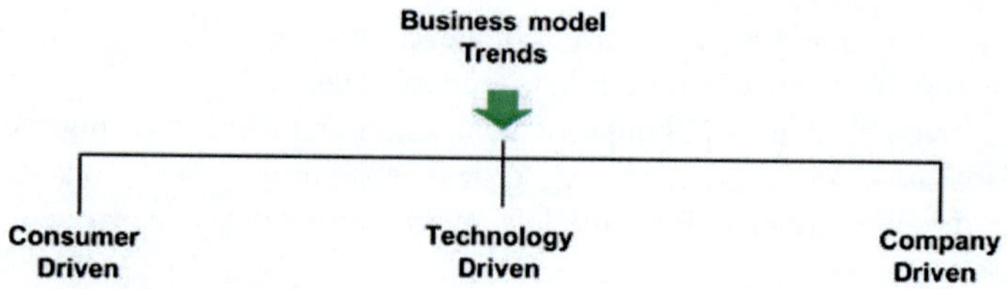

Fig. 11.2 Business model trends

business partners within the value chain. Customers expect faster business transactions, one-stop-shop solutions and transparency in the supply chain. This is only possible, if companies can digitalise information and data about products, customers, processes and services and thereby digitally transform their business model. With this new working method, a high amount of data is collected about business procedures and production processes, customer demands, as well as data about internal and external communication, requiring a high amount of management and data analysis [9].

Digital transformation means a reorientation of products, services, processes and business models towards the continuously digitalised world and results in faster transactions and more reliability through quality and security and therefore leads to higher customer satisfaction [10].

The digital transformation of business models can be implemented in three general phases (see Fig. 11.3):

Phase 1: Digitise the current business and build a platform for digital processes.
Phase 2: Integrate Internet of Things (IoT) functionalities into the platform and develop digital services.

Fig. 11.3 Digital transformation (according to [10])

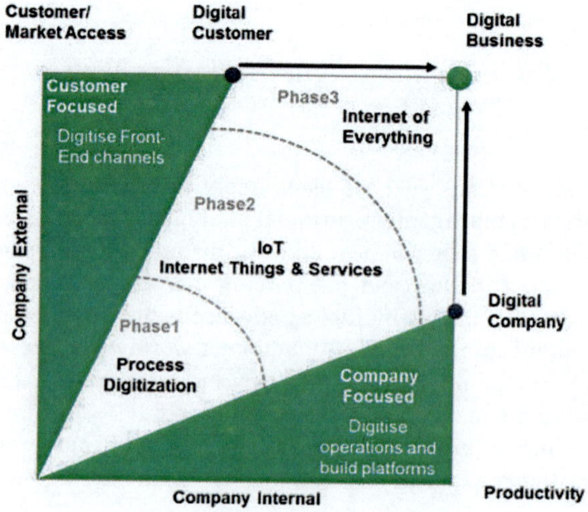

Phase 3: Close e2e loop of the entire business operations and modularize platform services [5].

Figure 11.3 depicts the journey from the traditional to the digital business. The model is divided into company internal and external elements, as the digitalisation of a business model can only work, if both the customer side and the own company can be "digitalised". This begins through the digitalisation of the channels and processes used to create or provide value. Afterwards, the digitalisation of products, services and other objects included in the value chain. Finally, full digitalisation of all transactions and procedures with a high automation level leads to a fully digitalised business model [11].

However, the amount of data and all communication processes, resulting from the digitalisation requires a high amount of management and analysis to ensure the correct interpretation of the data. A solution to the high maintenance of data quality and the management of all digitalised processes, services, machines, products and the network of partners and customers is a cloud platform. This allows a constant exchange of process data and an automated analysis throughout the whole supply chain as well as full value chain management through the platform owner [11] (see Fig. 11.4).

Cloud and IT-service providers can help with the setup of a suitable platform. Typical functions of a cloud platform are multi-client capability, scalability, availability and integration possibilities of external databases with and integrated development environment that supports different programming languages (e.g. Python, Java). A core aspect during the development of a platform, is to make sure that the collected data is processed in a correct and exact manner to ensure high output data quality, as all automated decisions and analyses are based on this raw data. Furthermore, new value adding services can be created and offered, such as: automated transaction procedures, invoice creation and download, monitoring services of products, machines and processes and even external services, offered by the service

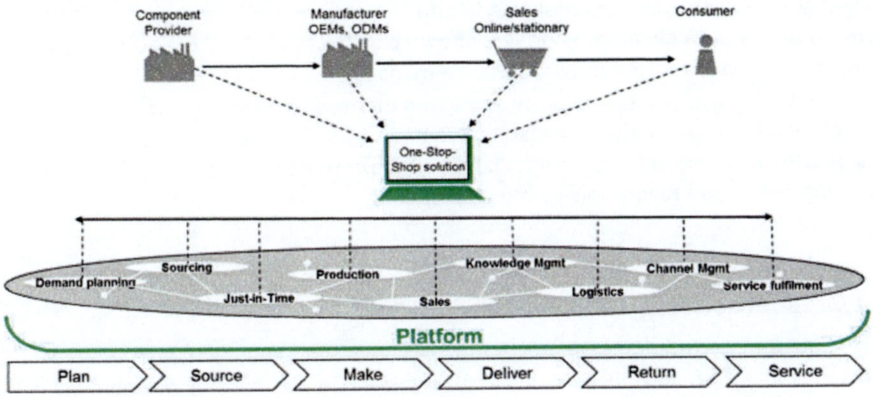

Fig. 11.4 Example of a platform (according to [12])

providers integrated into the platform. This means that the service portfolio can be extended without investing in an own service department, creating a one-stop-shop solution for customers [11].

To digitalise the whole value chain, not only must the products and processes be digitalised, but also the logistics, so that the digital image of all objects and their relevant data can be cross-linked to the physical world and also communicate with the external environment outside of the production site. Relevant information that can be collected is for example: status, condition, position, location and weight. This can be achieved through RFID chips (Radio frequency identification chips), GPS modules and scanning systems that can be placed into the products, the transport boxes and trucks and also used for the administration, quality or sales department [11].

Looking at an entire value chain, there are certain stakeholders: the supplier, the producer, the logistics companies, the service provider and the customer. By integrating all players, products and services into a cloud-based platform and by digitalising the value chain (products, services, trucks etc.) a transparent and fully automated manner of communication between the stakeholder but also between all objects in the value chain can be achieved. All this data can be collected through chips and sensors and sent to a cloud platform to which all stakeholders of the value chain have access to, in order to receive necessary information. A benefit of the platform is not only the automated communication, but also the integration of customers into one's digital infrastructure. A network of services can be provided, efficient and standardized processes and a transparency in the value chain, so that a customer "lock-in" can be achieved through a strategic integration into the platform systems. The key is to transform a straight-lined valuc chain into an ecosystem that encourages innovation and efficiency [11].

Many business model trends such as crowd based business models or for example Car2Go or Uber use such platforms to provide the services their business models promise. A simple example is the case of Uber: the service provider (car owner) is connected with the customer (traveller), who can provide the time and place he needs the service (transportation) and fulfil the payment all through the platform. The company does not employ drivers or own cars but provides the platform through which the service and the transactions are managed.

The platform business models have proven themselves in the digital age and have a high potential to remain as a sustainable model for the next years [10]. However, the platform is only the box in which the value proposition is served. The key is to develop a new and revolutionary business model.

11.2 Methodology

To develop a new business model and to transform the current business model, numerous methods exist to support the process. The methods that will be presented have been selected considering the factor of applicability in operational businesses and frequency of usage amongst enterprises. In the literature, one can find many different and specialized approaches on how the stepwise procedure should take place. To provide extensive applicability to any kind of business sector, the stepwise approach presented was based on a combination of methodologies from various authors.

A methodology is defined as "a system of broad principles or rules from which specific methods or procedures may be derived to interpret or solve different problems within the scope of a particular discipline. Unlike an algorithm, a methodology is not a formula but a set of practices." [12]. Table 11.1 explains the created methodology and shows the recommended methods that can be used.

However, this process and its tools must be followed and used by a competent team with a suitable set of skills. The selection of the team members is highly important, as the business model they will find could be crucial to success. Analytical and creative people should be part of the team which ideally consists of 5–7 individuals including a host or presenter, who oversees the process and leads the conversations into the right direction. The team members should come from different departments to ensure a holistic perspective of the business model elements. Furthermore, the path to new innovations or the or business model transformation, is often obstructed or prevented by pessimists in the company as they don't like change. Transformation is "a process of profound and radical change that orients an organization in a new direction and takes it to an entirely different level of effectiveness. Unlike 'turnaround' (which implies incremental progress on the same plane) transformation implies a basic change of character and little or no resemblance with the past configuration or structure." [12]. In this thesis, it refers to the change of the core of a business model which was formally in place in a company. The aim of the business model transformation is to improve the existing business or change it with the result of success and higher profitability.

Therefore, the project leader during the implementation phase should be a change agent, who is specialized in change management, as transformation is not always welcome. The project leader should be able to orchestrate the processes and make the team use the tools effectively and efficiently. One of the keys to success is not only the team, but also the dedication of the top management to increase innovation and change [8].

11.2.1 As-Is Analysis

Companies are a complex of various elements and interdependencies. Indicators and analytic data may point out deficits and faults, however their cause is a consequence of complex and interdependent influence factors and therefore not fully obvious. One of these elements is human behaviour, which is not easily measurable though quantitative models and data. A holistic diagnosis is the baseline for the preparation and sustainable redevelopment of the existing business model and the company.

Table 11.1 Methodology for the development of business models [5, 7, 8, 13–15]

Step	Explanation	Methods in use
As-is-analysis	The current state of the company is analysed to find out what issues exist and what the best practises are. The goal is to ensure common understanding on why a new business model is necessary and to find out what went wrong	1. SWOT analysis 2. Benchmark 3. Ishikawa
Goal definition	Before developing ideas, the enterprise must decide which direction should be taken and what the goals are	1. GAP analysis 2. Goal pyramid 3. Scenario analysis
Business model idea creation	Ideas are collected in groups and systematically developed. This phase lays the foundation for the further steps	1. Destroy your business 2. Empathy map 3. St. Galler business model navigator
Detailed business model design	The ideas are transformed into business models and described in detail	1. Canvas 2. SIPOC
Business model evaluation	The new-found business models are evaluated to enable a comparison and also to help decide on the most suitable one	1. PESTEL model 2. Porters five forces 3. Value benefit
Solution design	The aim of the solution design is to consider all relevant details of the new-found business model on which the planning and implementation phase is based	1. Detailed process design 2. Resource planning 3. Investment planning 4. Business case
Implementation strategy	When the new business model has been selected, the implementation strategy is developed. Furthermore, the setup of a transformation plan regarding change management aspects as well as some monitoring tools is important	1. Project charter 2. Action plan 3. RACI plan 4. Project plan 5. Transformation plan 6. Performance management

This is how the potentials, unused resources, opportunities and risks that may affect growth and profitability can be discovered [3].

An important point when analysing the own company, is not going too deep into detail, however deep enough to find the critical issues [16].

Three main parts of the company should be analysed: The own business model (customers, value proposition, value chain and profit model), the stakeholder (customer incentives, partners, competitors, own company) and the external influences on the business (ecosystem) [16].

Questions to be answered during the as-is-analysis of the business are:

1. What can my company provide to be attractive and accepted in the future?
2. What must be changed to ensure sustainable survival and competitiveness?
3. What mechanisms must be implemented to recognise opportunities, risks and the need for change early on? [3].

When the meaning and aim of the company and its vision is fully understood, and defined, the potentials and next steps to transform the business model can be seized.

The As-Is-analysis includes all functions and departments of a company, the product spectrum, technology, production depth, quantity framework, financial data, customer and supplier data, organization data, all methods and tools used, as well as the employees and their relationships toward each other and the company. During the As-Is-analysis deficits will be found, which are a result of various reasons. For example, wastage of resources or potential of the employees or not seizing opportunities by acting in a non-future orientated way [3].

The following methods are some of the many analytical tools found in the literature. These were estimated to be the most suitable and applicable in operational business.

11.2.1.1 SWOT

Explanation and Background

The SWOT analysis is an analytical instrument commonly used in strategic management and is the basis of many marketing strategies. It gives a snapshot of the company, departments, processes, products or projects and helps develop strategies on how to overcome risks and seize opportunities [13].

This method was developed in the 1960s by the military and then adopted by companies to enable a systematic situation analysis. It takes internal and external factors into account enabling a diverse perspective on the company. SWOT stands for strengths, weaknesses, opportunities, and threats. Strengths and weaknesses give the internal perspective of the company regarding resources in the form of image, motivation of the company, know-how of employees, financial issues and technology, while opportunities and threats regard external matters which influence the business model, company, or product (the analysed topic), for example customer expectation, trends, competitors, technology, or politics. The four areas of analysis are put in a matrix, which is to be filled out by a team. The team can consist of any number of people, however, the SWOT-analysis has the best and most accurate result if a diverse team of 6–10 people is working on it [13].

Procedure and Application

The procedure and usage of this method will be described in four steps. The premise of this method, is that the method is fully understood by the team and that the topic to be analyses is also defined. In the case of business model development, the current

Fig. 11.5 SWOT [13]

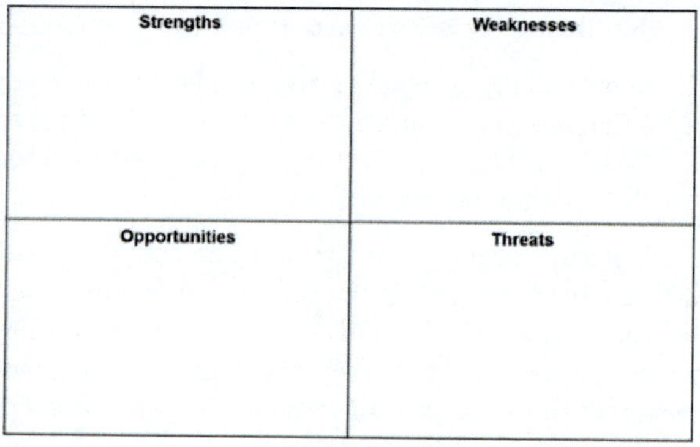

business model in place should be analysed and how the company is implementing it (Fig. 11.5).

Step 1: Describe strengths and weaknesses
The team members should collect all opinions of the detailed strengths and weakness of the company and then cluster them into categories. These can then be written in the according fields.

Step 2: External analysis
Opportunities and threats to the company and business model are collected here. These are all factors that cannot be influenced, as they are due to changing customer mentality, technology or politics. These are not to be filled into the SWOT-matrix yet.
As a guide line for the analysis, the following four dimensions should be considered:

- Main trends: regarding technology, regulations, social-cultural trends and social-economic trends.
- Market drivers: market segments, demands, market attractiveness, exchange costs.
- Macro-economic drivers: consisting of global market demands, capital market, economical infrastructure and other resources.
- Industry drivers: regarding stakeholder, competitors, substitutional products and suppliers [17].

3. Step 3: Derivation of opportunities and threats
By combining the strengths and weaknesses with the external analysis the significant opportunities and threats can be made out. A trend that can be seized by one of the strengths for example is an opportunity; a trend that cannot be seized due to a strength is a threat and so on. The result from the internal and external analysis is written in the SWOT-matrix. Then, they are clustered into topics and general titles. In a later stage strategies to seize opportunities and minimize threats can be worked on.
4. Step 4: Strategy and measures

	Opportunities	Threats
Strengths	Strength-Opportunities Strategies	Strength-Threats Strategies
Weaknesses	Weakness-Opportunities Strategies	Weakness-Threats Strategies

Fig.11.6 SWOT strategies [13]

In the last step, the opportunities and threats are analysed and measures to seize the opportunities and to minimize or prevent threats are worked out. This can be done so in an extended SWOT matrix as shown below (Fig. 11.6). Each field is combined with each other. This way, the company receives an overview of all internal and external factors that have led to the situation today and also of the opportunities that lie in the future [13].

11.2.1.2 Market Segmentation & Benchmarking

Market Segmentation

A thorough market analysis is necessary, as it can reveal insights that help generate new business models. The competitor analysis is one of the most important steps. Through this, demand niches can be discovered, new products can be developed and modifications or elimination of certain products or services can be implemented. The market analysis can be segmented into provider based market segmentation, product-based segmentation and demand based market segmentation. The market segmentation seeks to elaborate the complex interdependencies and relationships between provider, products and demand. Table 11.2 explains the main characteristics of the market segmentation [13].

For each of the defined segments a market and competitor analysis can be made.

Benchmarking

After the relevant market segments have been identified, the benchmarking process can begin. The goal of benchmarking is to compare capabilities within relevant fields or functionalities. Also, it can generate first ideas for new products or services by looking at the best-in-class. This analysis is made between the SWOT analysis and

Table 11.2 Market segmentation [8]

Type of market segmentation	Provider based segmentation	Product/Service based segmentation	Demand based segmentation
Description	Strategic groups of competitors	Goods and services provided in the market	Segments of consumer groups and their specific demands
Object of segmentation	Companies	Products/Services	Consumer
Characteristics of segment	• Company size • Company scope • Diversification level • Marketing methods • Financial strength	• Diversification between products and services, consumer and industrial goods • Product specific criteria such as usage possibilities, components, substitution rates	• Preferences depending on consumer groups diversified by age, gender, country etc. • Consumer behaviour

the GAP analysis, so that the strategic gaps and potentials can be found and realistic goals can be defined [8].

Benchmarking compares similar systems usually in the form of products or companies. This is how some strengths and weaknesses and the position of the own company can be found. From the findings, new solutions can be generated for the own company. The following phases are used to perform the benchmark [8].

Phase 1 Planning: The extent and scope of the benchmark is defined, benchmarking partners are identified and the relevant information is determined.
Phase 2 Implementation: Data plan creation and data collection.
Phase 3 Analyse: Data interpretation and derivation of concept ideas.

One must not forget that internal benchmarking in large companies can be very useful but can also involve high effort as not all departments are open to provide their data. Also, when using the benchmarking method, the comparability of the data and facts must be ensured [8].

11.2.1.3 Ishikawa Diagram

Explanation and Background

The Ishikawa diagram, also known as "cause-effect diagram" or the fishbone diagram, was created by Kaoru Ishikawa in the 1940s. It is commonly used in quality management as it shows causes for specific circumstances and helps find potential factors that are generating problems or inconveniences within processes, production cycles or companies in general. This method helps to understand the weaknesses that were found and what causes them [18].

This method became popular in the 1960s laying out the baseline for today's modern quality management. The diagram is shaped like a fishbone set up of arrows directed to a problem. All arrows end with factors that may have an influence on the problem. These are interdependent and can vary, depending of the analysed topic. Originally, the so called 5Ms were used (manpower, methods, machines, material, measurements) [18].

Now 6Ms are commonly used in practice, which are: [18]

- Machine (technology)
- Material (consumables, production materials and information)
- Methods (processes used)
- Man power (physical or non-physical work)
- Measurements (inspection, environment)
- Management (managerial decisions).

Procedure and Application

When it is decided, which problem is to be focussed on, it should be stated. The next step is to brainstorm potential causes of the problem. One can either develop new areas of problem influencing factors and decide on them in a team, or use the standard ones depending on the business segment. The key is to stay flexible. Each cause collected is added to the appropriate branch as the example shows in Fig. 11.7.

- Manpower: What influence do the people involved have (employees, managers, stakeholder etc.)? All aspects of the people can have an impact on the problem, for example education, well-being, health, motivation, know-how and so on.
- Management: What is the task of the manager in charge? What is his influence range?
- Methods: How and with which methods are the tasks done? Are these methods suitable or are used incorrectly?

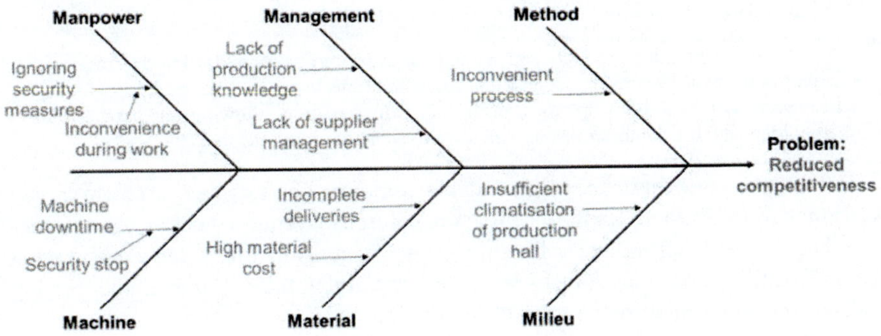

Fig. 11.7 Ishikawa diagram [18]

- Machine: Are the machines suitable for the tasks? Are they powerful and efficient enough? How old are the machines? Have all maintenance measures been done properly?
- Material: Are the tools, materials and consumables used in order? Are they used correctly?
- Milieu (Environment): What influences the system from outside? Are there regulations that constrict certain activities? [18].

After all influencing factors have been added to the diagram, a holistic overview is given, that helps to understand why the current business model is not working as it should. They can then be further analysed and the management board can decide which areas should be tackled and changed. The result of Ishikawa is that a problem can be thoroughly analysed and therefore, understood what the main causes of the problems are. For the development of a new business model this is very important, as it helps not to make the same mistakes again or to help solve the problem [18].

11.2.2 *Goal Definition*

The development of a strategic goal is essential when developing a new business model. It must be clear in which direction the company should develop and what its priorities are [13].

11.2.2.1 Goal Pyramid

Explanation and Background

The highest level of a company goal is the vision. The vision is the long-term goal of a company that describes the general purpose of the company [13]. An example for a vision from Procter & Gamble is the following:

> We will provide branded products and services of superior quality and value that improve the lives of the world's consumers, now and for generations to come. As a result, consumers will reward us with leadership sales, profit and value creation, allowing our people, our shareholders and the communities in which we live and work to prosper [19].

The next level down defines the strategic company goals, usually in form of key performance indicators (KPIs) like market volume, revenue or customer satisfaction. The lowest level indicates how the strategic goals can be achieved and what operational things must be done to reach and implement them. For example, pricing policies, introduction of new products, marketing, etc.

This goal structure can be depicted in a pyramid as shown in the Fig. 11.8.

When developing goals the company should decide what they want to achieve, where they want to stand in the future and in which amount of time the goal should be achieved. This should be done after the thorough company and market analysis

Fig. 11.8 Goal pyramid [13]

has been made. Goals are characterized as a desired status that is to be achieved and serve as a flexible guideline for further decisions. Only through explicit definition and understanding of one's goals, can activities be channelled and adapted to help support the goal. For this, strategies must be looked at, as they lay out the path to achieve them. They should be future orientated and potential orientated and specified for a medium to long-term period of time. The strategy should provide a certain continuity, however be established to be adaptable to changing surrounding conditions and allow creativity [13].

Procedure and Application

The goals should be measurable and include all possible dimensions to enable an exact explanation of the goals. There are four dimensions that should be considered: content, scale, time and scope. The content of the goal can be classified into economical and non-economic goals. Economical goals are driven by figures such as profit, revenue, market volume etc. Non-economic goals are not easily measurable and include for example a change in customer behaviour, brand image, name recognition and so on. The scale of a goal should define how much is to be achieved. This dimension limits the goal and sets borders. It goes hand in hand with the economic goals, as it is quantitively decided what is to be achieved, e.g. a growth in market volume of 20% or increase of revenue by 40%. The time limit should also be defined in which the goals should be accomplished. Finally, the goal scope should be specified where it is decided in which market segment the goals should be valid. For example, for which customer base, which department, geographical area etc.[13].

11.2.2.2 Gap Analysis

Explanation and Background

The GAP-analysis concentrates on strategic development graphs and is supported by quantitative values. It draws a comparison between the actual performance, the aspired and the possible performance which is realistically achievable. This method

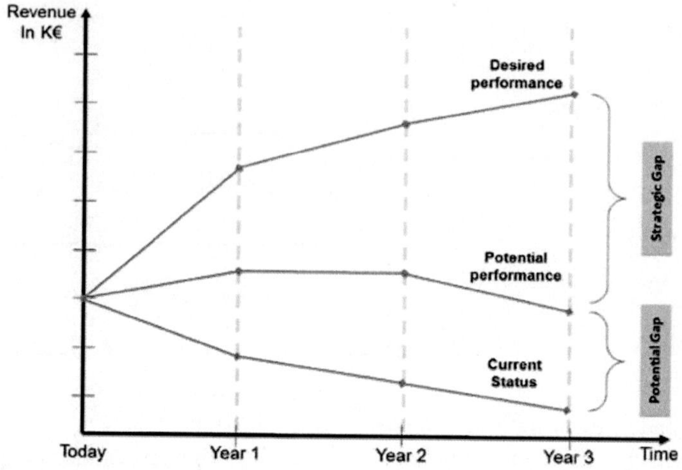

Fig. 11.9 GAP analysis

can be used additionally to the goal pyramid, as it concentrates on the strategic goals, while being more accurate and finds realistic quantitative values.

The aim of the GAP-analysis is to set a strategic goal. For this, a graph is plotted in which the x-axis serves as a timeline and the y-axis is used for a quantitative value that serves as an indicator to measure the success of the scenario. This figure can be profit, revenue, innovation level, sustainability or another KPI [13].

Procedure and Application

Figure 11.9 shows a finalised Gap-analysis.

Step 1 is to plot the graph and decide on the KPI of the y-axis and the timeframe starting from today, which is applied to the x-axis. Usually it is set to 2–3 years but a longer time frame can be selected.

Step 2 is to plot the current state of the business model or company and the estimated future development. For this, the chosen KPI from the last year can be used as the starting point. The development should be drawn keeping the threats and weaknesses in mind that were identified in the SWOT analysis.

Step 3 is to plot the desired performance throughout the years defined on the x-axis, whilst thinking of the opportunities from the SWOT analysis.

Step 4 is to plot the realistic and potential performance that could possibly be achieved. For this, the strategies defined in the SWOT analysis can be taken into account and the most feasible selected.

In the last step, the strategic and potential gap is measured and the goal for the next years is set. This serves as an inspiration for new ideas and to raise awareness of the need for a change in the business model.

Through this method, the aspired strategic goal is developed [13].

11.2.2.3 Scenario Analysis

Explanation and Background

A further method complementing the SWOT-matrix and the Gap analysis is the scenario analysis. The SWOT analysis, devoted to the internal and external factors of the business model give a snapshot of the current state of the company and indicate some opportunities and chances, as well as strategies to seize opportunities or condemn threats. The Gap analysis reveals the strategic goals and potentials whilst, the scenario analysis specialises on development scenarios that could occur and help identify measures on how the company can react to them. Furthermore, the analysis indicates ways on how to reach the set goals. Through this method employees are made to realize the necessity for change in their business and also think of new profitable ideas. It can be an eye opener for employees who have a negative attitude toward change. The Scenario Analysis originates from the military, where it was introduced to enable strategic planning and to leave no unexpected case observed during war. It was and is used to plan reactions and campaigning strategies. In the 1970s it was adapted by enterprises as a prognosis and an analytical instrument to evaluate business economic and political economic challenges and delivered a decision basis to estimate future development of products, business models or projects. The aim of this method is to find ways for an enterprise to survive or maintain their success in any future scenario that might occur [13].

For this, alternative situations that could take place in the future should be thought about for the business model or company, which are then formed into scenario specific strategies. Each scenario is to be described as a process and not as a result. These future scenarios can be developed by looking at trends, political decisions, competition and market development. After collecting possible outcomes, generic scenarios should be formed, two of which will be the extreme scenarios. The other scenarios can be more realistic either going in a positive or negative direction. A scenario should also be developed to estimate the outcome, if everything stays like it is now (market conditions, trends, company perspective). It is recommended to decide on 5–7 scenarios, one of which is the today scenario [13].

Procedure and Application

Before starting, a common understanding of the current business model should be in place within the development team and the SWOT analysis should be used as a basis. The scenarios are plotted in a graph with a timeline as the x-axis and level of profitability as the y-axis. The unit of measurement for the y-axis is percent, as to simplify the process. This way the development of a scenario can be depicted in detail with declining and rising periods of success. The following steps serve as a guideline to develop the scenarios: [13].

Step 1 Selection of timeframe (3–10 years).

Step 2 Identification of environmental influence factors and development trends (threats from the SWOT-analysis can be used).

Step 3 Projection of trends and influence factors onto the business model.

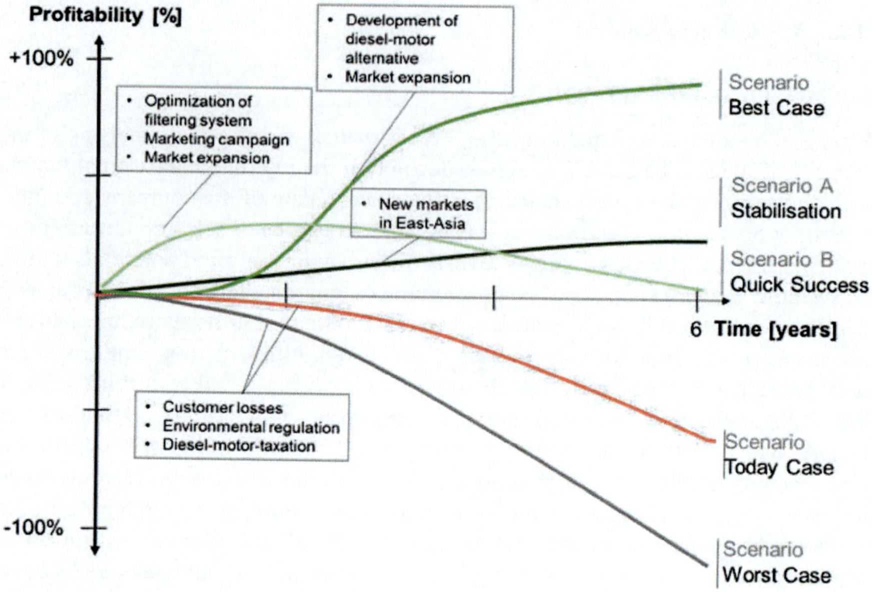

Fig. 11.10 Scenario analysis [13]

Step 4 Creation of worst, best and today case in the graph. Adding of realistic scenarios based on trends.

Step 5 Evaluation of success levels in percent during the periods of each scenario.

Step 6 Development of counteractive measures or measures to achieve a prosperous scenario.

Step 7 Define goal scenarios depending on feasibility.

Figure 11.10 depicts the outcome and result of this method. Different scenarios were developed and plotted into the diagram. This shows a fictive example of a diesel-motor manufacturer who has been facing losses in sales due to governmental regulations, diesel scandals and a more environmental friendly mentality.

First the "Today Case" was plotted. This shows a decrease in profitability due to customer losses and rising prices of diesel-motors. Depending on the impact of these factors, enforced by the government and depending on customer mentality, it could lead to an even worse case in which the company would face bankruptcy or similar. Then the "Best Case" was plotted in which profitability would increase dramatically and remain stable afterwards. For this, potential measures were thought about on how such a scenario could become reality. Moreover, two further scenarios (scenario A and B) were created, to show realistic outcomes and ways on how to achieve them were developed. After the graph had been finalised and possible measures to counteract the market changes and governmental regulation were formed, the company was able to evaluate the feasibility of the positive scenarios and set a goal, which was the basis for the next steps of the idea creation.

11.2.2.4 Business Model Idea Creation

The idea finding phase marks the beginning of the business model development process. The goals have been defined and the development direction set. Either an existing product or service is to be improved or a completely new idea is to be developed. In either case a key aspect is to take the customers into account. It is crucial to understand what the customer needs and where his inconveniences lie, as well as what the customer expects from the company. Another question to be answered is how the company can position itself in a way to satisfy the customers need in the best manner in comparison to the competition.

All methods to develop creative ideas and to emphasise creativity underlie the same principles: understanding of the challenge, loosening of transfixed stereotypes and assumptions, recombining existing approaches and solutions and refining of ideas through criticism and improvement [19].

This phase helps to create new ideas through creative thinking, without being influenced by existing business models and ideas, as well as the current business model in place [20]. The following sections will explain some methods to develop creative, innovative ideas. The most commonly used idea generation methods are mind mapping and brainstorming. However, these will not be looked at further, as they are very simple and well known. During the development process, using one of the following methods can suffice, if it generates enough ideas. However, the selected creativity methods all specialise on different perspectives and circumstances.

11.2.2.5 Destroy Your Business

Explanation and Background

This method is a rather radical method, resulting in creative ideas. The goal is to think of ways how one's business could become redundant. These ideas can be transformed into new business models. For example, if a DVD-player production company asks itself how their business model could become redundant, a possible answer could be: people do not buy DVDs anymore and only use streaming sites. From this conclusion, the idea of creating a universal device to create a portable digital library of all DVDs one owns, including all downloaded movies regardless of the site or provider it was downloaded from.

A group of 6 people guided by a presenter, who explains the rules and the purpose of the meeting, gather in a work-free environment. The group tries to think about whom and what could endanger their company and business model in general. By collecting all ideas and writing them down, innovative and new products or services can be created.

Procedure and Application

The following questions serve as a guide for this method:

- How could a financially strong company or a start-up copy the current business model and improve it?
- How could the competitors make the business model better?
- Is there a business model that would make the current business model redundant?
- What product or service would have that effect?
- What are reasons for customers to leave?
- What would a competitive price structure look like?

With these questions, threat scenarios can be developed and ideas for new services or products can be extracted [7].

11.2.2.6 The Empathy Map

Explanation and Background

This creativity method concentrates on the customer perspective and can result in the discovery of hidden potentials and may highlight a point of view or aspect of the business model that has not been regarded yet. Business models are made to satisfy a need of a customer. Enterprises should not ask themselves: "What can we sell the customer?" but "What jobs does the customer have to get done? What efforts does the customer face? What are the customer's needs? What is important to the customer?" Only then, can a successful business model be developed and designed properly [15].

The empathy map was developed by the company XPLANE. The tool highlights six aspects of the customer that according to Osterwalder "helps you go beyond a customer's demographic characteristics and develop a better understanding of their environment, behaviour, concerns and aspirations." [15]. Understanding the customer better helps to create a business model and value proposition and also ultimately answers the original question a company asks itself: "what can we sell the customer?" [15].

Procedure and Application

In the development team, all customer segments are defined. For the most promising customer segments the empathy map should be created and the following steps can be started. First the customer segment is humanised to one singular person and then, the six aspects of the customer are analysed by answering the questions in the empathy map depicted in Fig. 11.11.

Step 1: Humanize
Give the customer a name, an age and typical characteristics that suit the selected customer segment, for example a student, a mother, a middle-aged man etc. If the customer segment refers to an industry, think of the people who will use the service or product within the company. For example, the CEO of a company, the quality director, the cleaner etc.

Step 2: What does he/she see?

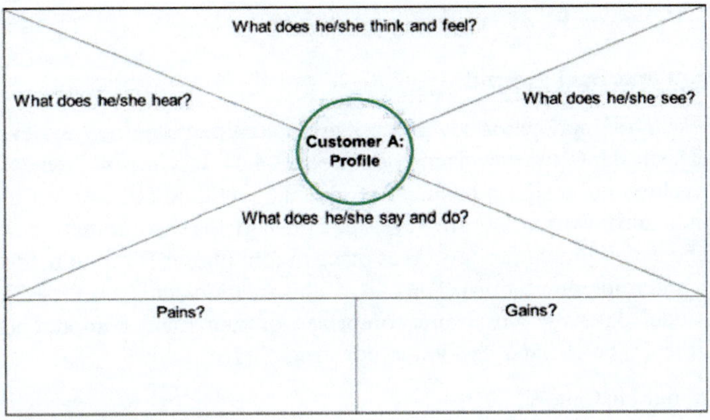

Fig. 11.11 Empathy map [15]

Describe the environment of the customer. Which people surround him/her? What market offers is he/she exposed to? What problems does he/she meet? What happens around her/him?

Step 3: What does he/she hear?
This specific question should answer how the environment influences him/her. For example, which people and media channels influence her? In a B2B case, this could be the boss complaining that the process takes too long, the media saying digitalisation is the key and will sooner or later replace workers or similar.

Step 4: What does he/she think and feel?
This elaborates what goes on in the customer's mind and finds out what their worries and aspirations are.

Step 5: What does he/she say and do?
What is the attitude, appearance and behaviour towards others? What are the conflicts between what the customers says and what he/she truly thinks or feels.

Step 6: What is the customer´s pain?
What is the biggest fear, frustration or obstacle? What risks might he/she fear taking?

Step 7: What does the customer gain?

What does the customer truly need and want? How does he/she measure success? Which strategies will he/she use to achieve his/her goals? [15].

This detailed map of the customer can help to generate customer specific ideas, as his exact needs are tried to be identified [15].

11.2.2.7 St. Galler Business Model Navigator

Explanation and Background

The business model navigator is a process orientated methodology to develop business models, similar to the canvas method (Sect. 3.4.1). It is one of the most common tools to develop and design a business model. The method is based on the realisation, that most business models are developed through the recombination of different patterns. The development process takes place in four phases (initiation, idea finding, integration and implementation). The core of this method consists of the 55 different business model patterns, which are comprised of four main components that are defined as the "Who-What-How-Value-construct." [16]

- Who are our customers?
- What is our value proposition?
- How do we provide/create value?
- How do we create revenue?

Customers and value proposition define the external perspective of the business model while the value chain and revenue model define the internal components. According to Gassmann, if two of the four dimensions are changed a business model innovation is created [16].

Procedure and Application

As described above the four phases are the initiation phase, the idea finding phase, the integration phase and the implementation phase. For the procedure, a team of maximum 7 people is suggested. A presenter is also needed, who deducts and explains the process [16].

Step 1 Initiation: A market and trend analysis is made to describe the changing environment of the company and the customers. This analysis should be made to find out and understand the future influence factors on business models [16].

Step 1 Idea finding phase: With the help of the 55 business model patterns, as described in Table 11.3 ideas to modify the current business model can be found [16].

Step 1 Integration: The adapted business models are evaluated in terms of realisation. For this, the ideas must be described briefly, discussed and evaluated within a group. The following perspectives are to be described for each business model idea:

- Need: What is the core customer need?
- Approach: What is the value proposition?
- Benefits: What is the customer benefit?
- Competition: Who are the competitors? [16].

After some ideas have been collected the business models can be developed in detail to ensure comparability.

Table 11.3 55 innovative business model concepts [16]

1. Add-On	2. Affiliation	3. Aikido	4. Auction	5. Barter By-product
6. Cash machine	7. Cross selling	8. Crowd funding	9. Crowdsourcing	10. Customer loyalty
11. Digitalization	12. Direct selling	13. E-commerce	14. Experience selling	15. Flatrate
16. Fractionalized ownership	17. Franchising	18. Freemium	19. From push-to-pull	20. Guaranteed availability
21. Hidden revenue	22. Ingredient branding	23. Integrator	24. Layer player	25. Leverage customer data
26. License	27. Lock-in	28. Long tail	29. Make more of it	30. mass customization
31. No frills	32. Open business model	33. Open source	34. Orchestrator	35. Pay per use
36. Pay what you want	37. Peer-to-peer	38. Performance based contracting	39. Razor and blade	40. Rent instead buy
41. Revenue sharing	42. Revenue engineering	43. Reverse innovation	44. Robin Hood	45. Self-service
46. Shop-in-shop	47. Solution provider	48. Subscription	49. Supermarket	50. Target the poor
51. Trash-to-cash	52. Two-sided market	53. Ultimate luxury	54. User designed	55. White labelled

11.2.3 Detailed Business Model Design

After some ideas have been gathered, they should be thought through systematically. All aspects should be considered, to enable a holistic perspective of the business model and to ensure its functionality. For this the most proven method used in practice is the canvas methodology. It highlights all dimensions of business models and gives a stepwise approach to systematically define the elements. Other methods have been looked at but have not managed to establish themselves as well as the canvas method.

This chapter will start with the canvas methodology, as it provides a high-level overview of the entire business model. Then the SIPOC chart will be presented, because it helps to understand and describe the internal processes that result through the new business model.

11.2.3.1 Canvas

Explanation and Background

The canvas method is a standard strategic tool for business model management and provides a basis for the development of business models. It helps to explain the core elements and logic, as well as the success factors. This concept allows its users to describe and systematically think about all relevant factors of the business model [15].

The Canvas method consists of a set of nine boxes in which elements of the business model are described. When the boxes are put together, a holistic overview of the business model is created and visualized, which can be used as a basis for communication, requirements definition and further development [10] (see Fig. 11.12).

Procedure and Application

The 9 Elements and their meaning are described in the following. These should be filled in by the development group for each business model idea that has been found. A detailed explanation of each box (see Table 11.4) can be found on the next pages:

1. Customer segments: Who are our customers? Who are important customers? Examples for customer segments are explained in the following table.

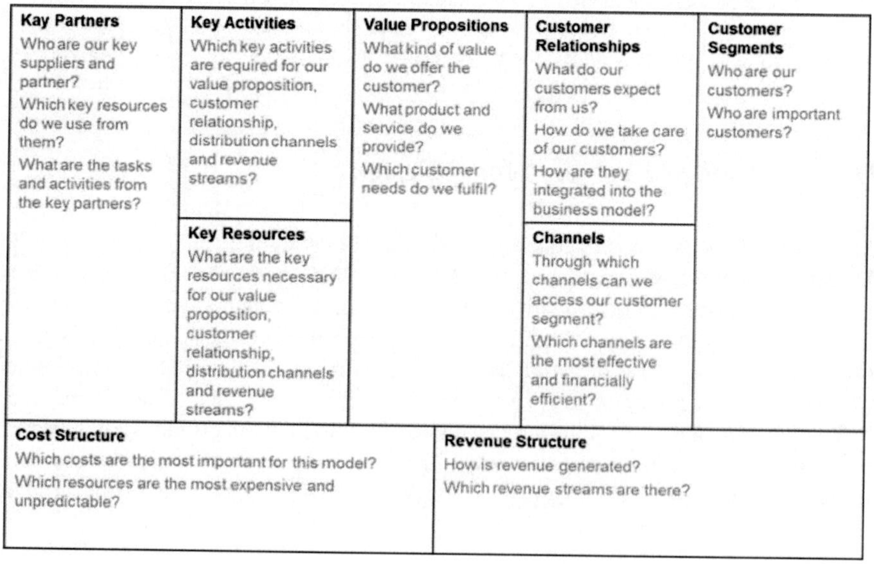

Fig. 11.12 Canvas [15]

Table 11.4 Customer segments [15]

Customer segment	Description
Mass market	The business model focuses on a large group of people who share a similar demand. The consumer electronics branch is an example for the mass market
Niche market	The business model provides a product or service to a specific customer branch. The customers and the supplier are highly dependent on each other as the business model is tailored to the exact needs of the customer. For example, when a car part manufacturer produces customised products for a car company
Segmented	The business model can provide solutions in form of a product or service that addresses different segments. The differentiation can be classified by income levels (e.g. a car manufacturer can sell normal cars but also luxury cars), age groups (e.g. cosmetics industry), etc
Diversified	The business model serves independent customer groups. Often this model is used to decrease the dependency on a specific customer segment. For example, Amazon.com not only serves the consumer side as an online retailer, but also provides cloud computing services to web companies, through the strong IT infrastructure it has built up over the years
Multi-sided market	The business model satisfies the needs of more than one customer segment. An electronics recycling company for example collects electrical waste form shops, manufacturers or consumers, extracts and refines the metals and sells these to production companies

Value Propositions: What kind of value do we offer the customer? What product and service do we provide? Which customer needs do we fulfil? [15] (see Table 11.5).

1. Channel: Through which channels can we access our customer segment? Which channels are the most effective and financially efficient? It is important to find the most suitable mix to make the customer chose the company's products or services. There are five channel types and phases: [15] (see Fig. 11.13).

2. Customer relationships: What do our customers expect from us? How do we take care of our customers? How are they integrated into the business model? The following categories of customer relationships can be explained: Personal or dedicated personal assistance, self-service, automated services, communities or co-creation [15].
3. Revenue Streams: How is revenue generated? Which revenue streams are there? [15] (see Table 11.6).

4. Key Resources: What are the key resources necessary for our value proposition, customer relationship, distribution channels and revenue streams?
   There are four types of key resources: physical (assets, buildings, machines etc.), Intellectual (brands, patents etc.), human (employees), financial (cash, credit, stock option pool etc.) [15].

Table 11.5 Value models [15]

Value models	Description
Performance	The performance of a product or service is continuously reliable and improved
Customisation	Products and services can be tailored to individual needs of customer segments
"Getting the job done"	The business model helps customers get certain jobs done. For example, Rolls-Royce produce and service their jet engines and sell the time the engine runs to its customers
Design	The product stands out from others due to an appealing design (e.g. fashion, consumer electronics)
Brand/Status	Displaying the brand is linked with certain characteristics that the customer wants to be identified with (e.g. wealth, fashionable, rebellious)
Price	Offering value to a lower price than the competitors
Cost reduction	Supporting the customers to save money through the usage of the product or service
Risk reduction	Value adding services or products that reduce the risks that can occur (e.g. guarantees)
Accessibility	Products and services are made accessible to a new customer base which couldn't access them before (e.g. Crowd funding, Car2Go)
Convenience/Usability	Providing more convenient solutions for consumers in a user-friendly way

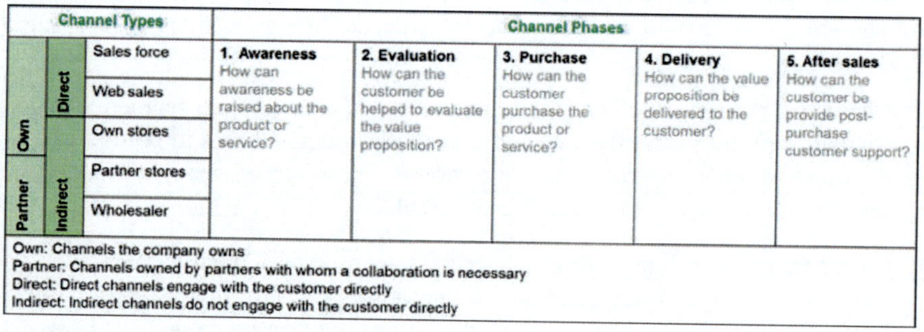

Fig. 11.13 Channels [15]

5. Key Activities: Which key activities are required for our value proposition, customer relationship, distribution channels and revenue streams? Examples for some key activities are production, problem solving or business models based on a platform or network (e.g. transaction platforms, bidding platforms, operating platforms etc.) [15].
6. Key Partners: Who are our key suppliers and partner? Which key resources do we use from them? What are the tasks and activities from the key partners? [15]

Table 11.6 Revenue streams [15]

Revenue stream	Description
Asset sale	Selling the right to own product
Usage fee	Selling the right to use a product or service
Subscription fee	Selling access to a service
Lending/Renting/Leasing	Selling the temporary right to use an asset
Licensing	Selling permission to use protected intellectual property
Brokerage fees	Receiving a fee by matching a buyer with a seller
Advertising	Receiving fees for advertising a product, service or brand

7. Costs Structures: Which costs are the most important for this model? Which resources are the most expensive and unpredictable (see Table 11.6)? [10, 15].

After filling out all boxes in a group an overview of all market and resource relevant aspects are visualised and serve as a communication basis for the further development and requirements definition [10].

11.2.3.2 SIPOC

Explanation and Background

SIPOC is a method used in the six-sigma methodology. The six-sigma philosophy uses DMAIC (Define-Measure-Analyse-Improve-Control) as a core process management phase model, to develop high quality process to the standards of six-sigma. In the measurement phase the SIPOC tool is used to map out processes and identify problems within them. Furthermore, it identifies inputs and outputs that that are necessary for the business model. In this thesis, it's purpose has been modified, to help create the first overview of the general process and to identify its stakeholders. This method complements the canvas methodology, as is goes further into detail on the process level of the business model.

The goal of SIPOC which stands for supplier, input, process, output and customer, is to align the process with the relevant internal and external target group, meaning internal (the company) and external (stakeholder) customers and suppliers [8]. With SIPOC a compact overview for all stakeholders of a process is created. Usually this method is used to optimise processes but in this thesis, has been chosen to create the first high level process of the newly found business models [21].

Procedure and Application

The development team should use the SIPOC method on all potential business models for which a canvas has already been made. As the customer side and the output has already been elaborated in the canvas method, they can be written under the according box (Fig. 11.14).

1. Customer: Not only the end consumers defined in the canvas can be the customers of the process, but also the executive board. They are interested in revenue and costs that are caused by the process. When developing a product, the quality department is a further customer of the process. If the process requires frequent reworking and repair or even results in high reclamations, unexpected costs may arise, which will not please the management.

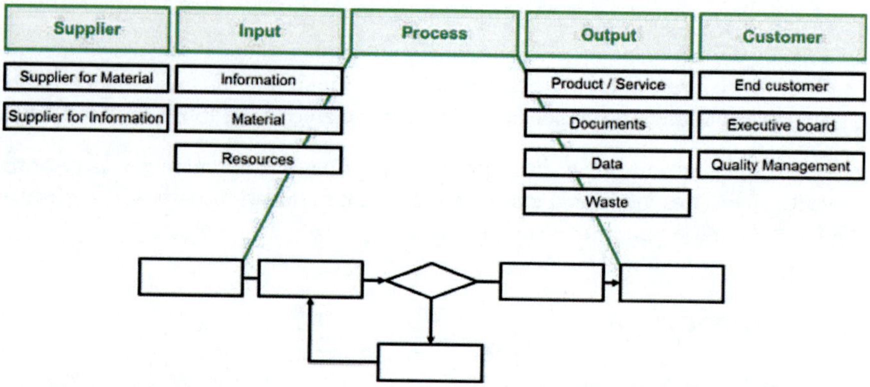

Fig. 11.14 SIPOC chart [21]

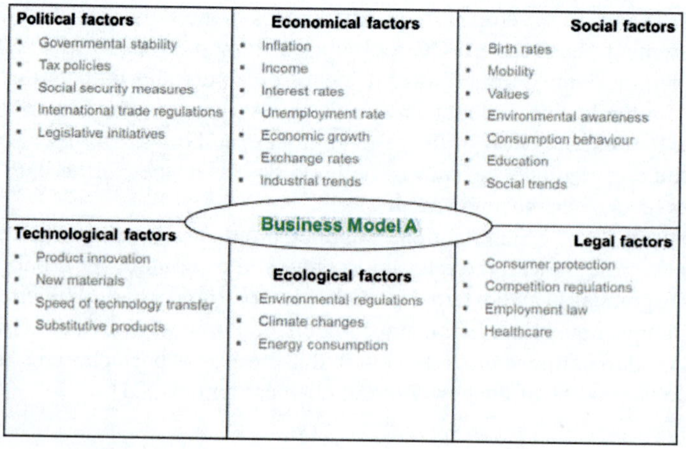

Fig. 11.15 PESTEL [20]

2. Process: The team should think of a process, which includes the suppliers and customers that shows how the business model could be realised. This process should only concentrate on the main internal process steps and not be very detailed. However, it should be designed to suit the customers and the company's interest by finding a cost efficient and secure manner of providing the service or product.
3. Output: The main output is the final product or service. However, other by-products or internal outputs can be generated. A product based business model could result in the production of waste and by-products, for which a solution or process must be made. A service based business model could result in high amounts of data or administration efforts that also need to be considered.
4. Inputs: The inputs consist of all materials, data, information and resources that are needed to make the process work.
5. Suppliers: When the inputs have been defined, the suppliers can be identified. Certain material or services may be required, that the own company cannot provide [21].

11.2.4 Evaluation and Decision Methodology

Having described the business models through the canvas and SIPOC method, the business models can now be evaluated. However, this can be a difficult task due to incomparability or through incomplete perspectives on the business models. Therefore, first of all an environmental analysis for each business model that is estimated to be promising should be made and then an objective evaluation with a systematic procedure and reasonable evaluation criteria should be followed.

11.2.4.1 PESTEL Model (macro-Economic)

Explanation and Background

The PESTEL model shows the macro-economic influence factors on a business model.

The macro-environment has six dimensions according to the PESTEL-Model, which are universally valid. The goal is to analyse all potential influence factors on the new-found business model.

- Political influence factors
- Economical influence factors
- Social cultural influence factors
- Technological influence factors
- Ecological influence factors
- Legal influence factors.

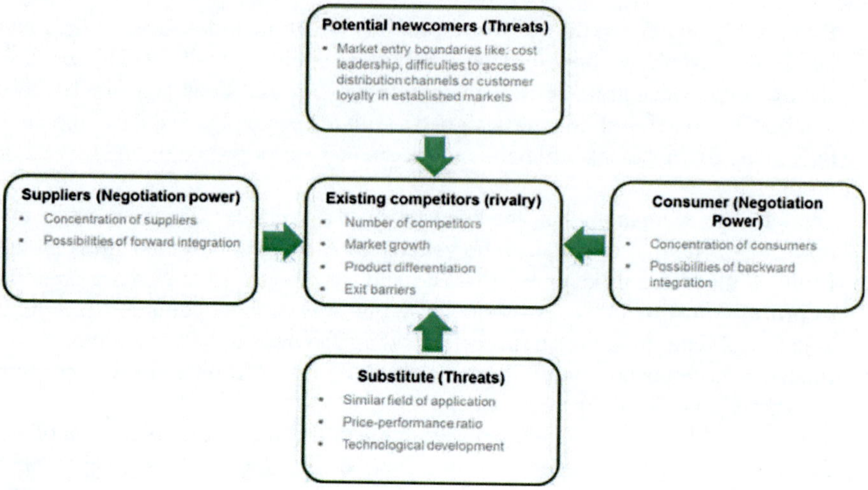

Fig. 11.16 Porter five forces [20]

For each of the described business models theses six influencing dimensions should be discussed and described [20] (Fig. 11.15).

Procedure and Application

The following table is an example for a filled-out table of the PESTLE method. It can be filled out by a team, which is familiar with the business models that have been described. It is also recommended, to introduce a new team member as to provide a neutral and new point of view (Fig. 11.16).

These factors can later be used as a basis for the value benefit analysis.

11.2.4.2 Porters Five Forces (micro-Economic)

Explanation and Background

After the macro-economic point of view has been analysed, the micro-economic perspective is to be considered. The micro-environment is important, to help understand the industry in which the business model is. For these Porters five forces can be used [20].

The dimensions that are regarded by Porter are:

- The threats through potential newcomers
- Threats through rivalry between competitors
- Threats through subsidiary products
- Negotiation power of consumers
- Negotiation power of suppliers.

Procedure and Application

Figure 11.16 shows the different dimensions and their content that is to be worked out for each potential business model that has been developed.

These forces are also to be considered in the selected criteria used in the value benefit analysis.

11.2.4.3 Value Benefit Analysis

Explanation and Background

The value benefit analysis is an instrument to make decisions. It comes into use, when a variety of aspects are to be incorporated into a decision and helps find the most suitable solution for the company and in this case, to find the most suitable business model. The holistic and complex problem is fragmented into different sub-problems. This allows to take all important factors into account and not be deceived by personal preferences. It also helps to be able to compare solutions with each other or business models that are not easily comparable [22].

This method prevents subjective decisions by individuals and manipulation of opinions. By separating the solutions into sub-problems, the emotional attachment or preference is taken, allowing a neutral evaluation and decision. The value benefit analysis is useful when many aspects come together, like for example evaluation criteria, a mixture of quantitative and qualitative evaluation criteria or different opinions of different people with different know-how. Furthermore, it is easily documented and presentable to stakeholders or board of directors etc.[22].

Procedure and Application

Before starting, some rules when working together using this method should be addressed. Firstly, the value benefit analysis should be done in a structured manner and be documented in every step. All steps are to be reconstructed later. A presenter should make sure, that all opinions are considered equally. Together the group should select reasonable criteria that suit the problem stated. The following steps explain the procedure in which the Value Benefit Analysis is made [22].

Step 1: Organisation of the work environment
A presenter is chosen, who explains the method and procedure to the participants. His task is to make sure all participants can express their opinion equally and that no one tries to manipulate the decisions of others. It is recommended to have 5–10 participants, however more or less people can participate. The time minimum is set to two hours and depends on the number of alternatives to be evaluated and the number of people discussing the topic. The process can be interrupted, because often more information may be needed to make proper decisions [22].

Step 2: The selection of alternatives

In this step, all business model ideas are collected and their suitability is discussed. Maybe all ideas have potential or some might be able to be ruled out beforehand, due to obvious feasibility reasons on unsuitability to the company's strategy [22].

Step 3: Collection of criteria

In the group, criteria are to be collected, that is relevant to solve the problem or to ensure success of a business model. Each company has a different emphasis. Some aim at value or profitability, others e.g. on sustainability. A list of 10–20 criteria is suitable. It is important, that the criteria fully grasp the key factors that are important to the company and its goals. For this, the goal pyramid (Sect. 3.2.1), the gap analysis (Sect. 3.2.2), the PESTLE model (Sect. 3.5.1) and Porters Five Forces analysis (Sect. 3.5.2) can be used as support [22].

Step 4: Weighting of the criteria

With the help of a cross-classified table as shown in Fig. 11.17, it is possible to weight the criteria by comparing each criterion with each other and deciding, if it is more important or less important than the other. This is done by going through each vertical column and deciding if the criteria in column A for example is more important than criterion in row B. If A is more important the digit 2 is written into the field and if it is less important than B a zero is written in the field. The digit 1 is given, if the criterion is equally important. After this has been done for each criterion, the values are added up at the end of each line. By calculating these values in percent

Criteria

Only fill white boxes

Instructions
The vertical criterion (colomn) is
- more imporant **(2)**
- equally important**(1)**
- less important **(0)**
than the horizontal criterion (row).

Criteria		1 Criterion A	2 Criterion B	3 Criterion C	4 Criterion D	5 Criterion E		Weighting Factor
1	Criterion A		1	2	1	1		15%
2	Criterion B	1		1	0	1		25%
3	Criterion C	0	1		0	1		30%
4	Criterion D	1	2	2		2		5%
5	Criterion E	1	1	1	0			25%
	Sum	3	5	6	1	5	*20*	100%

Fig. 11.17 Criteria weighting

Ranking	Criteria	Weighting	Alternative A Criteria fulfilled?	Value	Alternative B Criteria fulfilled?	Value
4	Criterion A	15%	5 Very Good	5	4 Good	4
2	Criterion B	25%	5 Very Good	5	2 Sufficient	2
1	Criterion C	30%	4 Good	4	2 Sufficient	2
5	Criterion D	5%	4 Good	4	3 Satisfying	3
2	Criterion E	25%	4 Good	4	3 Satisfying	3

Fig. 11.18 Evaluation of alternatives

of the total amount of points given, the weighting factor for each criterion is set [22].

Step 5: Evaluation of business model alternatives
Each business model alternative must be analysed for each criterion. For this, marks are given between 1–5, 5 being the highest value and 1 the lowest. It is to be decided, if this criterion is low or high in the analysed business model alternative. This is done for all business model alternatives and all criteria (Fig. 11.18) [22].

Step 6: Final result

The last step is to multiply the weight of each criterion with the value given for each business model alternative. Then the values are added up and divided by the number of criteria for each business model alternative and the one with the highest number is the business model most suitable to the goals of the company and most promising for future success (Fig. 11.19) [22].

11.2.5 Solution Design

The following chapter focuses on the business model that is planned to be used and helps to prepare and consider all aspects that are relevant for the implementation steps.

11.2.5.1 Detailed Process Design

Explanation and Background

A detailed process design is very important for compiling the business model, as it shows how it can be realised and identifies challenges, defines process times and is the basis to determine the costs within the business case. The goal of process design is to visualise all steps of the process between the beginning and the end. In the SIPOC method the first high level process was defined and serves as a basis for the further

			Alternative A		Alternative B	
Ranking		Weighting	Value	Value Benefits	Value	Value Benefits
4	Criterion A	15%	5	0,75	4	0,60
2	Criterion B	25%	5	1,25	2	0,50
1	Criterion C	30%	4	1,20	2	0,60
5	Criterion D	5%	4	0,20	3	0,15
2	Criterion E	25%	4	1,00	3	0,75
Total Value Benefit (in %) > 70% < 50%			88%		52%	
Ranking			1		2	

Fig. 11.19 Final result

detailed plan. This helps to develop the complexity and optimisation potentials of the entire process in a stepwise manner and also to define the responsibilities. There is a differentiation between the material or information flow process and the value stream. The material or information flow only gives information about the process steps, while the value stream also gives information about all process step data, like quantity, duration, frequency, availability, setup time, waiting time, etc. [8].

Procedure and Application

Material and Information flow:

Step 1: Usage of the SIPOC as a basis and definition of the start and end points of the process.
Step 2: Define all involved departments and responsible people.
Step 3: Define the individual process steps and their sequence and order.

The process and the departments can be depicted such as shown in Fig. 11.20 [8].
Value stream:

Step 4: All relevant data is added to the boxes. An example is shown below including the input, the output, the duration and its frequency (Fig. 11.21) [8].

If required, the processes can be simulated, for example with the program Witness to show bottlenecks, faults and optimisation potential [8].

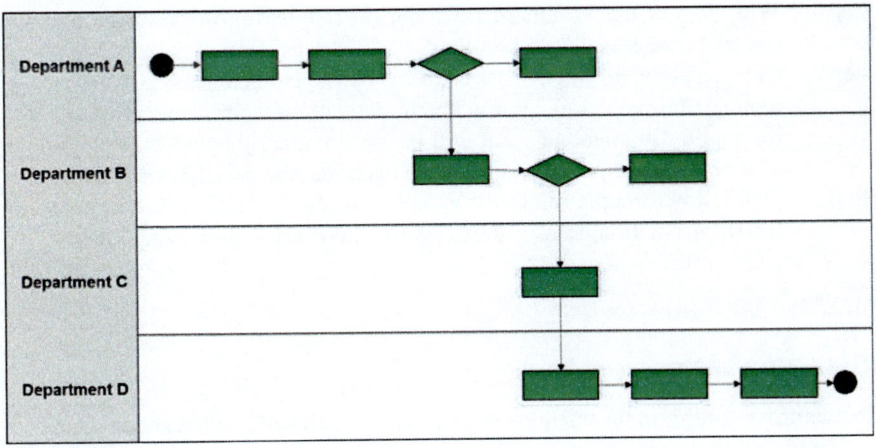

Fig. 11.20 Process design [8]

Fig. 11.21 Process box content [8]

Confirmation of customer order	
Input • Customer data • Order • Availability status	**Output** • Customer notificataion • Internal assignment
Duration: 25s	**Frequency:** 1

11.2.5.2 Resource and Investment Plan

Explanation and Background

When the processes have been defined the resources and necessary investments are planned. The goal of this plan is to identify what resources and investments are needed for the implementation and execution of the new business model [8].

Procedure and Application

Resource plan:

Step 1: Define competence groups or departments relevant for the execution of each processes

Step 2: Define the number of employees per department with help of the duration for each process step and the estimated number of cases or units processed per day, as described in the value stream [8].

Investment plan:

Step 1. With help of the SIPOC method and the detailed process design, the first internal investments can be found.

Step 2. Create a table, in which all investments can be collected and distributed over a selected time period. During the implementation phase, not all assets may be required immediately but will be systematically be added over time or after a few years, for example spare parts may be needed or similar.

Step 3. Define investments relevant for the business model, like for example for R&D, production, sales, infrastructure, auxiliary supplies etc. [23].

11.2.5.3 Business Case

Explanation and Background

To quantitively support the business model, a detailed business case should be calculated and different scenarios and outcomes should be integrated. A business case is a scenario to evaluate the effectiveness of a business model and to justify an investment. The purpose is to show the potential revenues, costs and the return on investment or cost savings, as to enforce a decision whether to make the investment or not [24].

The business case holds all necessary financial data about resources, profits, sales, personnel costs, investments and project costs and adapts the data to different scenarios: usually to a best case, worst case and a realistic case. The business case is set up from the project costs that are needed to implement the business model, the operating costs and the revenue achieved through the new business model [8, 24].

Procedure and Application

To create a realistic financial statement various elements must be planned. The key elements are the investments, the resources and the processes, which have already been developed in the previous chapters. These serve as the basic data that is transferred into the business case [24].

First of all, the time frame is set. This can be any amount of years, however it is recommended to look at 3–10 years, because a prediction that is too far in the future is less accurate, due to changing market conditions. The timeframe includes the implementation time and a defined operating time. Then the processes times ust be calculated into costs. For this, the amount of personnel and resources that is needed for each process is calculated. The number of transactions for each month is to be defined considering the implementation phase in which no customer orders will be received. Then the average costs for the operating phase are set for monthly or yearly basis [24].

Furthermore, the investment plan and resource plan can be transferred to the business case.

The following examples of KPIs have proven themselves in practice and are suitable for the evaluation of the business case's profitability:

- Total operation expenditures are defined through personnel, energy, consumables and maintenance.

- EBITDA (Earnings Before Interests, Taxes, Depreciation and Amortization): The focus lies on the profitability affecting payments of a company. Next to the taxation and interest expenditures also the depreciation is added. EBITDA is commonly used by companies, when large depreciations lead to a significant reduction of the EBIT result. It enables a direct comparison between companies or projects regardless of tax form or capital structure, like the EBIT (Earnings before interests and taxes). The difference is, that the depreciation policy is eliminated. It is calculated as shown below: [25]
  Annual net income
  ±taxes
  ±extraordinary profit/loss
  +interest expenditures
  =EBIT
  +depreciation of tangible assets (depreciation)
  +depreciation of intangible assets (amortisation)
  =EBITDA [25]
- NOPAT (Net Operating Profit After Taxes) describes the operational result after taxes and is calculated by subtracting the income tax form the EBIT. In practise the calculation of NOPAT is extended or differentiated. For example, the depreciation of intangible assets, differences in reserves or interest on leasing contracts are considered in the calculation [25].
- Net profit ratio shows the profitability of the revenue considering the financial structure, meaning the subtraction of interest on borrowed capital and is calculated be dividing the profit after taxes by the revenue. The result is given in percent [25]
- ROI (Return on Investment) shows the yield of the capital employed. It is calculated by multiplying the capital turnover (revenue/total capital) and the net profit ratio [25].
- Working capital is the difference between the circulating assets and the short-term liabilities and should be positive, meaning that the short-term assets cover the short-term liabilities. The higher the working capital is, the safer the liquidity of the company is [25].
- Cash flow is the surplus on operational payment receipt throughout a defined period. It is one of the most important financial figures for the analysis of the financial strength of a company. The goal is to estimate to what extent a company can finance its own investment and on the other hand how many financial resources for the liquidation of debt or for the dividend payment and for the liquidity planning are needed. There are different kinds of cash flow. For the example, gross-cash flow and free cash flow. The gross-cash flow stands for the surplus of operative revenue over the operative expenses: [25]
  Annual net income

    ±extraordinary expenses/profits
    +interest expenditure
    +depreciation of tangible assets
    ±input/output of long-term reserves

±latent tax expenses
+rent/leasing expenses
±inflation gain/loss of net-liquidity
=gross cash flow [25].

The free cash flow is also an indicator of the financial strength and flexibility of the company. It is calculated through the cash flow of the operative business activities plus net payments for investments in tangible assets and intangible assets. The free cash flow shows the withdrawal payments surplus, which could be paid to the outside creditors or investors. It can be determined using following calculation: [25]

- gross cash flow
- −/+ investments/disinvestments
- −/+ increase/decrease of working capital
- = free cash flow [25]

- WACC stands for Weighted Average Cost of Capital and indicates the required rate of return of a company, so that the capital expenditure is worthwhile. It is calculated as the following [25]:

$$\text{WACC} = i_b \times (1 - \text{tax rate}) \times \frac{\text{borrowed capital}}{\text{total capital employed}} + i_e \times \frac{\text{equity capital}}{\text{total capital employed}}$$

- i_b stands for cost of borrowed capital and i_e for the costs of equity capital
- ROCE (Return On Capital Employed) is a modified figure of the total capital profitability that measures the interest yield of the employed capital before taxes. This figure is particularly important for stock corporations and is one of the most published figures of DAX-companies for profitability. It shows the yield of invested capital (capital employed) independent of financing and tax systems within a period. ROCE is calculated by dividing the EBIT and the capital employed with each other [25].
- IRR (Internal Rate of Return) is the interest rate, that leads to the net present value equalling zero, when the surplus is discounted. An investment is beneficial, if the IRR lies above the required rate of return. It is calculated as the following equation shows [26].

$$\text{IRR} = i_1 - C_{01} \times \frac{i_2 - i_1}{C_{02} - C_{01}}$$

- Firstly, two interest rates are selected depending on internal company policies, i_1 being the lower rate and i_2 the higher rate. C_{01} and C_{02} are the Net Present Value figures that are calculated for each interest rate [26].
- NPV (Net Present Value) measures the profitability of an investment by subtracting the incoming and out coming cash flows over the period time of the investment. It is calculated as shown below: [23]

$$NPV(i, N) = \sum_{t=0}^{N} \frac{R_t}{(1+i)^t}$$

N: the total number of periods, t: the time of the cash flow, i: the discount rate and R_t: the net cash flow: [23]

- Amortisation describes the time needed for the profits to cover the investment and is calculated by dividing the investment by the average return per period [23].

All these figures and the relevant data is usually collected in an Excel-table which is highly complex to compile. Figure 11.22 shows an executive summary example of such a business case in thousand Euros.

11.2.5.4 Performance Management

Performance Management includes the selection of KPIs, service level agreements (SLAs) and other monitoring tools. Before the new business model is implemented, the framework of the performance management should be defined, to prepare what to expect from the implementation project and to help monitor the process. A typically used framework for performance management is the balanced scorecard (BSC). It is used by the executive board to help transform a strategy into operative measures. For this, different perspectives and strategic goals are defined, that are supported by quantitative figures, aims and measures, to ensure the performance throughout the strategy implementation. The BSC considers four perspectives, which are relevant for the performance: [27]

- Financial perspective: Financial goals and expectations
- Customer perspective: Customer requirement goals to reach financial goals
- Process perspective: Process requirements and targets to achieve customer and market goals
- Potential perspective: Goals regarding employees, potentials to enable the goals from the perspectives above [27].

The different perspectives and goals are dependent on each other and are monitored in the form of performance figures. Some benefits of using the BSC are: the definition of clear strategic goals to create motivation amongst the employees, constant performance improvement, identification of performance potentials and the communication and reporting potential from the management level. The key figures are defined with the help of key performance indicators (KPIs) [27].

Basis of the KPI is the performance guideline. There are four categories of performance success:

1. Financial success: Profitability and growth (for example revenue and EBIT)
2. Market success and customer satisfaction: Continuous improvement, reliability and quality (for example defined by new orders or market volume and CSI (customer satisfaction index)

Assumptions	WACC before Tax: 10,0%	Tax rate: 32,4%	**WACC after Tax:** 6,8%
Profitability:	**IRR:** 50,5%	**NPV** (as to 1st payout):	26.753 TEUR
	Amortisation (static):	3,5 years after 1st payout	
	Amortisation (dynamic):	3,7 years after 1st payout	

Years of launching	1	2	3	4	5
Business year	16/17	17/18	18/19	19/20	20/21
Cashflow overview(TEUR)					
Investment		-93	-724	-763	-802
Δ working capital	-280	-827	-1.331		-251
Gross-Cashflow	-271	1.919	5.594	6.274	6.950
Residual value					
Free Cashflow, nominal	**-7.067**	**999**	**3.539**	**5.510**	**5.896**
Free Cashflow, cash value per 1st payput	-6.843	906	3.007	4.385	4.395
Details as to Δ working capital [TEUR]					
Δ Liabilities	-135	-245	-126	-24	-34
Δ Accounts recievable	214	1.075	1.267	232	252
Δ Stock	201	-3	190	30	32
Δ Other					
Δ working capital (Aufbau=+/Abbau=-)	**280**	**827**	**1.331**	**238**	**251**
Details as to Gross-Cashflow [TEUR]					
Revenue through metal value	1.081	6.500	12.890	14.058	15.331
Cost savings					
Revenue	1.081	6.500	12.890	14.058	15.331
./. freight / other					
Gross profit	1.081	6.500	12.890	14.058	15.331
Personnel	14	675	832	836	840
Energy					
Consumables	252	1.507	2.967	3.206	3.463
Maintenance	21	251	257	264	270
Other	1.239	1.576	924	856	881
Total operation expenditures	1.526	4.009	4.981	5.162	5.454
Intermin result	-446	2.491	7.909	8.896	9.877
Administration & selling expenses					
EBITDA	-446	2.491	7.909	8.896	9.877
Depreciation	93	724	763	802	842
EBIT	-539	1.767	7.146	8.094	9.036
Tax	-175	572	2.315	2.622	2.928
EBI	-364	1.194	4.831	5.471	6.108
Plus depreciation	93	724	763	802	842
Gross-Cashflow	**-271**	**1.919**	**5.594**	**6.274**	**6.950**

Fig. 11.22 Example for business case

3. Process performance: Process stability, quality and efficiency for example defined by faults per device or reclamations
4. Employees: Competences and knowledge. For example, through employee productivity [27].

KPIs help enterprises to measure the performance and monitor the processes, quality and customer feedback. Each project and enterprise has a different set of performance indicators, as they vary depending on production processes, service processes or projects [28]. It is the management's task to define the KPIs and set

Key Performance Indicator / Project Element	Development Cost	Customer satisfaction index	ROI	Net Sales	Internal/External Costs
	Changes actual vs. baseline	Changes actual vs. baseline	Changes actual vs. baseline	Changes actual vs. baseline	Changes actual vs. baseline
Product A	0%	24,5%	28,7%	10,8%	-40,0%
Product B	0%	-0,6%	-0,8%	-3,6%	-16,1%
Servie A	0%	0,9%	0,6%	-52,3%	-18,6%
Service B	0%	-4,6%	-4,3%	22,3%	-4862,4%
Service C	0%	-1,1%	-1,1%	27,2%	112,8%

Fig. 11.23 KPI dashboard example

the strategic goals. This results in an overall transparency and understanding within the company and encourages the implementation and realisation of the goals [27]. Figure 11.23 shows an example of a KPI evaluation sheet.

Next to the definition of KPIs, SLAs are defined. If the new business model requires additional services, such as logistics etc. contractual agreements including KPIs and SLAs are necessary. These are agreements regarding the performance of the service provider, in order to realise the strategic goals. For example, that the logistics company delivers the product within 2–3 days after the order. It is also possible to create internal SLAs, for example when there is a system error, the IT department must resolve the problem within 24 h [29].

11.2.6 Implementation Strategy

When the entire framework has been set, the implementation strategy can be developed. The goal of the implementation strategy is that the current processes can be transformed without major delays and downtimes. Furthermore, the resource availability is to be considered and the influence on the running operations or departments before setting up a project. It is recommended to start a pilot first, to stabilise the processes and find gaps and potentials for improvement. The old and the new process should be operated in parallel, so that the new process can gain maturity and stability [8].

The implementation of a new business model within a firm should be done through a project. Beforehand, some planning must be done to define the framework. Therefore, project management elements will be used to support the transformation [20]. Beginning with the project charter, that defines all important elements of the project a second important document will be presented, that is used throughout the entire transformation phase: the project plan. Furthermore, a RACI plan is put together, to define the roles and responsibilities during the transformation project. Finally, a budget plan for the project is set up, which is usually done by the sponsor of the project and serves as a guideline and frame for the expenditures.

11.2.6.1 Project Charter

Explanation and Background

The project charter is a core document during the project. At the beginning of the transformation project all relevant information is collected and summarised in the project charter. The information can be updated during the project [8].

The Project Charter consists of the following elements:

1. Business case: Reason why the project is being done
2. Problems and goals: Explains the problems and opportunities, as well as the goals and defines the new product and/or service that is to be provided through the business model
3. Project benefit: Defines the estimated financial gain and non-quantifiable soft savings
4. Roles and responsibilities: The project team and their responsibilities are named
5. Project scope: It is defined what is in scope and what is not
6. Project management: Defines the key phases and milestones that lead to the project goal and presents the project risks [8].

Procedure and Application

The following example (Table 11.8) shows a completed project charter.

11.2.6.2 Action Plan

Explanation and Background

The goal of the action plan is to create an overview of all necessary tasks for the business model transformation. The action plan defines work packages, consisting of many smaller tasks and actions that are necessary for the realisation of the transformation project. It is a tool used by the project manager and by the team members, where information is given on how of the action has been fulfilled as a percentage. These way team members know about the progress of the project and know when they can start their next task, if it is dependent on someone else's results [8].

Procedure and Application

Step 1. Define work packages and sort them in a logical order (parallel tasks are also possible).

Table 11.8 Project charter [8]

Project Title:	Development of a full-service package for the XY Antenna systems	Date	01.08.2017	**Version**:	V1
Business case:		**Project scope**:			
Market studies have shown, that the Asian competition is producing similar products to similar quality. This will lead to a revenue decline of 30% in the next 5 years, mainly in Europe. To meet the market changes value adding services will be introduced to provide a closed loop value chain for the customer		IN	Setup of a service concept for the installation, maintenance and recycling of the antenna systems within Europe		
		OUT	Employment of installation workers. American and Asian sites are not in scope yet		
Problem/Goal:		**Roles and Responsibilities**:			
In the last 6 months, sales have gone down by 10% due to the Asian competitors. To provide customers with a closed loop solution and to create a lock-in, a service package will be introduced by the 10th of January. The service package includes the installation of the antenna systems, regular maintenance and recycling solutions		Sponsor	Dr. Franco		
		Project leader	Mr. Adams		
		Core team	Mrs. Smith (Marketing) Mrs. Choi (Sales) Mr. Watson (F&E) Mr. Evans (QM) Mr. Brown (Legal) Mrs. Jacquard (Operations)		
Financial Benefit:		**Milestones**:			
Amount to be sold 2018: 13,000, Service packages booked 2018: 7,000; Profit per Antenna system 10% of net-sales price, profit per service 20%. Net-sales price per system 2,500 €, Service package net-sales- price 350€. Revenue 2018: 34,950,000€; Net-Profit 3,740,000€		Requirements definition	01.08.2017		
		Process design	01.10.2017		
		Tools development	15.10.2017		
		Pilot	10.01.2017		
Risks:		Signatures: Sponsor project leader			
Reliability of outsourcing service partner					

Step 2. Brake down the work packages and define actions within.
Step 3. Decide on the earliest start, the duration and the latest end of each task [8].

11.2.6.3 RACI Matrix

Explanation and Background

The RACI matrix is designed to give an overview of the responsibilities within a project and shows, which activities belong to whom. Also, it is used as a communication tool, as it also reveals who must be informed or who can help with decisions or questions [8].

RACI stands for Responsible, Accountable, Consulted and Informed. These are the different rolls that the project team can be assigned to. The tasks of each role will be explained in the following:

Responsible: This person must fulfil the task and is responsible for its execution. The responsibility is decided through the "accountable" person. In a project, there can be many people who are responsible for certain tasks.
Accountable: This team member makes sure that the tasks are fulfilled and responsible and can be made accountable for the results. This person also has a veto right.
Consulted: This person can recommend final decisions and has knowledge and experience on the topic of the project.
Informed: This team member must be informed about decisions and results.

Procedure and Application

Step 1. Identify the tasks and the people involved in the project. This is usually the project team but can also be supported by experts who are not in the operative team of the project
Step 2. Decide upon the roles and responsibilities of each person and task. Only one person can be made accountable for each task
Step 3. Remove overlapping of responsibilities and responsibility gaps.

The chart (Fig. 11.24) shows an example of a RACI chart and is used to give an overview of the tasks and the roles and responsibilities during the project.

11.2.6.4 Project Plan

Explanation and Background

The project plan helps to maintain an overview over the timeline, tasks and project phases. This is often the most important instrument of the project controlling, as it fully describes the project tasks. The tasks are grouped to task packages with subtasks.

Tasks	Position / Name	CEO	Head of Marketing	Sales & Marketing	Legal	Quality Management	Key Account Manager	Director of Operations	F & E
		Mrs. Grey	Mrs. Smith	Mr. Taylor	Mr. Brown	Mr. Evans	Mrs. Choi	Mrs. Jacquard	Mr. Watson
1. Marketing plan development		I	A	R	C				
2. Supplier Management		A	I	I			R	C	
3. Process developmemt		I		C	C	C		A	R
4. Legal verification		A		I	R			I	I

Fig. 11.24 RACI chart [8]

There are different forms of project planning instruments, for example milestone-trend-analysis, appointment schedule, Gantt-diagram and the network planning technique. In the following the Gantt-diagram will be introduced as it is commonly used in practice. The Gantt-diagram, named after Henry Laurence Gantt, gives an overview of the chronological order of activities and tasks in form of a bar chart. In comparison to the network planning technique, the task duration is visible [29].

All tasks are considered in the Gantt-diagram on different levels of detail. The highest level describes the overall task and functions as a subproject. The latter is then broken down into task packages (these can be copied from the action plan), which are made up of smaller subtasks, which fill the lowest level. For each task package, specific goals can be defined and a subproject leader is selected, who will be responsible for the execution of the tasks and their results. When the tasks have been set, they are combined with a timeline and sorted into a logical order, considering the interdependencies between the tasks and task packages [29].

Procedure and Application

Step 1: Create sub-projects concerning overall tasks or goals that are necessary for the business model transformation

Step 2: Define tasks that are relevant for the sub-projects and group them into task packages. Depending on the level of detail, only the task packages can be plotted or the task packages including the subordinate tasks. For example, if the subproject is "Pilot preparation" and the first task package is to prepare the "Kick-off Meeting", then subordinate task would be to create a presentation, to inform the members of the project, align it with the project owner, schedule a meeting and so on.

Step 3: Select a timeline and sort the task packages in a logical order, considering their chronological dependency and then estimate the time needed to realise the task.

Step 4: Assign subproject leaders to the different task packages (Fig. 11.25).

Figure 11.26 is an example for the usage of this method.

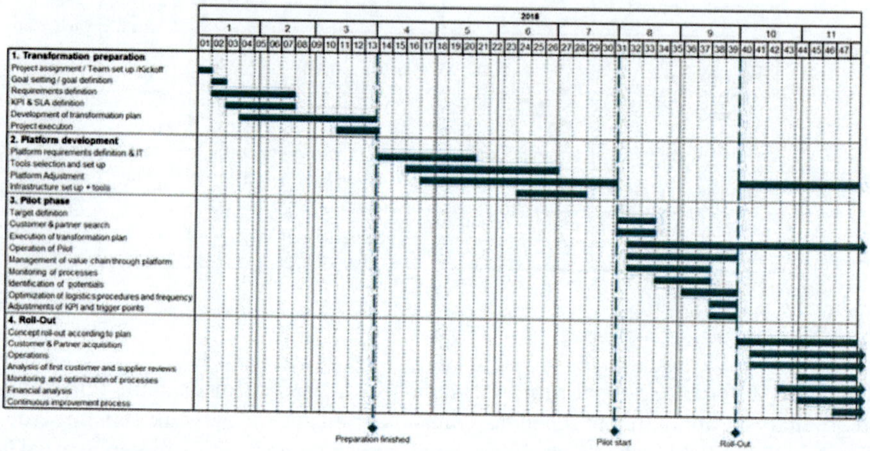

Fig. 11.25 Project plan

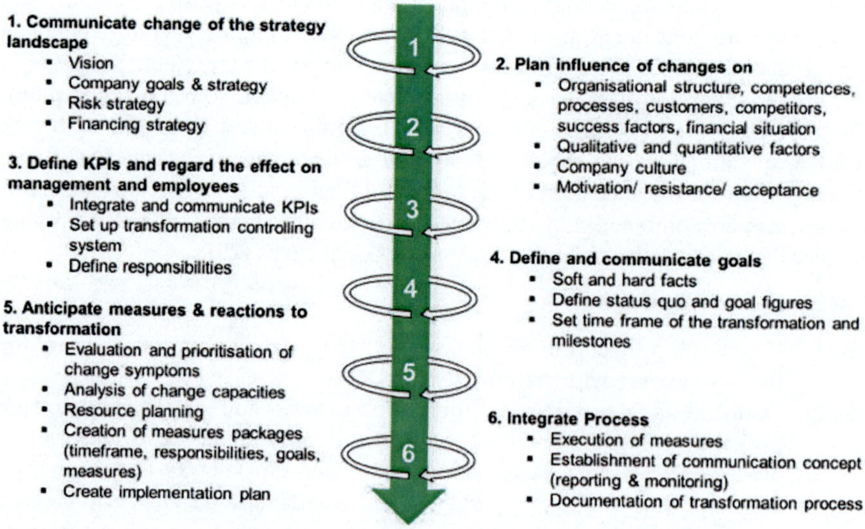

Fig. 11.26 Six loop change management [27]

11.2.6.5 Transformation Plan

Before implementing the new business model it is recommended to have a transformation plan in place [7]. The reason is, that often the culture and employees of a company can be the main obstacle when exposed to major changes. Companies, whose culture is driven by entrepreneurship, creativity and innovation are more likely to successfully transform their business model, rather than companies with a traditional and closed-minded mentality. A study of McKinsey from 2008 with top managers and executives showed that a company's culture is the biggest obstacle and at the same time the main driver of innovation and change. On the one hand, company culture influences the internal aspects, such as the behaviour of the employees while on the other hand it reflects the personality and reputation of the company [30].

Therefore, before changing the business model and all related processes, the company culture should be analysed.

Schein developed a model to describe the 3 different layers of company culture. The first layer is the outside perspective. It is visible through the clothing and language of the employees but also through the logo and company concept. The second layer contains the company values and norms, which are fulfilled by the employees. This is a mutual feeling and behaviour of how things should be and are handled. The third and deepest layer of the model is shown through subconscious behaviour, which is provoked by fundamental influences and is the most difficult layer to influence, as it was formed by generations of employees [31]. A company with a strict corporate dress code, a very dominant hierarchy and a conservative mentality for example, will be less susceptible to change and less flexible when adapting, than a company with no dress code and a casual informal environment between the executives and the employees. The corporate environment can also have a high impact on the values and corporate identity as well as on open or restricted communication [30, 31]. The key drivers of corporate culture and identity are the executives of a company, as they exemplify the set values and behaviour.

The "Six Loop" Change Management tool can be used to actively design the transformation process (Fig. 11.26).

In the six loop concept supports the continuous change and improvement process, while focussing on the company culture, the employees and the mentality [27, 32–35].

1. To start, the vision and the strategy of the new business model must be communicated to the employees. If it is unclear in which direction the company is moving, any further attempts of convincing the employees of the new business model will fail. The vision should motivate and be challenging, but reachable within time.
2. All internal changes that are caused by the implementation of the new business model should be considered, for example the changes on the organisational structure, competences, processes, customers, partner, competitors, success factors and the financial situation. Also, the change in the company culture is to be analysed and the possible resistance and acceptance factors that come along with it.

3. The KPIs represent and quantify the strategic goals in terms of the financial perspective, market and customer perspective, process perspective, training and development perspective. These should be communicated to create common understanding on how the success of the new business model can be measured. For this, a controlling system can be introduced, to support the change.
4. Together with the KPI introduction, the goals for each department should be defined. These can be in form of quantitative or qualitative goals. The current status quo is to be communicated and a timeframe in which the goals should be achieved, including milestones, is to be set up.
5. Measures should be defined on how the processes will be changed or newly introduced with the help of an implementation plan. Before the implementation however, also the resources should be evaluated and possible change symptoms and change capacities per department. Then individual measure packages can be set up to address each department differently.
6. In the last step, the implementation plan can be executed and the transformation can start. The controlling system and standardised documentation during the transformation is essential to keep track of all factors within the process and should be continued well after the implementation project has finished [27].

References

1. Reichwald, R., Piller, F.: Interaktive Wertschöpfung. Open Innovation, Individualisierung und neue Formen der Arbeitsteilung. 2., vollständig überarbeitete und erweiterte Auflage. Wiesbaden: Gabler Verlag/GWV Fachverlage GmbH Wiesbaden (2009)
2. Glauner, F.: Zukunftsfähige Geschäftsmodelle und Werte. Strategieentwicklung und Unternehmensführung in disruptiven Märkten. Berlin, Heidelberg, Springer Gabler. https://doi.org/10.1007/978-3-662-49242-0. Accessed on 10 June 2017
3. Nührich, K.P., Hauser, A.: Unternehmensdiagnose. Ein Führungsinstrument zur Sicherung der nachhaltigen Existenzfähigkeit von Unternehmen. Springer, Berlin (2001)
4. Böhmann, T., Warg, M., Weiß, P.: Service-orientierte Geschäftsmodelle. Erfolgreich umsetzen. Springer Gabler, Berlin.https://doi.org/10.1007/978-3-642-41625-5. Accessed on 19 June 2017
5. Schallmo, D.R.A.: Design Thinking erfolgreich anwenden. So entwickeln Sie in 7 Phasen kundenorientierte Produkte und Dienstleistungen. Springer Fachmedien Wiesbaden, Wiesbaden (2017)
6. Rawitzer, H., Hefti, J.: Ansoff-matrix reloaded. Wachstumsstrategien für die digitale Zukunft. In: ZFO—Zeitschrift Führung und Organisation **85**(06) (2016)
7. Kohne, A.: Business development. Kundenorientierte Geschäftsfeldentwicklung für erfolgreiche Unternehmen. Springer Vieweg, Wiesbaden (2016). https://doi.org/10.1007/978-3-658-13683-3. Accessed on 04 June 2017
8. Staudter, C., Hugo, C., Bosselmann, P. Mollenhauer, J.P. Meran, R. Roenpage, O.: Design for Six Sigma+Lean Toolset. Innovationen erfolgreich realisieren. 2., vollst. überarb. und erw. Aufl. Springer Gabler, Wiesbaden (2013)
9. Niemann, J.: Die services-Manufaktur, Industrielle Services planen –entwickeln – einführen. Ein Praxishandbuch Schritt für Schritt mit Übungen und Lösungen. Aachen, Shaker Verlag (2016)
10. Kreutzer, R.T.: Konzeption und Grundlagen des change-managements. In: Wirtschaftswissenschaftliches Studium 1 (2017)

11. Dehmer, J., Kutzera, A.-A., Niemann, J.: Digitalisierung von Geschäftsmodellen durch plattformbasiertes value chain management. In: ZWF - Zeitschrift für wirtschaftlichen Fabrikbetrieb 112 (4) (2017)
12. Business Dictionary "Transformation": Transformation. Hg. v. Inc. WebFinance. https://www.businessdictionary.com/definition/transformation.html. Accessed on 25 June 2017
13. Herrmann, A., Huber, F.: Produktmanagement. Grundlagen - Methoden - Beispiele. 3., vollst. überarb. u. erw. Aufl. Springer Gabler, Wiesbaden. https://doi.org/10.1007/978-3-658-00004-2. Accessed on 14 July 2017
14. Gerberich, C.W.: Praxishandbuch Controlling. Trends, Konzepte, Instrumente. 1. Aufl. Wiesbaden: Gabler (2005)
15. Osterwalder, A., Pigneur, Y.: Business model generation. Ein Handbuch für Visionäre, Spielveränderer und Herausforderer. 1. Aufl. Frankfurt am Main: Campus Verl. (2011). https://search.ebscohost.com/login.aspx?direct=true&scope=site&db=nlebk&db=nlabk&AN=832895. Accessed on 30 May 2017
16. Gassmann, O., Frankenberger, K., Csik, M.: Geschäftsmodelle entwickeln. 55 innovative Konzepte mit dem St. Galler Business Model Navigator. München: Hanser (2013). https://www.hanser-elibrary.com/action/showBook?doi=10.3139/9783446437654. Accessed on 14 July 2017
17. Homburg, C.: Marketingmanagement. Strategie - Instrumente - Umsetzung - Unternehmensführung. 6., überarbeitete und erweiterte Auflage. Springer Gabler, Wiesbaden (2017). https://doi.org/10.1007/978-3-658-13656-7. Accessed on 05 May 2017
18. Munro, R.A.: Six Sigma for the office. A pocket guide. Milwaukee, Wis.: ASQ Quality Press (2003)
19. Hoffmann, C.P., Lennerts, S., Schmitz, C., Stölzle, W., Uebernickel, F.: Business innovation: das St. Galler Modell. Springer Gabler, Wiesbaden (2016) (Business Innovation Universität St. Gallen, Profilbereich Business Innovation).https://doi.org/10.1007/978-3-658-07167-7. Accessed on 08 July 2017
20. Schallmo, D., Brecht, L.: Geschäftsmodell-Innovation. Grundlagen, bestehende Ansätze, methodisches Vorgehen und B2B-Geschäftsmodelle. Zugl.: Ulm, Univ., Diss. (2012). Springer Gabler, Wiesbaden (2013)
21. Melzer, A.: Six Sigma—Kompakt und praxisnah. Prozessverbesserung effizient und erfolgreich implementieren. Springer Wiesbaden Gabler, Wiesbaden (2015). https://search.ebscohost.com/login.aspx?direct=true&scope=site&db=nlebk&AN=1050548. Accessed on 03 June 2017
22. Kühnapfel, J.B.: Nutzwertanalysen in Marketing und Vertrieb. Springer Fachmedien Wiesbaden, Wiesbaden (2014)
23. Heesen, B.: Investitionsrechnung für Praktiker. Fallorientierte Darstellung der Verfahren und Berechnungen. 3. Auflage. Springer Gabler, Wiesbaden (2016). https://search.ebscohost.com/login.aspx?direct=true&scope=site&db=nlebk&AN=1087143. Accessed on 20 June 2017
24. Brugger, R.: Der IT Business Case. Kosten erfassen und analysieren, Nutzen erkennen und quantifizieren, Wirtschaftlichkeit nachweisen und realisieren. 2. Aufl. s.l.: Springer-Verlag (Xpert.press) (2009). https://site.ebrary.com/lib/alltitles/docDetail.action?docID=10297037. Accessed on 01 July2017
25. Wöltje, J.: Finanzkennzahlen und Unternehmensbewertung. 1. Aufl. s.l.: Haufe Verlag (Haufe TaschenGuide—Band 00381, v.381) (2012)
26. Olfert, K.: Investition. 13., aktualisierte Auflage. Herne: Kiehl (Kompendium der praktischen Betriebswirtschaft) (2015)
27. Gerberich, C.W., Teuber, J., Schäfer, T.: Integrierte Lean Balanced Scorecard. 1. Aufl. s.l.: Gabler Verlag, 2006. Online verfügbar unter https://gbv.eblib.com/patron/FullRecord.aspx?p=748707
28. Gabler Kompakt-Lexikon Wirtschaft.: 11., aktualisierte Aufl. Springer Gabler, Wiesbaden (2013)
29. Bergmann, R., Garrecht, M.: Organisation und Projektmanagement. 2. Aufl. 2016. Springer Gabler, Berlin, Heidelberg (BA KOMPAKT) (2016). https://doi.org/10.1007/978-3-642-32250-1. Accessed on 19 June 2017

30. Mervelskemper, L., Paul, S.: Unternehmenkultur als Innovationstreiber? Ein Einblick in die Praxis. In: Zeitschrift für das gesamte Kreditwesen (15) (2016)
31. Schein, E.H.: Organizational culture and leadership, 4th edn. Jossey-Bass, San Francisco, CA (2010) (The Jossey-Bass business & management series). https://www.esmt.eblib.com/patron/FullRecord.aspx?p=588878. Accessed on 14 July 2017
32. Niemann, J., Pisla, A.: Sustainable potentials and risks assess in automation and robotization using the life cycle management index tool—LY-MIT. Sustainability **10**, 4638 (2018)
33. Niemann, J., Schemann, T., Erkens, J.: Servitization—pathway of transformation from product manufacturer towards a service provider. In: 2018 International Conference on Production Research—Africa, Europe, Middle East 5th International Conference on Quality and Innovation in Engineering and Management, July 25–26, 2018. Cluj-Napoca, Romania
34. Stöhr, C., Janssen, M., Niemann, J.: Smart services. In: 14th International Symposium in Management, Challenges and Innovation In Management and Entrepreneurship, 27–28 Oct 2017. Timisoara, Romania
35. Niemann, J.: Eine Methodik zum dynamischen life cycle controlling von Produktionssystemen. Heimsheim: Jost-Jetter Verlag, 2007IPA-IAO Forschung und Praxis 459). Stuttgart, Univ., Fak. Maschinenbau, Inst. für Industrielle Fertigung und Fabrikbetrieb, Diss. (2007)

Part II
CAD/CAM/FEA/PDM and Robotics: Factors of PLM Implementation

I don't know what your destiny will be, but one thing I know:
The ones among you who will be really happy are those who have sought and found how to serve.

Albert Schweitzer

Chapter 12
The Next Digital Age

"Digitalization is shaping the new manufacturing"
Siemens PLM.

12.1 Introduction

Do you want to become a PLM-Engineering Manager?

You May Clarify Yourself:

IF Yes

Well, This is a Good Decision.

If not—then

You will find some hints to become, in the chapters of this book.

Are You Ready for that?

For your best decision let's start from the ***JobDescription***, releases by a company looking for an Engineering Manager [1].

EngineeringManagerWanted!

"Professionals focused on technical sales that provide technical direction and business guidance to the engineering team".

Three sorts of demands are simply specified in the job description:

1. ***Are required*** a lot of ***technical skills and technical understanding*** before the: **business, sale** and **communication skills**;
2. ***Drives gross profit****through* ***account planning, resource planning*** *and* ***allocation.***
3. ***Proactively developments*** *that leads a team to high-performing: Pre-sales Engineers and Post-sales engineers, continuously* ***seeking for innovative methods****to* ***improve team performance*** *and* ***profitability***.

But, what really company needs is: **THE PROFIT**.

In this case the "good part" is that the company indicates exactly the meanings, to obtain profit. By exercising:

J. Niemann and A. Pisla, *Life-Cycle Management of Machines and Mechanisms*, Mechanisms and Machine Science 90,
https://doi.org/10.1007/978-3-030-56449-0_12

Fluxes management:

"cash flow"—account planning;
materials, energy, and parts (acquisition, storing, cost and availability evaluation)—resource planning;
logistics—allocation;
marketing—pre sales;
development, manufacturing, service—post sales;
Human Resources Management—improving team performance.

Work Frame:

measurable responsibilities;
capabilities improvement;
necessary skills; and technical competences.

Probably you understand already the same thing. Based on the expressed description the profile of the **Engineering Manager** is designed.

If you apply for, this will be **YOUR Responsibilities.**

- **Build** and **Hire** a professional in the **Systems Engineering Organization** (SEO);
- **Build** the **engineering structure** to scale and grow with the strategy of the company;
- **Work jointly** with the sales organization to create a **territory plan** utilizing the sale engineer feedback and adding own knowledge of local market demand:
  - o **Create** a **technical resource plan** by vertical management and technology
  - o **Collaborate** with Sales VP to prioritize and target **team opportunities**
  - o **Understanding** the mechanisms to build **team capacity** and improve team performance
  - o **Align** resources to deliver on commitments and **drive results**
  - o **Lead team** to technical account strategies that align to customer business requirements and goals; **assigning resources properly**
  - o **Monitor and approve customer requests** focused on pre and post-sales resources
  - o **Assist** in determining domain/solution focused resources, understanding and developing the **requests for these resources**
  - o **Provide** timely and appropriate feedback that focuses on those things that will make the biggest difference in performance; **reinforces efforts and progress**
  - o **Track and report** team metrics for a given opportunity. Selling Professional services is a must in understanding how to **make money**!

12.2 It a Helping Tool and a Nightmare

Now that you are aware about the importance of your position "***PLM Engineering-Manager***", you probably know that is nothing on this job without IT (Information Technology).

Good or bad we face in every sector the human evolution towards: globalization and digital instruments connection. In this "environment" an advantage is to appreciate which are the right digital components for you, and to be aware about what other existing options or applications.

Every research institute and research group has its own vision about the efficiency and the opening in using the IT instruments.

From early 70 we are prepared that the computers and the robots will be part of our lives and the machines and the mechanisms are the elements that will surround us to make our life possible and effective. This image is a part of an element that from the beginning of the **Homo Faber** existence (Man the Creator) defines the humanity: ***The technological culture***.

Homo Sapiens (the Wise Man) is considered as a unique example of a "different animal":

the wise one = an entity with self-conscience, feelings and intelligence.

Having the doubt about the intelligence of the "Homo sapiens", great philosophers and political theorist like Johanna ARENDT, Henri-Louis BERGSON or Max Ferdinand SCHELER, made the reference to the Homo Sapiens more likely as a Homo Faber "***Humans as controlling the environment through TOOLS, as a result of the ORIGINAL SENSE of the intelligence***.

This approach leads to the idea that: **what we are** and **how a social group reacts** is a ***cognitive function based on the existing technological culture*** and one's evolution and status is defined by the power that tools can do.

The **NEXT digital age** is the result of almost 50 years (in 2020) of evolution within computer science and IT.

At the beginning it was a struggle to create powerful processors and microprocessors, larger memories and very performant computers. This was the road for the apparition and the evolution of personal computers (PC) and the machine-tools computer numerical control (CNC), soon followed by the industrial and non-industrial robots development.

This evolution is continuing, but now the balance is moved toward creation of stronger algorithms and more powerful software applications, parallel computing and efficient user interfaces.

For the next digital age, essential elements are the user approach and the detailed definition of the applications processes. Components like Computer Aided Design (CAD), Computer Aided Manufacturing (CAM), Computer Aided Engineering (CAE) or Product Data Management (PDM) are of most importance, being valid for both industrial and non-industrial applications. The complex of these major components forms the CAC (Computer Aided Control) enabling the processes and resources management within a Product Lifecycle Management platform (PLM).

These components, one's considered useful only for engineers, are including now biological, medical, social, political and environmental elements to be modelled, simulated and economically considered as technological means for professionals. In the next digital age the Homo Faber must, and will, **engineering** and **manage** all the aspects related to life.

The tool-intelligence and the combination recognition, starts in 1950 with the Alan Turing famous test, presented in the paper "Computing Machinery and Intelligence", for machine capability to exhibit equivalent intelligent reaction that are indistinguishable from a human one. In the Alan Turing test, a human judge engages in natural language a conversation with "a partner" (human or a special designed machine).

If the human judge cannot reliably tell the machine from the human, the machine is said to "have passed the test", even the test does not check the ability to give the correct answer to questions but how closely the answer resembles typical to the human answers.

The IT users every day problem is the continuously increasing of the Data Bases (DB). In every **second** are generated in the world more data that one can read in a lifetime. Now, the challenge is for the search capacity (***query capacity***—the expression of questioning expressing a doubt).

If one uses any Internet browser looking for a term like "technology" in less than 0.3 s you will receive over 760 million references, and if you are looking for "PLM" in the same 0.3 s you will receive "only" over 7 million references.

Communication and knowledge transfer between mentor and follower reach a different level: ***the capacity of basic data selection and interpretatio***n.

Even more, the communication and knowledge transfer is not limited between human subjects but evolving at the communications between machines M2M (machine to machine) and between devices IoT (Internet of Things).

In many applications the interfaces (pages, texts, images, sounds, collage, etc.), are not made any more by peoples but are automatically generated by computer programs (the follower of the Turing special designed machines).

At the global level, one of the famous inter-servers communication languages (the second within the ORACLE applications) is the XQuery designed by the Romanian computer scientist **Daniela FLORESCU**, since early 2007.

The scientific activity, in testing, simulation or process forecasting is based on the EPP (Electronic Prototyping Platforms). From the logistic point of view the rules and the philosophy that made works the EPP already exist within the PLM (Product Lifecycle Management) and therefore is a natural evolution that EPP to be direct implemented within the PLM.

The forecast of EPP evolution is a strong element in recognizing (once more) the PLM concept not only as leading element within a company activity and evolution but as crucial in reaching the success. ***ThereforePLMmust be used as the "Driving Element".***

Regardless the branch of activity, in the digital age context, the universality and the operability of PLM is given exactly by the elements that define the next digital age: the capacity to generate knowledge as half-structured data bases, the query capacity and the capacity to keep under control this type of activity.

Why is considered that PLM can be applied to any branch of activity?

How is this possible?

The answers are not simple but the core element that made understandable the phenomena is that PLM is using modular components that are digitally dealing with virtual reality: virtual products, virtual processes, virtual machines or mechanisms, virtual tooling, virtual persons or virtual projects (scenarios).

The virtual reality (VR) is getting closer and closer to the "real" reality through the ***features***.

Features: are distinctive attributes that can be used as standalone or in correlation. One feature may be a characteristic or a limit (boundary or condition) for other features, contributing at the overall appearance and the final shape of the represented item.

Considering mainly the industrial activity, the PLM implementation capability is closer to the companies that are using already the CAD and the CAM or "only" the CNC (Computer Numerical Control) for their processes.

Traditionally CNC is related with the numerical controlled machine-tools, but the CNC concept is extended also with other valences in connection with:

- CAD (Computer Aided Design) where parts may be generated by numerical control, independently from the designer but under his supervision;
- CAM (Computer Aided Manufacturing) the extension of the initial numerical controlled machine-tools with as many as possible additional features;
- CAPP (Computer Aided Production Planning) where (again) entire sequences can be automatically generated based on the existing features.

The CAD/CAM/CAPP integration is possible by having another specialized module named Product Data Management (PDM). The PDM easy the features representations through data, and so are enabling computers to store and process the DATA.

Using together the PDM and the CAE (Computer Aided Engineering), results a broader operational environment and more functionality is added, as a frame for the integrated manufacturing process systems (IMPS), namely the PLM Platform (PLMP), Fig. 12.1.

As long as PLM looks like a panacea is justified the rise of the next question:

How did appear PLM?

The PLM Appears in the Mid'90, Within the Automotive Industry and then Expand to All the Other Industrial Activities.

This answer was easy, but a more interesting question, to justify the success of the PLM and the billions invested every year in PLM research and implementation is:

Why did appear PLM?

Our answer is: as a **NECESSITY!**

Fig. 12.1 The PLM Platform. [*Source* ADA Computers srl]

12.3 PLM Functionality in the Next Digital Age

From Descartes and his "Discourse on Method" the "**human common sense**" was the main instrument for decisions as a basic ability to judge and understand things. The common sense is reasonably expected by nearly all people without debate, as long as the complexity and the challenges in developing new products are limited to local or regional aspects.

Not surprisingly in the "modern times" and especially in the western technological civilization, that the "common sense" term becomes frequently used for rhetorical effects and sometimes with pejorative sense as result of the increasing complexity and engineering challenges of developing new products for the global competitive markets.

In the new global environment, the human common sense (crystallized in instincts) is not enough anymore to adapt at change. **Therefore the PLM appears!**

PLM appears as a system in form of a "***common sense tool***" needed by the organizations to help them within in operation process, having practical functionalities and wider boundaries, in much broader scopes, using and define the most advanced technological culture for the company.

PLM foster innovation in a product development lifecycle. ***Appropriate change in the early stage is good***, and in fact, should be encouraged. The more performant products and process simulation is experiencing a lot of changes in the early stage when the change is less expensive. In the same time the PLM functions enables an effective way for managing the evolution of product-related information and the product development process action flows, capturing the actions that must be done, revealing the affected components and the demand for resources.

Within industrial process the PDM is not just a way to store complex and multiple DATA or to possess performant query mechanisms but is also a multitasking multi user digital tool (DT) that manufacturing engineers needs in selecting sequences and part design to work together with each other and also simultaneously with other partners.

This way of activity requires permanent notifications when any change occurs. Now, the PDM is used to control information: documents, and work processes in order to design, build, support, distribute and maintain products.

Within the PLM platform the PDM is the essential core consisting in 3 important elements:

- **Parallel engineering;**
- **High complexity;**
- **Synchronous technology.**

The parallel engineering enables engineers from all around the globe to work in the same time and on the same product.

The single engineer or the engineers' team, which subscribes to a portion of a design, is not alone but needs to cope each other and always to notify or to be notified when changes occur. This is fulfilled automatically by the parallel engineering.

The parts complexity result from different sources: the geometry, the tolerances, material characteristics, assembly conditions, surface engineering requirements, manufacturing and properties portability, assembly characteristics, tracking coding, etc. all together representing a product and must be embedded in the same file.

Synchronous technology (ST) captures ideas quick in a structured technical manner, using data from different CAD systems, enabling change in an implicit interactive way and is storing the parameters change in a log that is updating all the connection without the human intervention. That is connecting everyone in the product lifecycle and in the product lifecycle management.

The PDM ensures the interface with all the other design and manufacturing (D&M) components and fluxes creating the **data source** for modelling, optimization, visualization and simulation.

A product is created in a virtual world and represents a virtual object, considered an error free version of the hardware product and therefore PDM is effective needed for the product-related process development (manufacturing flows, history analysis, actual decisions, predictions, etc.).

The PDM system is used to control data, files, documents, and work processes grouped in the form of **information clusters** required by the Product-Process (PP) in: design, optimization, simulation, manufacturing, packaging, distribution, service and maintenance. Therefore PDM is an integration tool connecting many different areas and placing the right available information in the right form to the right person at the right time in the entire project network, and that from the very beginning [2].

The new generation, the web based PDM (WPDM), is designed to support the entire product lifecycle and to offer the best time-to-market with certain resources and defined quality level, from the initial concept (**Styling**) until the end (**Recycling**).

Not forgetting the triad **GCS Good-Cheep-Swift**: **G**ood quality, low price (**C**heap) and **S**hort manufacturing time, from which we may always afford maximum two components.

The PLM power comes from the readiness to manage the information throughout the entire lifecycle, with the capability to handle hundred thousands of users, from multiple domains, with different business processes and different representations of the multitude of native product-models, resource-models and process-models [3, 4].

This functionality creates an automatically collaborative activity—**diagnose oriented company**—using synchronous technologies for a real time activity or for speed up a process scenario.

12.4 PDM Methodology Evolution

In the 70s are the beginnings of the CAD (Fig. 12.2a). Originally the CAD systems provided an electronic pencil that leads to electronic 2D drawings that can be stored, printed, copied, transferred and updated within an informatics format.

In the 80s the 2D design evolved to 3D in the shape of wireframes and surface shading to give the 3D ***sensation*** and for the CAD users the ability to build 3D virtual prototypes and simulate interactively the product performance by checking products interferences (Fig. 12.2b).

This system provides low capacity in:

- assembly modelling;
- versions generations;
- products comparison;
- products configuration;
- components relationship;
- components management;

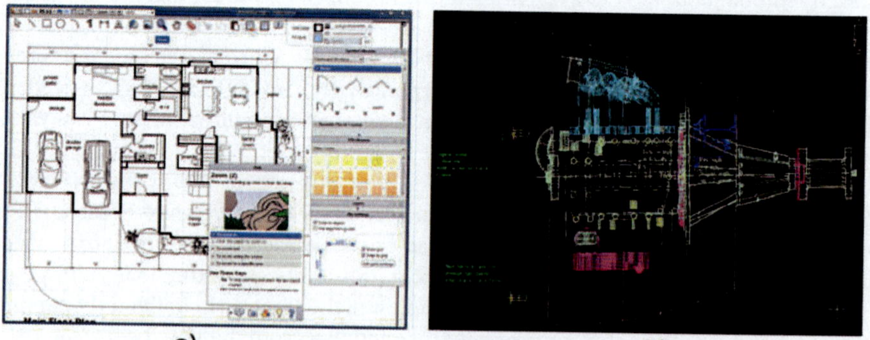

a) b)

Fig. 12.2 The 2D and 3D early CAD systems. [*Source* Autodesk]

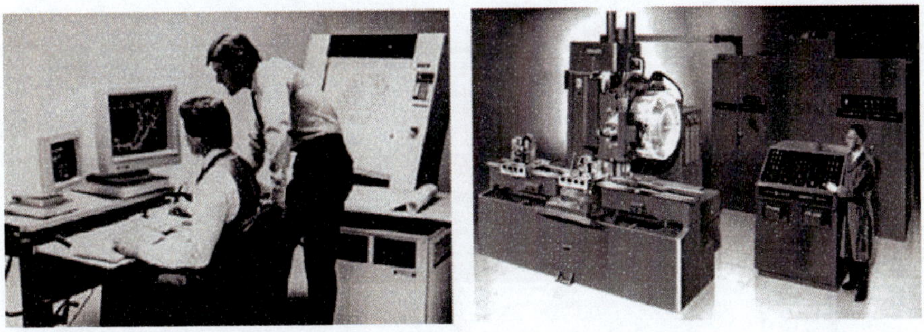

Fig. 12.3 Early CAD/CAM systems. [*Source* Bobcad]

- cost and bill of materials generation;

Together with no data link with other processes or company general databases, made the leaders in manufacturing industry, to reconsider the **CAD/CAM** technology. The conclusions about the severely downgraded efficiency came from the missing of a **PDM** system to provide a **secure location** for universally access **product data**, and the **support** for **structured workflows** in order to evolve a product design through its lifecycle, enabling the **data sharing** between design engineers, manufacturing engineers, economic engineers and other legacy applications.

At that time, early 90s, with no commercially available systems each large company started to develop its own **DMS** (Data Management Solution). In parallel a number of software companies started to realize the problem associated with the so-called "***automation islands***" (CAD and CAM tools) and the potential market of efficient data management methodologies, Fig. 12.3.

This is the first generation of commercial PDM systems, as products offered by the vendors already involved in the CAD/CAM/CAE software market, but adding the new developed the **Data Management** solutions, forming the PDM of their product lines.

The companies that realize the advantages of the PDM successfully implementation, achieved multiple benefits in terms of productivity and competitiveness. Now just imagine an updated CAD and a modern CAM without the PDM capabilities! The result is a missing link, Fig. 12.4, [Source: HP, Autodesk, and Dawson-Shanahan].

12.5 PDM Capabilities

In terms of PDM's capabilities, there are five basic user functions that should be supported by a PDM system:

1. **Data vault and Document Management**

   Provides storage and retrieval of product information.

Fig. 12.4 CAD (left) and CAM (right) No PDM (middle) systems. [*Source* HP, Autodesk, and Dawson-Shanahan]

Data Vault attempts to solve the problem of dealing with change in the business environment by separating the **business keys** (business entity identifier) and the associations between business keys, using **descriptive attributes** of those keys.

The business keys and their associations are structural attributes, forming the skeleton of the **Data Model**.

One of the Data Vault main axioms is that: real business keys change only when the business changes, being considered as the most stable elements from which derive the structure of a historical Database.

The keys are stored in tables, with few constraints on the structure, tables that are called **Hubs**. Choosing the correct keys for the Hubs is the prime importance to ensure the stability of the model [5, 6].

Hubs contain a list of unique business keys, with low susceptibility to change, containing also a surrogate key for each Hub item and metadata describing the origin of the business key. The descriptive attributes on the Hub (such as the key description—possibly in multiple languages) are stored in structures called **Satellite tables.**

Data Vault Modelling is a Database modeling method, designed to provide long-term historical storage of Data coming from multiple operational systems. It is enabling also the looking at historical Data with issues such as auditing, tracing, loading speed and resilience to change.

Data Vault Modelling focuses on several things:

First, is the need to trace of where all the Data in the Database came from? This means that every row in a Data Vault must be accompanied by record source and load date attributes, enabling an auditor to trace values back to the source.

Second, it makes no distinction between good and bad Data ("bad" meaning not conforming to business rules), the Data Vault stores "***a single version of the facts***" (also expressed by Dan Linstedt as "all the Data, all of the time") as opposed to the practice in other Data warehouse methods of storing "***a single version of the truth***", where mechanisms "cleansed" that is removing the Data that does not conform to the definitions, is active.

Third, the modeling method is designed to be resilient to change in the business environment where the Data being stored are explicitly separating structural information from the descriptive attributes.

Finally, Data Vault is designed to enable parallel loading as much as possible, so very large implementations can scale out without the need for major redesign.

2. **Workflow and Process Management**, controls procedures for handling product Data and providing the mechanisms to drive a business with information.

Workflow is a process oriented resources allocation sequence, as a digital form of the real work necessary to obtain a product (material processing, Data processing or service offer) expressing a series of operations and requirements, made by a single person or a team, with defined meanings, within a define interval of time within an organization.

Process Management, is related with the physical activities as result of the process definition, workflow management and the human interactions.

The **Workflow Management**, is an important element in the Process Management, from the static control of the material and Data buffers to the flow speed, flow volume and flows interconnections, that in the beginnings instinctively leaded to the just-in-time and just-in-sequences philosophy and procedures for the job shop control.

In other types of activities the workflow management creates the necessary discipline and the necessary level of redundancy facilitating the implementation of the document-driven workflow (i.e. banks or insurance companies); enables the continuous work around the globe (follow-the-sun methodology); the manipulation and search unstructured Data (as in health care); the coordination of the service-oriented applications (where specific geographically distributed facilities are available); the competent and reliable activity within informatics (getting-things-done philosophy); and also in the most nonconventional environment like scientific research by avoiding repetitive loops (i.e. DAG- Directed Acyclic Graph model).

3. **Product Structure Management** handles bills of material, product configurations, associated versions and design variations.

The Product Structure Management (PSM), appears in applications since 1998, starting with the CIM concept (Computer Integrated Manufacturing) and CIM data. PSM is not a static defined component, it represents the moment status but comparing different considered moments offer the dynamic of the change and therefore also the configuration management is to be considered. Usually the PSM is represented in an organigram shape, fundamentally being a hierarchical tree structure starting from the core (root) of the Product, as an element integrated within a functional system.

The PSM is putting into evidence the: relations, the Product data structure, the part modelling parameters, the rules, the Product items (group, users, host access, transforming tools, world representation, etc.), all connected references, the Product changes history, attachments, files and directories, associated geometrical CAD,

CAM CAE models, papers, documents, certifications, etc. This is practically a breakdown of the product to recreate partial functionality of the product defining subsystems, advisable to follow the manufacturer assembly procedures. The common PSM consideration is reflected by the BOM (Bill of Materials) directly applied within the manufacturing and connected activities process. In most industry this is the first (some times and only) component directly understand and accepted by both the engineering and economic (management) staff. If the company realize to go over to the BOM are added other product meta-data.

This approach facilitates new functionalities for the product structure definition and for the process workflow management.

The product structure point of view:

- The product definition data are linked to the structure data;
- All departments may visualize simultaneous specific views of the product;
- The creation and the management of the product is facilitated for every department;
- Every product is considered as a unique self-standing product, with no versions, the PSM enables an effective tracking of the design variations;
- Ensure the bidirectional communication and data transfer between the product structure and the enterprise resources availability, leading to the next step, the work flow management.

The product workflow management point of view:

- Authorization of individual in connection with actions over the product data;
- Triger actions;
- Notify automatically individuals over the triggered actions;
- Defines the changes on the product data;
- Control the changes on the product data;
- Tracking the actions and changes on the product data.

The need for PSM is to create a modular structure for the components of a product, named the Part, where the Part is seen as a system with many interrelated components. Modifying a component is leading to the updating of the entire Part Structure [7, 8], integrating all product related information.

The machines and mechanisms lifecycle management is referring to the projects topology within mechanical engineering, and therefore the PSM refers to this type of applications.

The PSM is defining the modelled physically relationships among the Product components integrating all related data from design (CAD models), manufacturing (CNC programs), functionality (Product early versions, functions, requirements and subcomponents), restrictions (standards, norms and regulations), materials and suppliers availability, time to deliver, marketing, cost and risk analysis, warranty and post-warranty service and maintenance plans.

What is of outmost importance is the knowledge behind the PSM not only for the components description and data structuring but also for the relations and interfaces between the components.

The result is a virtual Product as **reference model** for the company, that may include own or assimilated best practices for the product processing, promotion and derivate version development. The different expectations from different departments of the company lead to a decomposition of the product structure and functionality in the used information system. Redundancy ensures the continuity of the process if same systems fail but if the same data are stored in different locations these data must be updated in the same time as corresponding to a single product and not to product versions. Therefore the PSM may work simultaneous with different IT platforms considering intensive active changes in the product development and that way is needed to handle in an appropriate way the information, the IT systems and the organizational (hierarchy) structure.

There are proposed different strategies in generating and using a PSM, the solution in selecting the adequate combination is based on the facilities and methodology in storage, search and retrieval of the data for a better data management, reducing the data processing and the data search of the *product structure* below the manufacturing process times, avoiding any information availability delay.

Once established an adequate PSM the next step is to implement within the company a systematically manufacturing process improvement environment. The improvement is based on the company capacity of innovation as knowledge intensive process in the direction of new technological achievements (including reengineering capacities, workflow support, processes monitoring and control and automated notifications) as well as in the personnel training.

The improvement based on the well done PSM will lead in short time to lower costs, quality satisfaction and reduction of the lead-times.

As mentioned the PSM is using different IT systems, commercially available or developed in house. One of the well-known systems is the ERP (Enterprise Resource Planning) that is managing data about the product that must be manufactured, offering structured data and process support. The ERP is not containing data about the parts characteristics and process technology and therefore another well-known system is to be added, the PDM (Product Data Management) dealing with the product data and documents as drawings and text specifications, supporting the product development process but not the manufacturing one. The use of both leads to a better information management process but ads also associated problems, in the end being a great help in the process administration and product development but not necessarily (directly) in the manufacturing process.

Between the design, planning and manufacturing is entering a new player to spans the gap, the Engineering Change Management (ECM) a different matter from CAE (Computer Aided Engineering) that is actually not only the name of a new entry IT platform but also the task of the Lifecycle Manager: solve conflicts derive from the decomposition of the product structure, use different information from different sources, make compatible and synchronise the functionality of different IT platforms, data transfer, data translation and data update between the systems, use every internal workflow modules to control the complete workflow.

To be valid in the real life, when the PSM is put into practice, relays on the ECM as a process with a great influence on the products manufacturing. Considering an example from the aerospace industry that relocates a company from Central Europe to Eastern Europe.

Usually the design department and the manufacturing workshop are not having close locations and two main situations may be encountered during the ECM: the change is made during the design process and company internal influences of the change from the change initiation until starts the manufacturing process; and the second situation is the change in the product structures and their handling within the company departments.

The change is imposed by change orders as starting point of change within the design and development department. That made from the change orders an instrument of control for the product structure and the corresponding documents. In this sense it is well structured and known:

- Which Roles are involved (departmental roles and individuals within the departments is allocated to each role;
- What information will be handled during the change;
- What tasks and activities must be performed during the change;
- The information system dedicated to this change.

Regardless of the characteristics or the importance of the Product (part) that is suffering the change the influenced levels are the same. In the design phase the change may consider the materials or the geometry of the product. The new version must continue to ensure the conformity with all the considered standards.

The ECM process must be totally documented regarding the captured characteristics data, the roles distribution and the information system. As long as the manufacturing process is general based on numerical controlled systems (CNC—Computer Numerical Control) or more the ICAM (Integrated Computer Aided Manufacturing), the documentation and the decisional process is based on a functional methodology that enables the analysis, development, integration and reengineering of the product based on the capability to describe manufacturing functionalities. In this class can be mentioned methodologies like: SADT (Structured Analysis and Design Technique) describing systems as a hierarchy of functions, or IDEF0 (Integration Definition for Function Modelling), used to show the functional flow of lifecycle processes and the system control, graphically represent a wide variety of business, manufacturing and other types of enterprise operations to any level of detail.

The changes of the product geometry and eventual material may induce the involvement of 15–20 roles design, accounting, CSR (Corporate Social Responsibility), environmental, human resources, maintenance, manufacturing, purchasing, spare parts, technical documentation, technology elaboration, tooling, etc., as result of geometry change in elaboration of a new part number. The problem is that this large number of roles and activities must be managed (controlled and coordinate) in a given time and with given resources, usually very limited, and considering that the product is reflected in different product structures depending on the number and type of the involved IT platforms.

(a) If the product structure is considered the master structure and can be accessed by the entire company, decomposition is oriented versus the functionality of assemblies and subassemblies or/and towards a particular structure useful for dimensioning, modelling or simulation, used to make parts classification and update product structure data in any other systems, this is a product structure customized for **design**;
(b) If the product structure is based on the product specifications, customer orders variant and how that affect the consumptions of parts in each manufacturing workshop, this is a product structure customized for **forecasting**;
(c) If the structure is oriented versus the assembly, operations, materials and BOM, with no practical meanings for division in assemblies and subassemblies but very connected with specific manufacturing variations, this is a product structure customized for **manufacturing**;
(d) If the product structure is based on the customer order for a specific manufacturing and after the order manufacturability validation is proved and allocated to a specific manufacturing facility, to be broken down based on the order specifications, this is a product structure customized for **order management**;
(e) If the structure is oriented versus the assembly, operations, materials with no practical meanings for division in assemblies and subassemblies, but the manufacturing locations and the parts number necessities are important for the forecasting process to estimates the quantitative needs for each component, this is a product structure customized for **purchasing**;
(f) If the product structure is oriented to satisfy the needs for service by continuously maintain the stocks to ensure the spare parts and the spare parts kits to ensure a complete service operation, this is a product structure customized for **spare parts**;
(g) If the product structure is on the storing capacity of all service data are structured enabling to create instructions and procedures on how the service activity to be carried on, this is a product structure customized for **service**;

These are the tasks and how are affected different departments when the change is made in the design. The next crucial activity is the transfer of the structures and the adjacent systems to the manufacturing plant or workshop. In real cases multiple product structures are needed in accordance with the specificity of each manufacturing workshop, what system is managing the product structure and the manufacturing workflow.

In order to provide an efficient ECM it is advisable that all product structure data to be stored and managed in a centralized data base the continuous monitoring of the changes may offer the appropriate workflow control. As long as the cost, quality or the time factor is more or less important for each company, the strategy in defining the PSM. The decision in building a single or multiple structures results from the customers' requirements on the number of the product variants, how the product structure to be managed and how the product to be documented; different departments within the company may have different demands regarding the product structure and the IT system functionality; the cooperation with suppliers, certification bodies and partners needs quick and reliable communication system, parallel and synchronized activities; never the less the most important is the product where the number of components, variants and complexity creates specific demands over the product structure and the IT platform.

In order to respond at such long list of criteria, demands and specifications an optimization process must be conducted. That means that a good compromise has to be done.

4. **Parts Management** provides information on standard components and facilitates re-use of designs;

A product may consist in a single part or multiple parts that may be organized in systems and subsystems. The Parts Management (PM) is connected to the PSM by an activity originate from the military and space applications the CM (Configuration Management) [9], based on a series of standards, considering that the complex products are built after detailed specifications, the accuracy of the product design, manufacturing and maintenance being the main task of the CM.

As long as PSM is a set of functionalities will include also the CM as a set of activities described by the considered set of standards.

Imagine a helicopter as Product and the product systems, Fig. 12.5 [10] main rotor, cockpit, fuselage, airframe, hydraulics, landing gear, tail rotor system, transmission, avionics, power plant, etc.

The Parts Management considers if the parts in the Product are standard parts that will enter via a supply chain, are parts from a previous project existing in the

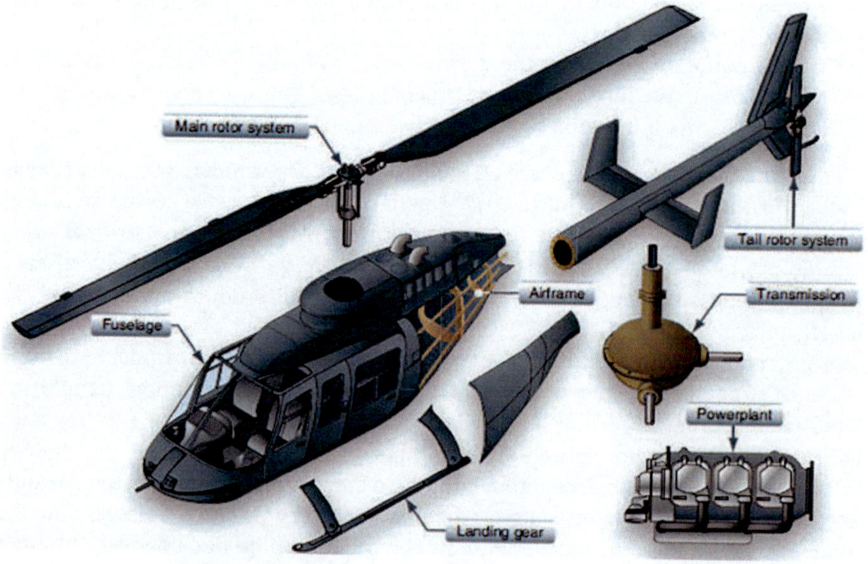

Fig. 12.5 The Helicopter main systems. [*Source* FAA]

company storing facilities or must be manufactured within the company or at third part suppliers.

These four situations are considered in the PM module, and must be able to provide the adequate Data in accordance with the encountered situation. This is the moment when CM enters with 4 components: the Configuration identification—tag and register the Product items, the Configuration change control manage the changes and corresponding documentation during the entire Product lifecycle, the Configuration status accounting—reports for instance the items change status, revisions and versions, and the last Configuration audits and reviews—audits and meetings to verify and validate the product–specifications conformity.

The program management is an important component being very much connected with the company structures, facilities and personnel skills, reflecting in a much customized way the real (possible) process evolution vis-à-vis of the considered Product.

The Program management result is: the work breakdown structures (WBS). The WBS is showing the total scope of the activity together with the links and dependencies between different departments and different planned tasks.

Some other utility functions can enhance a PDM system, such as **communication capabilities** that provide information transfer and event notification.

Data transport functions track Data locations and move Data from one application (location) to another.

Data translation functions exchanges files in a proper format as an online image of handle storage services, access and viewing the product Data.

Administration functions, control and monitor system operation and security.

The all encompassed functions and management subsystems enables the authorized PDM user to search company's Data, the actual searching process being handled by the PLM platform server, using meta-database search engine.

The files are stored in the managed files or **Data Vault** and transferred to the users, in a proper format, Fig. 12.6.

In conclusion the PDM manage product and process data in a central single system, ensuring change management capabilities by correct and quick data finding; lifecycle visualization without special technical knowledge and visibility for business decisions; accessibility for multiple applications and multiple teams collaboration; expend to a full PLM platform, secure configuration and data management; reduce costs, cycle-times and errors.

As the product structure must be adapted to the company departmental demands also the PDM can go beyond and be considered as an integrated or/and extended PDM, meaning to increase the number of people that are currently use the managed data, pushing downstream the use of data (from design and development to manufacturing), collaborating in generating product data, managing intellectual property and improving part reuse. The PDM support processes that cross departmental boundaries, share engineering and management data in any cross functional processes like change management, reviews/approvals or revision control.

The functionality of extended PDM can be described by the effect of the 5 M:

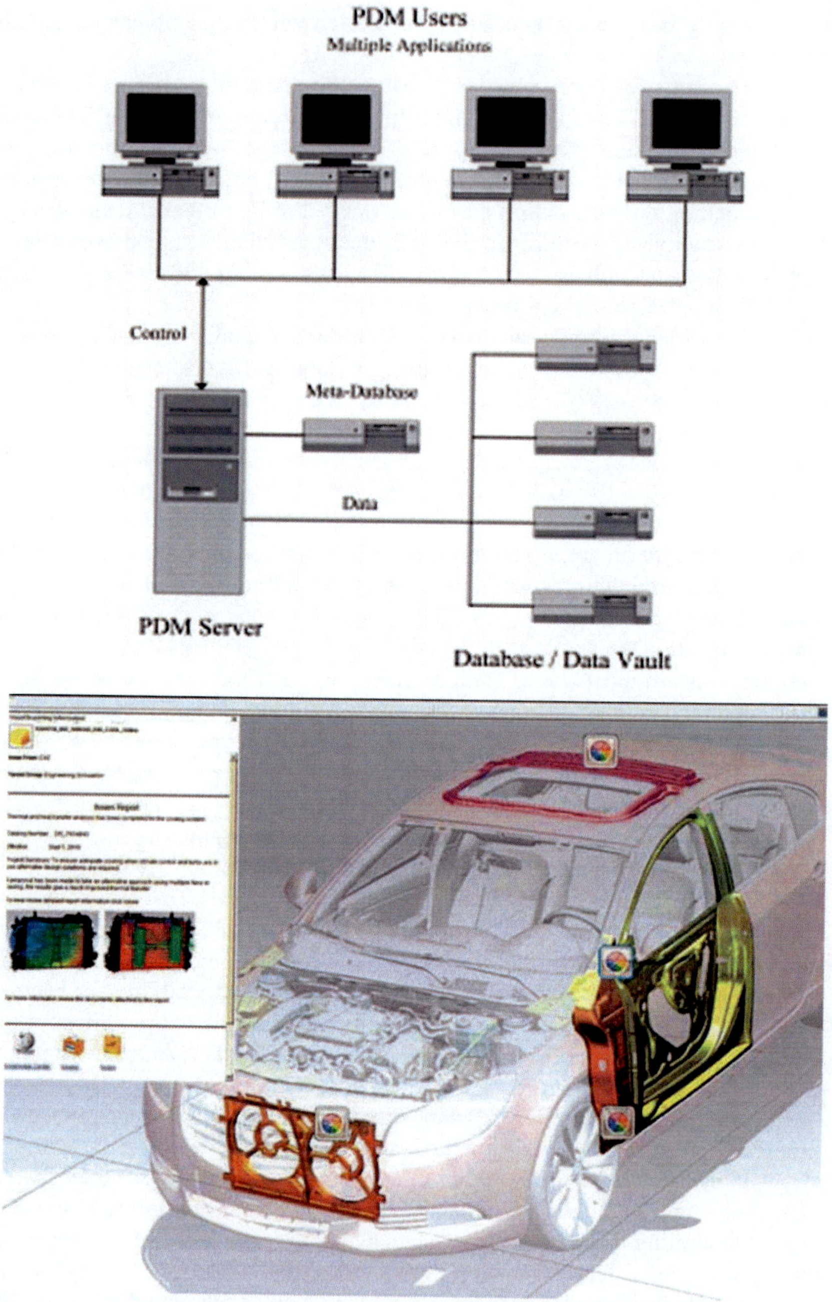

Fig. 12.6 PDM—data transfer. [*Source* Siemens PLM]

- More departments;
- More locations
- More people (roles);
- More processes
- More product details

The PDM, PSM and PLM affects practically any department in a company, sometimes based on their own demands but also on the existing IT platform.. Even small changes are generating extensive activities so the PSM will be a guide before any change (restructuring) even is a change intensive activity difficult to perform but is considering in the same pot the product structure the IT platform structure and the relations between the two.

Therefore this is a hot topic for industry but also an interesting topic for academia that is getting more and more attention. From the product and process description to the information management systems and data exchange are practical problems to be solved that in the meanwhile are big issues for industry, and that can be seen in the worldwide investments PLM and PDM systems, staff and services with an systematically yearly growing rate of approximately 20%.

References

1. Engineering Manager Job Description https://www.highpoint.com/docs/engineering-manager.pdf
2. Peltonen, H., Pitkanen, O., Sulonen, R.: Process-based view of product data management 1996 Comput Ind Elsevier **31**, 195–203 (1996)
3. Colin, M., Matthew, S.: What do investors look for in a business plan? A comparison of the investment criteria of bankers. Venture capitalists and business angels Int. Small Bus. J (2004). https://doi.org/10.1177/0266242604042377
4. Jonas, R.: Federated through-life support, enabling online integration of systems within the PLM domain. In: 1st Nordic Conference on Product Lifecycle Management—NordPLM'06, Göteborg (2006)
5. Hans, H.: Data Volt Modelling Guide. Genesee Academy, LLC (2012)
6. Roland, K.: Data Vault Rules. Data Vault Community (2014)
7. Schuh, G., Assmus, G., Zancul, E.: Product Structuring—the Core Discipline of Product Lifecycle Management. RWTH Aachen University, Laboratory for Machine Tools and Production Engineering WZL, Germany (2005)
8. Daniel, S., Johan, M.: Strategies for product structure management in manufacturing firms. In: Proceedings of the 2000 ASME Design Engineering Technical Conference 10–14 September, Baltimore, Maryland, USA, 10 pages (2000).
9. Buckley, F.J.: Implementing Configuration management—Hardware Software and Frameware. IEEE Press, New Jersey (1996)
10. FAA, https://www.faa.gov/regulations_policies/handbooks_manuals/aviation/helicopter_flying_handbook/media/hfh_ch04.pdf

Chapter 13
Machines and Mechanisms in the Digital Age

"When mechanics stops the world rust"

13.1 Different Models

In the beginnings, the most popular toys were purely mechanic motion solutions enhanced in our times with "***electric powered mechanisms***". Every system kinematics is made with mechanical or mechatronic mechanisms. Everybody meet mechanics from the early stage of its life (Fig. 13.1).

In the everyday life we are not notice anymore the huge number of cases in which mechanisms is used. Practically our "life in motion" is based on mechanisms. From office devices, transportation vehicles and sports, to kitchen facilities, manufacturing, science, jewellery and guns, all are mechanical and mechatronic structures, examples in Fig. 13.2.

Within mechanical systems, **Mechanisms** is a group of specialized machine parts, assembled to work together in order to ensure a certain kinematics and perform a complete functional motion, considered (by definition) to be a piece of a **Machinery**, group of machines. Starting from this knowledge machinery could be defined as a system made from the combination of hardware, mechanisms, apparatus that is kept in action to obtain a certain result.

A **Machine**: is a group of fixed and moving parts that transforms mechanical energy and transmit it in a useful form as machine work.

Using the design and the manufacture of so many mechanisms, it is very important to optimize all the aspects related to them: concepts, design rules and standards, materials, manufacturing processes and nevertheless their lifecycle as Mechanism and Machines PLM.

Considering the machines and the mechanisms in the digital era, every simple search on Internet reveals that the digital era is not a revival for the machines and mechanism but the Digital era was created by machines and mechanism.

Quote a Google compilation from "History of computing hardware" that state:

J. Niemann and A. Pisla, *Life-Cycle Management of Machines and Mechanisms*, Mechanisms and Machine Science 90,
https://doi.org/10.1007/978-3-030-56449-0_13

Fig. 13.1 Mechatronic toys and simple mechanisms [*Source* Meccano & Google free pic]

Early mechanical tools to help humans with digital calculations were called "calculating machines", by proprietary names.

The first aids to computation were purely mechanical devices which required the operator to set up the initial values of an elementary arithmetic operation. Later, computers represented numbers in a continuous form. Numbers could also be represented in the form of digits, automatically manipulated by a mechanical mechanism.

A tally (or tally stick) was an ancient memory aid device used to record and document numbers, quantities, or even messages. Tally sticks first appear as animal bones carved with notches, in the Upper Paleolithic (a late Stone Age 50,000–10,000 years ago). A notable example is the Ishango Bone, exhibit at the Royal Belgian Institute of Natural Sciences, Fig. 13.3.

The first considered mechanism with arithmetic tasks is the Abacus. Used by Babylonians in Mesopotamia (2400 BC), the Chinese, the Greeks (400 BC), the Egyptians (Herodotus time), the Romans, the Turks, the Persians, Indians and Japanese.

Interesting is that the Abacus (Abaci—Turkish, and Soroban—Japanese) was found in decimal form and in binary form, Figs. 13.4a and b [*Source* Google—Wikipedia]. When in China from the XIV Century until around 1950, the hexadecimal weight calculation was common.

Fig. 13.2 Everyday mechanisms applications. [*Source* Siemens PLM]

A little bit later, in the medieval times, as a need for astronomical calculus, in different regions analog computers were invented. Chronologically, in ancient Greece (200–100 BC) the "***Antikythera mechanism***", Fig. 13.5a, a clock work mechanism designed to predict astronomical positions and eclipses, together with Astrolabe, an inclinometer used for spherical astronomy to predict Sun, Moon, Stars and Planets positions, Fig. 13.5b and ***Torquetum*** analogue computers that converts measurements made in three sets of coordinates (horizon, equatorial and elliptic), [*Credit* Google—Wikipedia].

Fig. 13.3 The Ishango Bone, a tally stick used as memory aid device in the Late Stone Age. [*Source* Google—Wikipedia]

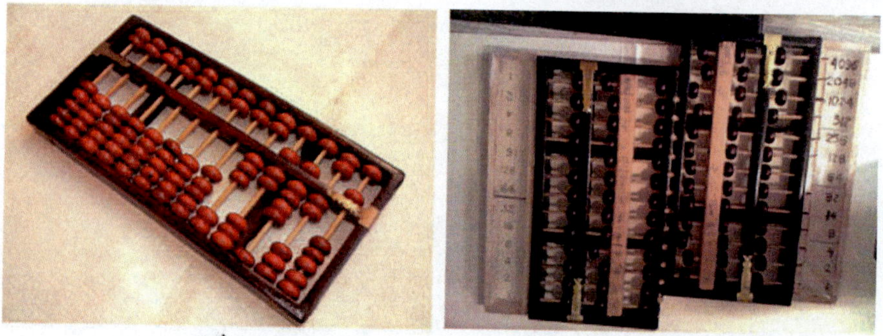

a) b)

Fig. 13.4 The Abacus in decimal (**a**) and binary (**b**) weight construction

The Antikytheramechanism is composed from 30 meshing bronze gears among the 82 components of the system. That reveals a high knowledge and technology that surprisingly disappears for about 1600 years since the middle age when watch craftsmanship rise again at very high standards preserved since today.

In Alexandria (10–70 AD), the Greek **Hero of Alexandria** that was working in the Roman Egypt is known even today from his mathematical discoveries: the Heron Formula states that the area of a triangle and the iteratively computing of the square root.

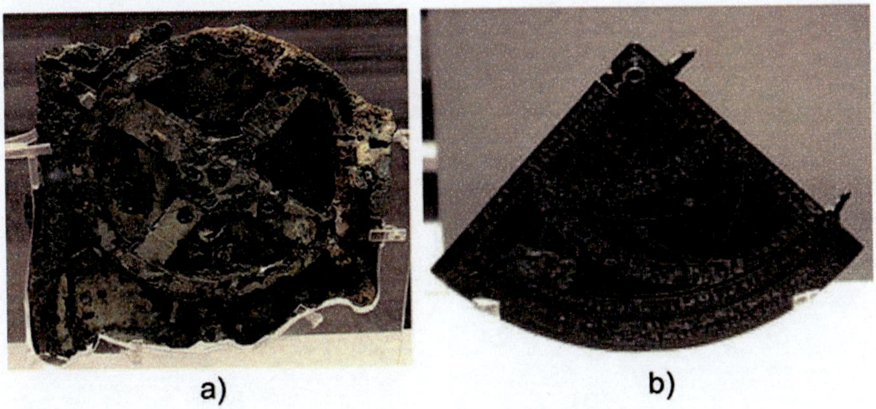

Fig. 13.5 The Antikythera mechanism (**a**) and Astrolabe England 1388 (**b**)

$$A = \sqrt{s(s-a)(s-b)(s-c)}$$

$$s = \frac{a+b+c}{2}$$

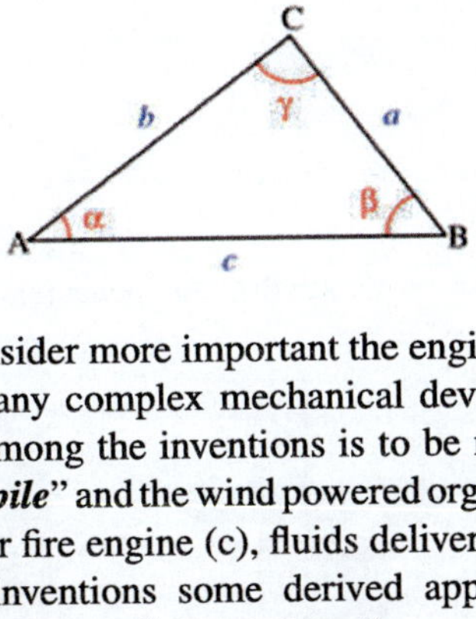

At this moment, we consider more important the engineering activity of the very same person. He made many complex mechanical devices presented in his book "Mechanics & Optics". Among the inventions is to be mention the steam reactive powered engine, the "***aeolipile***" and the wind powered organ, Fig. 13.6a, b, automated devices like force pomp for fire engine (c), fluids delivery control with syringe (d).

And from the initial inventions some derived applications are presented in Fig. 13.7, a coin holly water dispense vending machine (a), a hydrostatic energy self-containing fountain (b), and a programmable cart powered by falling weight and controlled by strings wrapped around the drive axel (c), [*Source* GooglenoPara_Without_NoWikipedia].

All these studies and experiments represents some of the first formal research into **cybernetics**, the science of communication and automatic control systems in both machines and living things.

That demonstrates once again the long history and the importance of the:

Engineering Manager

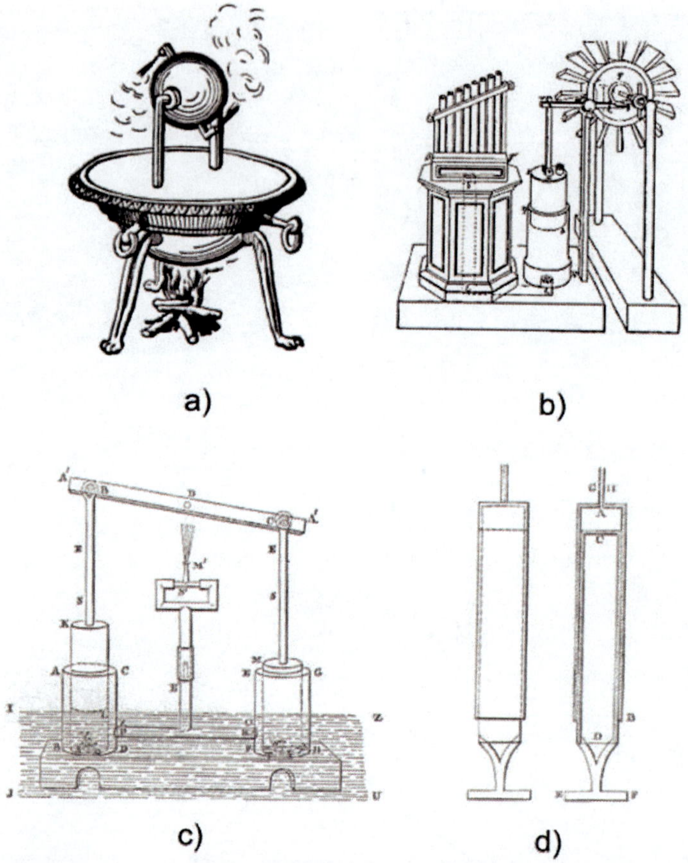

Fig. 13.6 Aeolipile (**a**), wind powered organ (**b**), force pomp (**c**), syringe for fluids control (**d**)

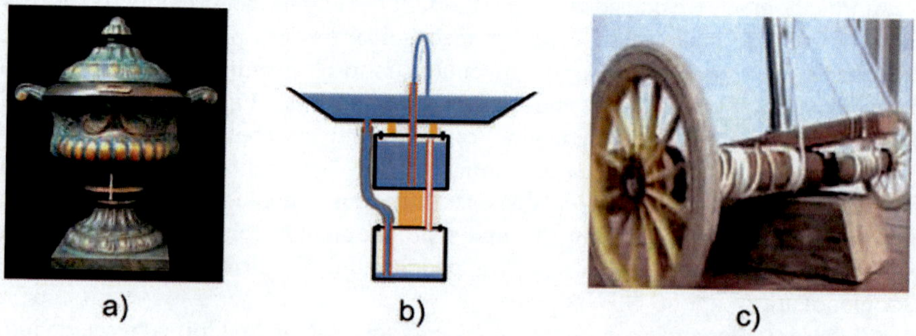

Fig. 13.7 Hydrostatic energy self-containing fountain and the programmable cart

and opens the way for what is more and more claimed in our days the:

Lifecycle Manager

as a specialist and representative of the Engineering and Management domain with a broad spectrum of knowledge, essential skills for management and the capacity to be deeper specialized in an industry or a company profile.

Why is not enough to be "just" a manager, because is missing the subject characteristics?

And why not enough to be "just" an engineer, similar with the situation when the 2D people try to understand the existence of 3D people?

Simply: because even in the 3D people world without the right tools and the right management approach we will face the situation of the "blinds and the elephant". What is that?

1. Imagine the Lifecycle Management (LCM) like an elephant: a big complex, self-motion but potential controllable structure.
2. Now imagine that you are in a small group of blind people different located around the elephant.
3. For all of you the elephant is a totally new structure, with no clue or representation of this structure.
4. You may use only the tactile sense no other senses being available, but at the end you must: Define the elephant.

All the descriptions will be true based on a tactile limited experience, you only may touch one part, partial or totally separated from other experiences. In the end the result definition may be an opaque fuzzy one, certainly affected by relativism or inexpressible matters, as you may also deduct from Fig. 13.8.

Now you must know that people in a company like to relay on the "Master of the Elephant". At various times and usually in the moment of crises or simply in decisional moments they do on the ostrich to escape from certainty sensing the

Fig. 13.8 The elephant as (LCM) and the "blind" people that have to deal with. [*Courtesy*: 27. July 2018, patheos.com]

possibility that something might be happening that is not right so they will do very conscientious his job in his small area.

The Master of the elephant is the Lifecycle Manager (Rara Avis) if we consider that locally even exist such an expert. Is a commonly inaccessible situation, with deficit of information that requires a lot communication, adequate IT infrastructure, respect different perspectives and all the activities around. That means a lot of time and no certain solution.

Considering the Fig. 14.8, the conclusion is: that it will be relatively easy to become the PLM Expert on your Side and that may lead to violent conflicts when all the Sides PLM Experts are gathered in an assembled product.

The second version is to totally forget about to decide alone and start a collaborate activity that will lead to "see" the assembly, and this is the real Lifecycle Manager.

In the brutal reality neither of the two situations makes us happy, we will still remain blinds, but with a Certified level of certainty.

> What we observe is not nature in itself, but nature exposed to our method of questioning.
>
> Werner Karl Heisenberg.

Theoretical physicist and pioneer of the quantum mechanics.

In the Hindu genuine form, one of the story versions is introducing the number of 6 blind men like the 6 Sigma, a set of techniques and tools for process improvement developed by Motorola in 1986 and introduced as central business strategy at General Electric in 1995.

In spite of the fact that is used in many industrial sectors, the Six Sigma offers partial views (like the blind men) and from here is suffering from a lot of criticism: lack of originality, just adding new terms (expert J. Juran); the overselling of the Six Sigma by too great number of consulting firms with rudimentary understanding of the involved tools and techniques; potential negative effects; over reliance on statistical tools; sifting creativity in research; lack of systematic documentation and academic rigor; sigma shift is "goofy" do its arbitrary nature (Donald J. Wheeler).

If ten, fifteen years ago PLM was largely immature, the present looks differently. The best cars manufacturers are the ones derived from the aerospace industry, the most complex one, and here we have: Austin Motor, BMW, Hyundai, Honda, Mercedes, Mitsubishi, Porsche, Rolls-Royce, Saab and Subaru.

You may recognize that these are the most competitive cars and so is no wonder. In the same way the PLM concepts and applications are coming from the aerospace industry. In the beginning it was a single vendor opportunity in Aerospace & Defense. Now is applicable everywhere until the extensive and most interesting industry of CPG (Consumer Packaged Goods), things that get used up and have to be replaced frequently in contrast to items that people usually keep for a longer time, such as cars and furniture.

Ready or not, as future managers and better lifecycle managers, you have to deal in any case with data. So the first close step is to the PDM—Product Data Management.

Machines and mechanisms are design to transfer motion using or performing a certain kinematics. The MM (Machines and Mechanisms) are also design to transfer forces and torques of certain intensity, with a certain orientation, duration and frequency, in reliable conditions when the structure that forms the MM is neither cracking nor deforming.

Never the less the MM design must consider the weight, size, fixing or positioning capabilities, manufacturing and assembling capabilities, manipulation and ergonomic operation, environmentally friendly, safety and cost efficient, clear documentation, easy maintenance and repairing, effective service etc. all together are aspects that must and are considered when a new MM design start to be on time to market. And that is engineering.

To ensure all this characteristics is required a lot of experience and even so, when we do not have it, engineers are creating models to express the performances of the MM and to assess the MM performances.

Modelling and assessing performance is connected with every mechanical industry, actually where the MM are developed. From automotive industry, constructions, hydraulic static or dynamic systems to toys or choppers, everybody is testing.

In this activity there are two major aspects: how to represent the MM and how to assess the MM performance.

Considering the MM representation it can be real, the MM (prototypes) or real scaled MM; statically passive mock-ups or functional dynamic mock-ups, usually at a smaller scale but could be also larger. A general definition is given by Gains since in 1979 [1] is mentioning:

> The largest possible system of all is the universe. Whenever we decide to cut out a piece of the universe such that we can clearly say what is inside that piece (belongs to that piece), and what is outside (does not belong to that piece), we define a new system.
>
> A system is characterized by the fact that we can say what belongs to it and what does not, and by the fact that we can specify how it interacts with its environment. System definitions can furthermore be hierarchical. We can take the piece from before, cut out a yet smaller part of it, and we have a new system.

In engineering from all considered categories [2] the most common are the: Schematic Mathematical and Physical models expressed in form of mock-ups, to not to be confused with the mathematical models in the Physics field. So, when we cannot resume to a schematic or mathematical model we are creating mock-ups, but also in the case when is find out that schematic model is too simple and the mathematical one is to complex. In all cases the research and testing is based on empirical observations as exploratory or causal research.

13.1.1 Schematic Models

The classical schematic models are designed in form of charts, drawings, flow charts, graphs or maps, using colours and codes in an abstract and structured way to help understanding the construction, relationships and the MM functionality based on quantities, schedule, timing and positions.

The usually 2D representations, via computers evolution and digitalization are changing into 3D representation design that made difficult to keep the classical definition of the schematic model generation as long as by including the capability to insert the MM evolution based on mathematical relations, may result a realistic virtual model.

13.1.2 *Mathematical Models*

To create forecasting possibilities the generation of mathematical models is much appreciated. The mathematical model uses behavioural equations that may represent a single part or a single process but can be extend to enhance also phenomena from different other parts and a multitude of processes form industry. The equations may encapsulate the interaction between mechanical, physical, chemical, attributes and also capturing physical parameters like temperature, pressures, forces, vibrations or material flows together with economical, biological or legal influences.

13.1.3 *Physical Models*

The Mock-ups are considered the physical models for engineering. Usually they are made to understand and to put into evidence parts behaviour, from the conceptual look to the recycling procedures.

Combined with the Schematic models or Mathematical models the Physical models may lead to an Analytical model, that may have references also about processes considering the technological implications, manufacturing scheduling (like planning) or specifications for a drawing detail. The Analytical model is not improving the part properties, characteristics or performance but may have an influence on the company structure and organization or even over the evolution of a Project, implying iteration that shows in due time the evolution of the activity (i.e. business) helping in the decisional process of selection and evaluation of the system conception and the system layout.

Modelling and models can be completed before the day one of the research or can be developed in parallel with a continuous higher degree of complexity, as long as is considered and known that a model is a "simplified version" of the reality. The models and modelling are needed because of time and cost reducing facilities and in some cases to the impossibility to work with the real components. The disadvantage of the modelling activity is that the weakest link determines the modelling process quality. But isn't so in any process?

Within the engineering and management activity, we must consider the benefits of modelling, never forgetting the danger and the implication of the advantages but with a constant focus on transforming a disadvantage in an advantage.

With digitization, the modelling activity becomes in many cases a purpose per se, just look around about the evolving VR (Virtual Reality) solutions, together with the AR (Augmented Reality) and the ER (Enhanced Reality) that copes with all possible simulation models to create an App or a product.

For MM industrial environment the models and modelling impose to the users to provide a systemic and systematic approach in solving the problem. The systemic approach is considering the company modelling possessing an intrinsic value, as requiring the organization and the quantification of the data and information in the process. This creates the capability to indicate where data is missing or additional data are needed and, of all the benefits, provides within an industrial environment an "industrial like development environment".

Models and modelling, by increasing the capacity of problem-understanding, requires a higher specificity about the final objectives and the tasks objectives of a process, being capable to offer the adequate tools to serve consistently in the evaluation and validation process.

Nevertheless, with digitalization and accepted implementation of the I4.0, modelling offers a standardized manner for the problem-analyse and the problem-solving.

Summarizing the advantage of having a model:

- it is easy to change MM design;
- a simulation can be scheduled any time;
- the MM may not exist yet, is just a concept;
- the real experiment can be too dangerous;
- manufacturing the real MM is too expensive;
- a new simulation can be performed any time;
- it is easy to change experimental parameters;
- perform experiments that may destroy the MM;
- offers variables that are not accessible in real experiments;
- the time scale of the system may be extended or shortened;
- any variant can be evaluated separately or in combinations;
- the experiment can be rewind and analyzed playing forward and backward the process;

As a model is a simplified representation of reality, in the MM performance assessing they are different stages depending on the product achieved TRL (Technology Readiness Levels). The accepted level of abstraction and the desired conformity level is depending on that.

The TRL is a method for estimating the maturity of technologies during the acquisition phase of a program or a project, in this class is included a product and the technology that leads in achieving the product. The TRL concept was developed at NASA during the 1970s, enabling consistent, uniform discussions of technical maturity across different types of technology.

As long as conformity level and conformity standard that must be fulfilled represents a very specific situation, with very specific tasks, it is of interest and useful to consider the level of abstraction and in this category enter the model complexity, the simulation procedures (conditions, restrictions, duration, measured parameters, etc.) to run the simulation process.

In many situations the MM performance assessing stops here, but that is enough only if the so called virtual certification is accepted. In the other cases "the Rig", a testing stand, is required to certify the assumptions and most important the performance of the considered MM previously on any real life application of the manufactured MM. That implies the use of the Rig test, within the testing standardized or accepted procedures.

Actually a Rig is a piece of machinery or a testing environment to conduct a basic or a complex test, to assess the capability and performance of components for industrial use. A Rig is containing hardware, support elements, instrumentation, but also software tools and simulators.

As long as different type or tests are run there also different testing techniques and different structure and configuration of the rigs. Among the testing procedures and techniques we may have Structural test design technique base on the analysis of the internal structure of a component, mechanism or machine, known also as the "white box test design technique".

The Equivalent class technique is considering a portion of an input or output for which the behaviour of a mechanism component or the entire system, based on the specifications is assumed to be the same.

The design, the concept and the construction of the rig can vary greatly. Exits rigs for bearing, testing, gear testing, agricultural mechanisms testing, machines testing or shoe testing. Usually the rigs can be found within specialized research institutes and universities. Within universities is a sort of dedicated activities implying rigs: the material science, machine parts, mechanisms, tribology, hydraulics, and aerospace, robotics, high and low voltage components, PCBs (Printed Circuits Boards), automotive or thermodynamics.

Some examples of the used rig types are selected and pictured with the courtesy of Mechanical Engineering Faculty, Purdue, USA, DLR (Deutsches Zentrum für Luft- und Raumfahrt), German Aerospace Center, NASA (National Aeronautics and Space Administration) as independent agency of the United States Federal Government responsible for the civilian space program, aeronautics and space research, the MTU Aero Engines and the Siemens PLM.

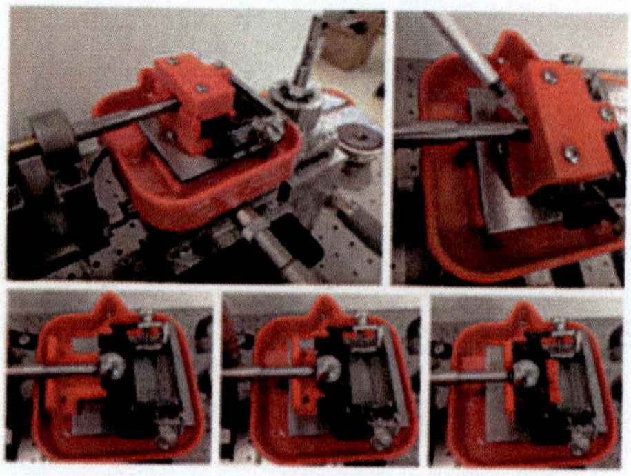

BCFTR (Bearing Cage Friction Test Rig) designed to directly measure the friction between the cage and rollers of a variety of ball bearings. The X–Y stages for positioning is combined with a high precision load cell measures forces and torques in all 6 DOF, to determine the friction coefficient of the ball—cage contact in a variety of slide to roll ratios.

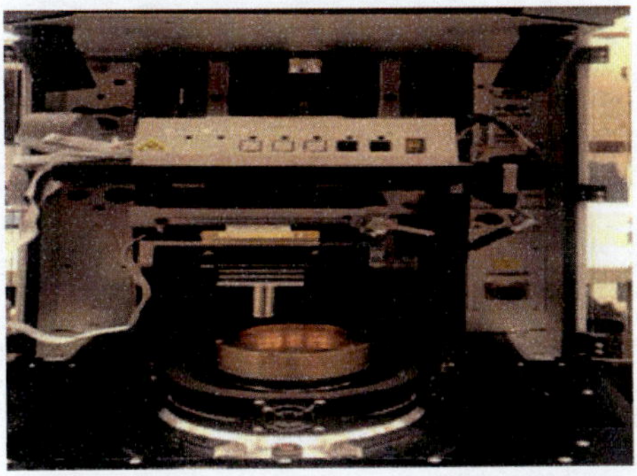

POD Pin on Disk, Bruker UMT Tribolab, test rig is used for fretting wear, sliding wear, and friction measurements of different configurations such as ball-on-flat, flat-on-flat, and cylinder-on-flat. Several dual-axis force sensors are available to measure the friction force and the normal force simultaneously. The force sensors can measure a load in the range of 0.5–200 N. There is a reciprocating linear drive with an adjustable stroke length ranging from 30 to 25 mm with a frequency of up to 60 Hz for room and high temperature tests. An ultra-high temperature chamber allows testing up to 1000 °C.

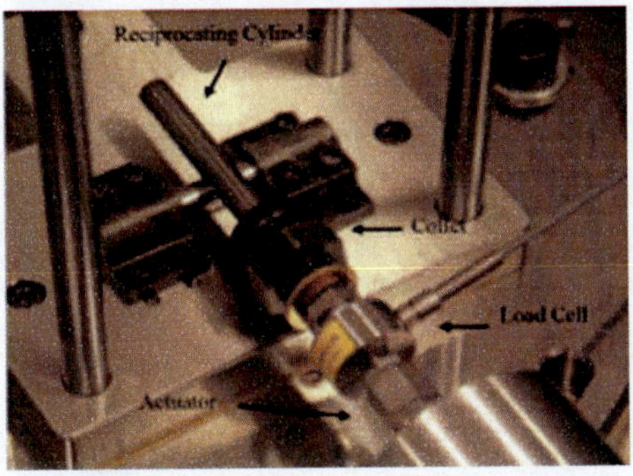

FWTR (Fretting Wear Test Rig), modular fretting test rig designed to measure friction and wear for various contact geometries. The test rig is mounted on a machine base and operated using a computer equipped with a data acquisition card and the LabView virtual instrument software. A microscope equipped with a high speed camera can be used to observe the fretting contact of the ball on flat configuration.

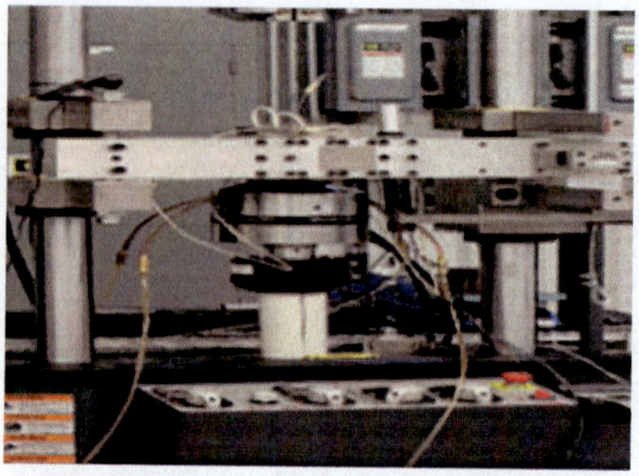

MTS (FWTR) designed to measure friction and wear at the interface of transition inlet ring and spring clip in the Gas turbines material, up to 500 °C. The actuator of MTS axial fatigue tester is used to induce small amplitude reciprocatory motion at the contact.

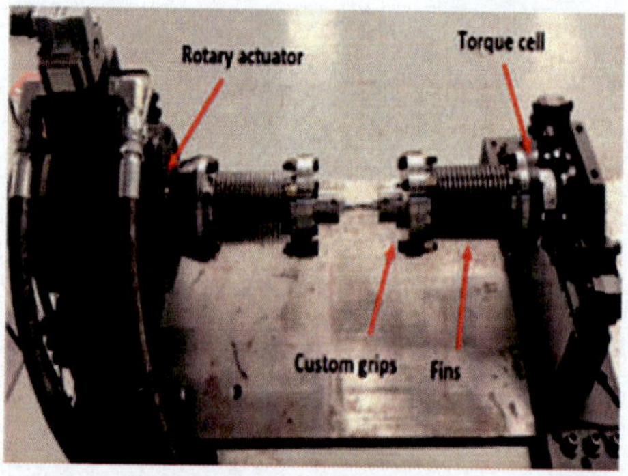

The torsional fatigue rig, beside initial destination can be used to experimentally measure the S–N behavior of different alloys at elevated temperature. A customized heater can be used to perform the test at high temperatures up to 1000 °C.

MTS Axial Torsional Fatigue Test Rig is a hydraulic linear and torsional actuator capable of applying loads and torques. Investigation on the effect of hydrostatic pressure on bearing steels are conducted.

EHD (Elasto Hydro Dynamic) Test Rig, measures thin films produced during elastohydrodynamic lubrication through the interferometry. Measure the coefficient of friction as a function of temperature, load, speed, and sliding.

MPR (Micro Pitting Rig) monitor specimens subjected to RCF (Relative Centrifugal Force), investigating the effects of sliding on surface deterioration, and may measure temperature, wear, and friction during the test.

TTR (Turbocharger Test Rig) is a cold gas turbocharger test stand developed for characterization of the bearing system in ball bearing turbochargers. Ball bearings are replacing journal bearings in turbochargers due to their low friction characteristics which not only increase overall efficiency of the powertrain but also reduce turbocharger lag. To support the rotating assembly of a turbocharger, two opposed angular contact ball bearings are integrated into a single cartridge and supported on a squeeze film damper.

VTR (Valve Test Rig), the cam and follower valve train test rig is designed to measure the individual angular speed of the intake and exhaust rollers.

APPTR (Axial Piston Pump Test Rig), the axial piston pump test rig was designed to operate pumps and measure the lubricating gap between the valve plate and cylinder block. The hydraulic circuit used is based on a general design for steady state testing of axial piston pumps commonly used in industry. The APPTR is instrumented with flow meters pressure transducers and thermocouples to measure and control the desired pump.

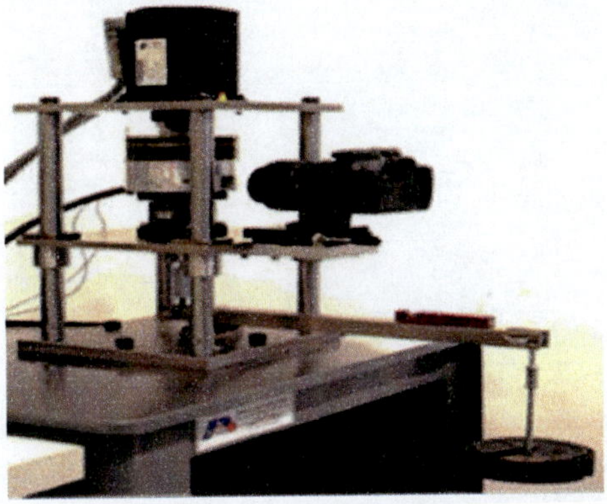

TWTR (Thrust Washer Test Rig), a pocketed thrust washer test rig is designed to experimentally measure the cavitation and shear driven cavity flow inside of surface modifications using optical measurement techniques such as Particle Image.

Velocimetry(PIV). Any desired lubricant can be used in the reservoir as well as any desired pocket geometry or orientation. An encoder is used to measure and control the servo motor speed.

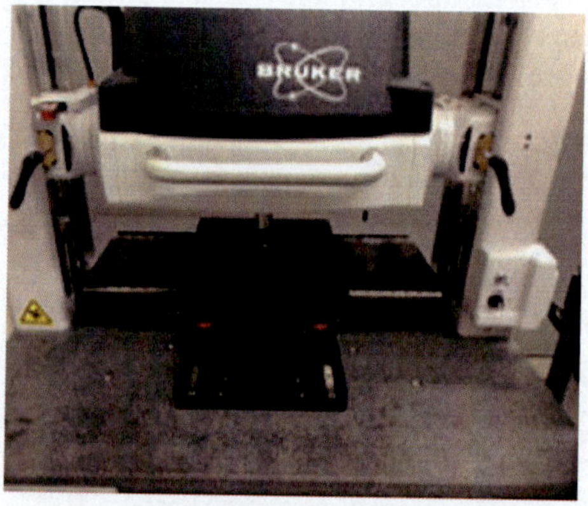

Optical Surface Profilometer is used in the examination of tribological surfaces. Especially in lubrication, we commonly encounter film thicknesses that range from 1 μm all the way up to 10 μm.

The BFTM (Bending Fatigue Testing Machine) examines the classical bending fatigue phenomenon. Once a cylindrical test specimen has been placed into the machine's bearing, load is applied to the specimen. This rig utilizes hanging weights to place the cylindrical test specimen in bending.

The High-pressure Combustion Chamber Test Rig 3 at the DLR Institute of Propulsion Technology in, provides an economical test environment for single and multi-sector combustion chambers used in aviation.

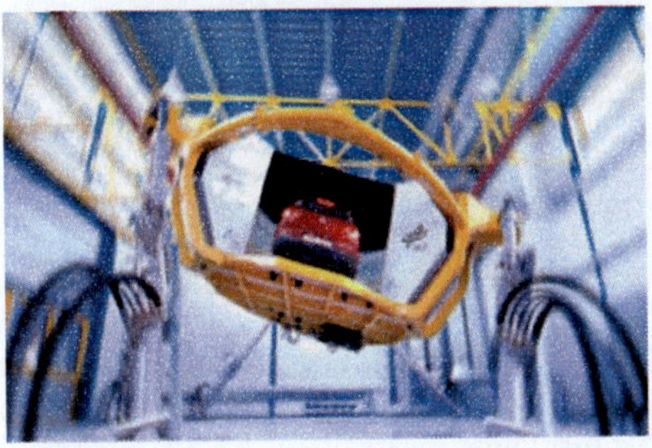

Part of the DLR car testing rig is dedicated to the car overall behaviour testing but also to the pilots drive testing, electro mechanical suspension setting, video and audio comfort, rolling stability, and dynamic feedback control.

The DLR Institute of Materials Research develops new materials and associated process technologies and applications in aerospace, automotive and renewable energy.

The Tiltrotor Test Rig (TTR) is the NASA developed new capability for testing the prop-rotors up to 7.9 m in diameter at airspeed up to 555 km/h. In November 2019 the testing achieved 505 km/h, the highest airspeed ever achieved.

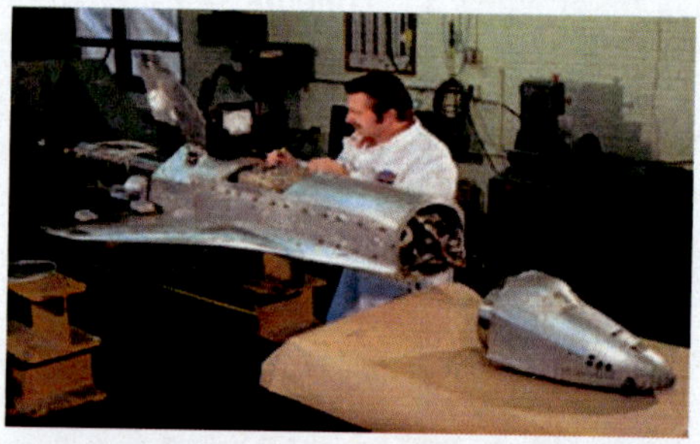

The NASA rig for the space shuttle Endeavour payload bay, facing the Airlock to Middeck. The windows on top peer into the flight deck with the astronauts control panels, and the thermal insulation.

The new MTU Aero Engines high-tech centre with testing rig for compressors, turbines and components like blades, disks, rings, but also engines for aviation industry development and certification tests.

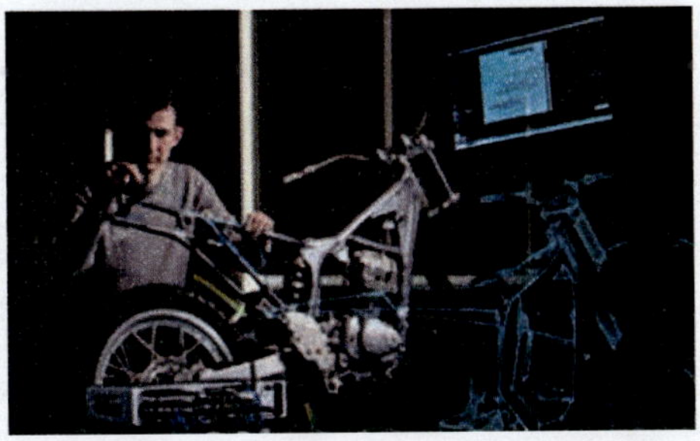

Besides configuration and monitoring in mechanism testing, using testing rigs, knowledge management is of utmost importance. Therefore the upgrading to digitalization and PLM brings great deal in parts optimization and validation, including the influence of more constrains in shorter testing cycles, with the conflicts solver in required performance achievement.

The rigs will continue to be used a lot, basically because the first element every manager body looks at is the cost reduction. Is coming naturally once with the engineering flexibility by enhanced collaboration and multi-physics interconnected simulation to use as much as possible rigs and simulation techniques.

The mentioned combination enables the alternative design adjustments, analysing stress, heat, collision or airflow capabilities for systems that are made of thousands of parts in motion, that works at temperatures around 1000 °C, at rotations over 8000 RPM, in vibration conditions like gas turbines or jet engines. Prototypes for these conditions are hard and costly to manufacture with no evidence of insurances for elimination of most of the problems during the design, or during the prototypes manufacturing process.

13.1.4 PLM and MM

Within the relation between the PLM and mechanism there are some aspects that PLM helps to take care on a mechanism offering more features and better solutions at lower costs.

Usually in literature can be found frequently that more represent aver double 200% of the considered data and the achieved results, better is almost 100% ensure of the expected quality (99.99%), cost is with minimum 40% cheaper and in the same amount 35–40% is reduced the time to market, regardless if the market is the selling of a product or the validation the results in a research project.

Synthetic it looks like in Fig. 13.9.

Why is so important?

Because in an increasing competitive market, the mechanisms as a product are more and more complex than ever and the companies must not only to ensure innovative solutions to the market with a proper mechanical design, but also to manage the complexity of this products concurrently and across multiple disciplines like errors

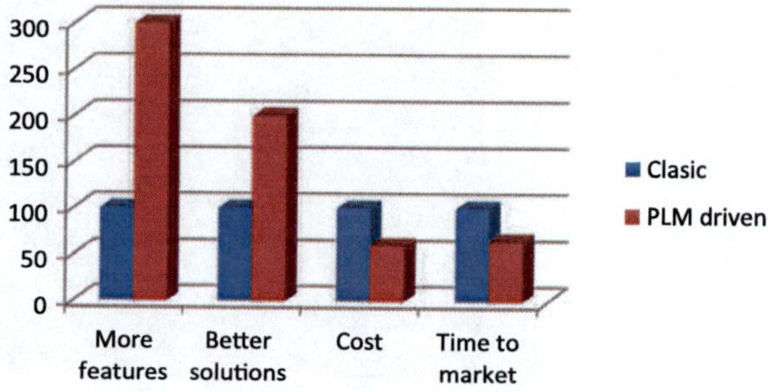

Fig. 13.9 Synthetic advantages of PLM implementation in MM

management, cycle times development and monitoring, combine mechanical with electrical, automation and software issues and digital validation.

In this moment is coming the PLM solution to assist to: build right the mechanism and also to build the right mechanism. Only doing right based on a wrong decision is very costly. What about is actually PLM taking care in the mechanisms development?

- In the first stage is assisting the concept to understand and insert the customers' requirements, that became the guiding lines and the restrictions frame for the entire lifetime of the product and the entire lifecycle of the product development;
- Secondly is not just setting the "working frame" but is also assisting dynamically and continuously the work team in the design and development process with considerations that the design parts:
  - Must be available for all the participants in the process;
  - Must be manufactured;
  - Must ensure the compliance;
  - Must benefit from secondary products from suppliers;
  - The purchasing tasks needs time;
  - All the activities must be scheduled in a synchronized way;
  - For the design mechanisms must be considered the packaging and the delivery conditions;
  - The developed product will suffer maintenance and service activities.

How is this possible? By over 70 specialized modules integrated in the same platform. All starting from the "Modelling Technology Platform" that generates direct, explicit or parametric solid and surface modelling, benefiting from:

- Cloud connection;
- DTCSDigital Twin Close-Loop System;
- MBSEModel-Based Systems Engineering;
- Integrated materials and composites characteristics;
- NGEDNext Generation Electromechanical Design;
- Generative Engineering—based upon meeting design constraints;
- Equation driven modelling—creates functional surfaces and solids easily from the mathematical equations definition;

- Syncron technology enables to modify 3D geometry intuitively by mouse push-and-pull methods, without the requirement to understanding how models are generated;
- Adaptive user interface—in an adaptive UI panel the tracked user actions are displayed as predicted commands based on the currently actions as result of AI and Machine Learning process;
- Topology optimizations—creates a new generation of products by taking performance requirements into account from the start and optimize geometry by automated calculations iterations;
- Convergent modelling—benefiting form a point cloud mesh giving the ability to perform faceted-based modelling without data conversion, combining the three representations facet, surface and solid modelling in a single integrated environment.

Additional to the mechanisms generation are the modules wit supplementary tasks that help in the mechanisms manufacturing, improvement, and connection with other components or systems, connection with their future functionalities or connections within the company environment processes like:

- Digitally transformed manufacturing to program CNC machine-tools and control robotics activities;
- Mechanisms are usually seen as assemblies so the fastening and join design and processing are essential. Therefore within the mechanism 3D definition is also included the assembly definition generated throughout iterative processes to cover the fastener management, visualization and verification. This will be communicated and latter monitored within different manufacturing stages;
- The company assets and product portfolio is worth to be considered and therefore they are used both in the design and in the manufacturing process in optimizing structural solutions, leading and response time;
- Through the interoperability visualization and collision detection modules is ensured the risk decrease for the mechanisms construction and functioning;
- Modules to digitalized products and processes and modules generating data for IoT;
- Modules that are managing the data generated by and for IoT, coupling the machine-tools with the robots, systems and plants;
- Digital planning and digital automation services;
- Service and maintenance management of the functional system and of the manufacturing facilities that contribute to the mechanism fabrication.

As it seems the PLM cover the assistance over the entire lifecycle of the product. In the "commercial" lifecycle consideration of a product where there are practically 4 stages: the product launch (L), the product maturity (M); the product revival (R) and the product end of life (EoL).

In the mechanism PLM assisted lifecycle, we may identify also 4 stages and we may couple the PLM software module that are mostly used during each of the stages, considering how the PLM modules are initiating and maintain the collaboration, handling features, monitor and manage the engineering workflow, beside the primary aspects mentioned for the company the improvement of the data quantity, quality and availability, involving all kind of team from developers and administration to suppliers and maintenance.

In the lifecycle of a robust mechanism exits a launching time, not for selling but the beginning of the product lifetime as a development phase with all the design, optimisation and simulation procedures that can be called Creation of Life (CL). In this stage the PLM role is to:

- Ensure an effective product development process;
- Design and designate the assignments and milestone for avoiding impediments in product development;
- Generates a single data store administrating the admittance and the priority for every involved personnel;
- Creates permanent concurrent access to the available design data;
- Reduce design times by managing the assets reuse and creating automated reports generation.
- The second phase is actually the manufacturing phase when the mechanisms must be obtained that can be called Birth Level (BL).
- Generates templates for data generation and parts modification;
- Generates automatically pre-defined workflows for reviews;
- Creates a direct connection between the notification and approval process;
- Ensure a fully electronically communication.

The BL is not possible without manufacturing facilities the PDM and suppliers.

- The existence a of a single data warehouse containing the product and production data, facilitate the manufacturing process;
- No spoiled or waste files are preserved;
- The manufacturing operation are directly linked with the contractual forms, resources allocation;
- Automatic reporting operations;
- Connection of the product launch and product updating with the manufacturing capabilities and processes;
- Supply chain management;
- Supply chain associates documents management;
- Interoffice communication, order documents and repetitive modifications generation;
- The MES (Manufacturing Execution System) is included.

The validation stage can be considered as the next level to proof the functionality, performances and compliancy of the mechanism, or, can be made also for and on the prototypes, on the testing rig previous manufacturing. This stage can be called Validation Level (VL). Finally will be also an end of life (EL) that can be shaped in different forms, all manageable by the PLM system. In the final the product can be stored, packed & stored, delivered, distributed, destroyed, recovered or recycled.

- Validity of all protected data;
- Withdrawn modifications can be resubmitted and revised;
- Access to various users and specialists for reviews and testing elaboration;
- Dispositions assignment is correlated with the departments destination and their workflow of review.

As a conclusion, a mechanism is a group of components, assembled in such a way that may transmit motion and force following a certain lawfulness.

The mechanical systems drawing is a technical drawing that represents the dimensional and constructive details of a system as a relational part of the system for construction, optimization, revision, certification, conformity. There are different levels of technical drawing representation: schematic drawing, details drawing, assembly draws, fabrication drawing, and installation drawing.

The mechanical systems design means how these components are going to work in collaboration. It is an iterative process starting with the representation of real structures and components, the system model; on the system model are applied loads to calculate the force and moment distribution, mechanical reaction and share moments; the calculus leads to the stress estimation in the mechanisms components, due to the combined loads and accordingly to the deflection estimation and critical points identification; adding the materials properties now static failure theory may be applied for brittle and ductile materials components to predict failure loads and estimate safety factors for critical points, geometry variation and loading conditions; for the new geometry are determined buckling predictions for statically loaded compression components; once defied the cyclic loads on different system components the alternating and the mean stress can be calculated for the defined cyclically loading and predict the safety factors for the estimated mechanism lifetime (S–N curves, Weibull distribution, Goodman diagram), establishing the reliability of the system when different components works together; in an and structural and ergonomic problem solving conclude the analyse of a well design mechanism.

A machine can be formed by multiple mechanisms and by definition has the role to transform energy in mechanical work. An engine is not a machine, is "just" an energy transformer that can be a machine-tool, an automobile, a plane or a fax machine, all can be incorporated in the same definition.

Different from a mechanisms or mechatronic systems, the machine has an operation system, a command system, a control system, the working conditions, the mechanical work and the energy consumption management.

Like in mechanism design in the machine design is considered a design lifecycle formed by cyclical, iterative, processes that leads in the end at refined design of the machine. All start with the Contract or the Engaged project ending in the RMMD (Refined Mechanism-Machine Design). In Fig. 13.10 is depicted the PLM assisted MM design. As can be seen there are 3 major Milestones: Project engagement—Experiments & Testing—RMD.

The project engagement is well defined by PMBOK (Project Management Body of Knowledge) a standards terminology and guidelines for the project management, resulted from the overseen work of PMI (Project Management Institute): "*Project*

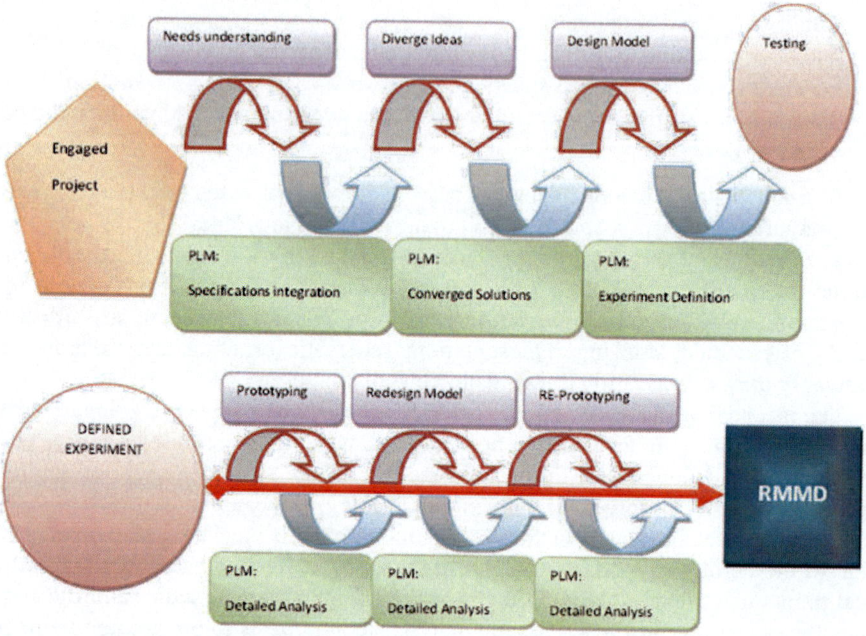

Fig. 13.10 PLM assisted MM design

engagement, refers to each stakeholder's investment of energy, skill, ability, effort, and eagerness in a defined project."

The departmental involvement in the project is primarily defined by the company management and later by the conclusions resulted by applying the PLM mechanisms. Therefore the participations and the commitment in the various actions defined in PMBOK are varying, as is specified:

1. Attending to task details
2. Commits to project completion
3. Involves self in special sub-projects
4. Speaks highly about the project to others
5. Innovates effective processes and procedures
6. Demonstrates personal/professional development
7. Initiates problem-solving and/or conflict resolution
8. Communicates willingly and effectively with others
9. Balances support and challenge with inquiry and advocacy
10. Building intra- and inter-departmental strong connections and relationships.

In the initial phase, exits the conceptual and design activities leading to the experiment is a sinuous trajectory between a series of engineering tasks resulting from one or more departments, doing calculations, creates drawings, define physical layouts, evaluating, selecting and reviewing the adopted solutions, handle existing equipment and the equipment proposals, generating components, materials and subassemblies lists.

The managerial tasks benefit of the PLM support generating the Project Planning, and the Project management in parallel with the assistance for the communication and data collection, validation and transfer and for the Project engineering.

The second phase is starting with the experimental activities, where the established "red direction" will lead to the finish-line through a complete iterative correlations and correction engineering activity, having the constant and continuous support of the PLM driven analyses. At a certain point the RMMD will reach the end, the accepted final form.

Considering the MM destinations there are 2 revolutionary approaches for the next applications of the PLM: in correlation with the ML (Machine Learning) and in correlation with the DTT (Digital Twin Technology).

The development of "data mining" and "machine learning" leads to a logical connection with the PLM as long as PLM was developed to collect and manage huge amounts of data about the products, materials and processes attributes, together with the data regarding the company behavior and the company assets, including staff, suppliers and clients.

Machine Learning and learning models will better capitalize the use of the data adding new perspectives and insights to the MM lifecycles like Adapted User Interface that change its functions, keys, colors, position and the windows dimension accordingly with the user that placed at that moment at the active PLM module console.

Supplier allocation or alternative supplier's recommendation is based on their location, costs, historical performance, compatibility with the company digital platform, conformity in compliance standards, reputation, solvability, etc. all will lead to changes in the Adapted User Interface.

Nevertheless machine learning will lead to the identification of most valuable activities, identification of the product lifecycle moment on the global market, assist the managerial process based on the contractual and customer behaviour historical data impose policies regarding the use of the company insights and assets, define benefits and turnover based on virtual prototyping and VR business scenarios simulation.

Actually the so called "best in class" companies are really embracing these 2 revolutionary approaches, the essence of I4.0 industrial revolution within the PLM. In Fig. 13.11 is represented the implementation evolution of the two mentioned technologies, within the BC (Best in the Class) and the other type of companies.

To have a complete view of the evolution and the impact of these new technologies in combination with the PLM is not only the idea that really oriented companies, the best in the class, are double oriented in the new implementations than the others. Of course the availability of implementation resources must not be ignored, but the second important element considers that if the classical PLM implementation based on CAD/CAM/CAE is capped as a constant monthly platforms implementation within the global industrial environment the same is with the machine learning

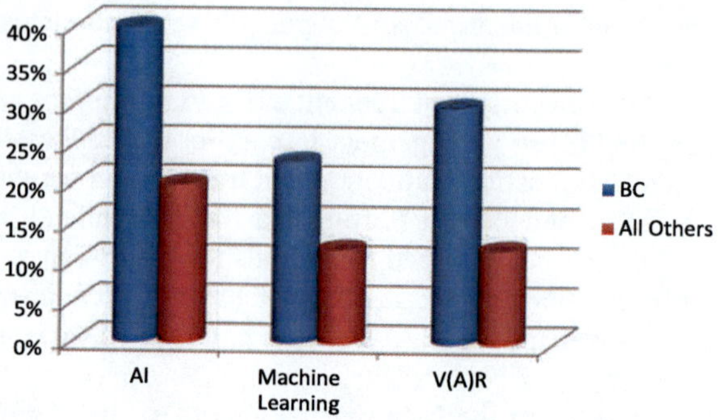

Fig. 13.11 PLM assisted MM design

platforms, but the number is 10 times higher. So if there are 10 classical PLM implementations in a period of time, will be 100 Machine learning implementations in the same period of time.

That leads to the new connection is to rely on coupling PLM to Machine learning in parallel with adding Machine learning capabilities to existing PLM implementations.

On other hand the Digital Twin, a concept introduced in 2002 [3], is a strong singular tool as a result of the combination between MM systems synthesis, IoT and VR.

Even if it is a young technology in the past years and mostly in the last year's surveys continue to confirm. Digital Twin was of great implication in the management and decisional process of the product design, development, manufacturing and implementation, to evaluate performance, identify errors or "acceptable" improvements.

In Fig. 13.12 is depicted the survey results made by the marketing firm "Reboot Online" and published in Machine Design by Stephen Mraz in February 2019.

Regarding the findings we just like to consider "the worst component" with only 19% support considering the DTT capability as "Unique Physical Asset" is still in the class of the "most necessary components of digital twin technology", and surprisingly "the best component" with 71% support considers the DTT capability of (Physical Asset). In our opinion, considering that the investigation was well conducted, these two items must be considered together so we have actually a 90% agreement that DTT is well dealing with the Physical Asset for a new product design and development.

Even so, the DTT is a tool firstly destined to the non-creative teams, in creative teams there are highly trained people with a lot experience and with a trained imagination demonstrated in already made successful products. They will find in this technology a way to easy their activities and get quicker (but not necessarily better) to the market.

There are some other departments like the ones showed in Fig. 13.13 with the Siemens courtesy, departments that can benefit more: the manufacturing (especially assembly), selling and implementation at customers' site, training, maintenance and service.

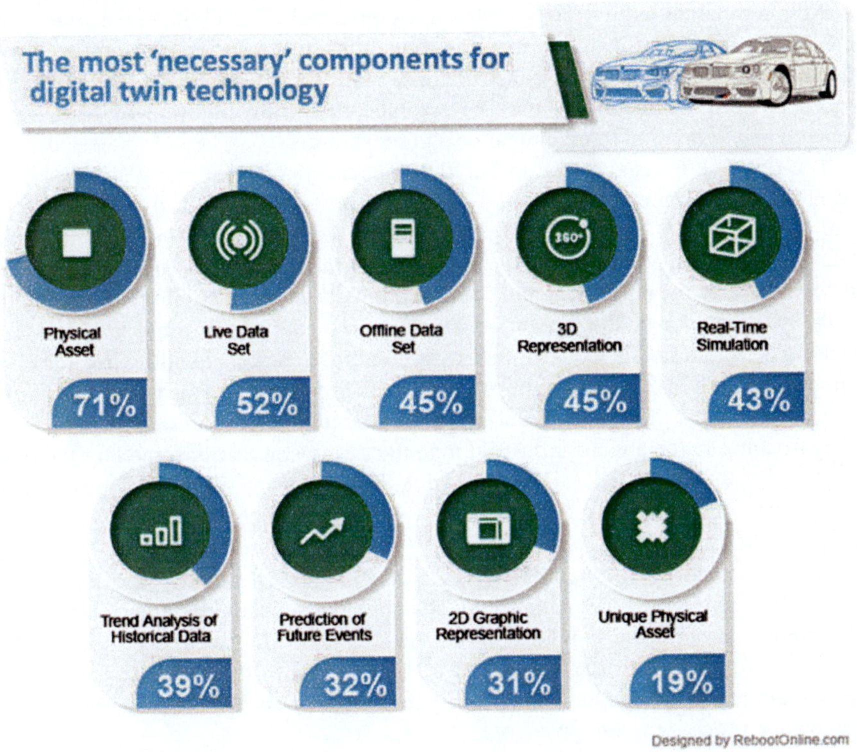

Fig. 13.12 PLM assisted MM design

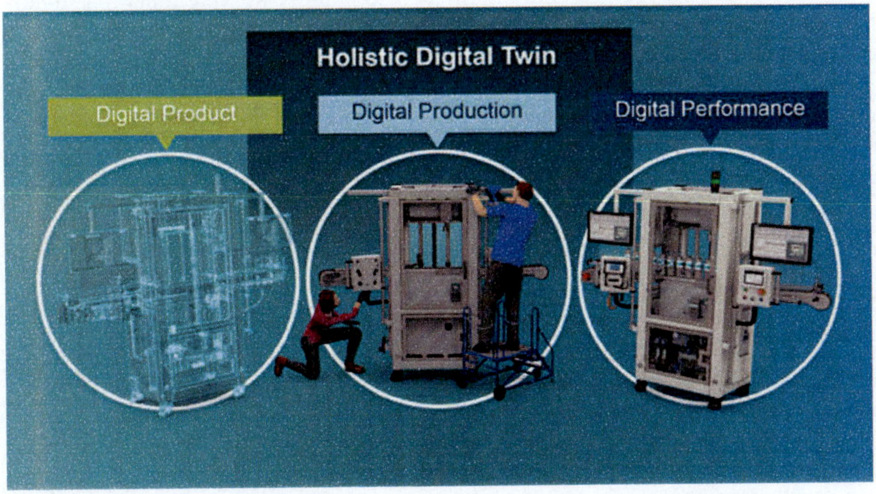

Fig. 13.13 PLM assisted MM design

At this moment Digital Twin is strongly connected with robotics, machine-tools and MM, with a significant number of applications in aerospace, automotive, heavy machineries, agriculture and constructions, real estate. Enabling MBSE (Model-Based System Engineering), may be spanned the entire lifecycle of MM, from concept and validation to maintenance and recycling.

The opening for real estate and civil engineering could be really successful by extending the number of applications at this moment dedicated to manufacturers. The social, legal and traffic data can be linked with the structural and materials data in creating an architectural project connected with the actual property location that may lead to the placement and configuration decisions, including the evaluation of the maintenance or the "green behavior" of the energy consumption (HVAC—Heating Ventilation/Air Conditioning), in combination with facilities for families with many children or aged people, or people with impairments, as a digital alternative in establishing the value of the real-estate and the in time probable evolution.

Remaining in the area of industrial manufacturing applications, the Digital Twin is used for the fault and failure identification, repairing solutions, or scenarios generation, all will apply for different kind of products or products versions differentiations.

The real trend for the manufacturers is in the area of remote surveillance and control for remote MM maintenance by changing the schedule-based and the proactive maintenance in the condition-based maintenance. A research project in this sense was running with the support of a Ph.D. student of our University and the Airbus Company in Bremen, ending in a patenting process and the start for airspace acceptance procedures and protocols.

Identifying the overtime different operation modus and their implication in the overall MM lifecycle the Digital Twin may not only save money and reduce the dead time in the MM functionality but may also identify missing functions, redundant functions or components and even the usefulness of a certain product and the consistency in unwanted functionalities.

As "technical matter" will continue to be developed in larger and better solutions, the "legal matter" will raise new problems to be solved like IPR (Intellectual Property Rights). The extensive share of data, not only internal within the company staff but also with customers and suppliers, certification and authorization bodies etc., is not clarifying in the same reliable and structured manner the sharing of the responsibilities. Who is to be blamed and charged in case of failures and even worth in case of accidents?

The idea of Digital Twin implies also the feedback from the operators at the customers' site.

- The new founding and the new data sharing regulation will be regulated by whom?
- Who is the owner of the Digital Twin generated data over the system lifecycle, sometimes as the single source of lifecycle data?
- And probably are many other questions the will require an answer in the near or far future [4].

Do not forget that embracing the digital transformation is not a single seat shuttle. Exist a global community that is striving with digital transformation across their company and you may build meaningful connections with thousands of people. One way is to embrace and enjoy a virtual experience with “Realize Live” events or specialized webinars.

References

1. Gains, B.: Quo Vadis. General Syst. Res. Yearbook. **24**, 1–9 (1979)
2. Stevenson, W.J.: Operations Management, 7th ed., Chapter 1. McGraw-Hill/Irwin, Boston (2002)
3. Madni, A.M., Madni, C.C., Lucero, S.D.: Leveraging digital twin technology in model-based systems engineering. Systems7(1), 7 (2018)
4. Kraft, E.: *Challenges and Innovations in Digital Systems Engineering*, NDIA 20th Annual Systems Engineering Conference. Springfield, VA, USA (2017)

Chapter 14
Challenging PLM

"Companies have to find ways of growing and building advantages rather than just eliminating disadvantages"
Michael Porter.

Now it is opportune to consider the PLM by definition.

There are so many discussions regarding the PLM solutions, benefits, and implementation models. But, following the discussions on product lifecycle management, all ends in the description of the **Data Management** using the PDM modules. Therefore it is important to have a clear description about the definition apparition and functionality of the PLM.

14.1 PLM—Definition

A considered successful attempt is by putting together the conclusions of different considerations, analysis, or other definition attempt versions and combined them with the personal experience. In conclusion we may state that:

PLM is a sort of systemic concept that represents the entire lifespan of a product.

By product we will understand: a part, a process, or a project, depends on the topic. regardless of the topic the lifecycle management functionality is the same.

Managing and document all the **Product Data**, from concept and marketing phase through the design and product processing stages, PLM is helping and influencing the production planning and manufacturing processes. In parallel, PLM is assisting the documentation, packaging and delivery tasks, finally ending with the warranty and post warranty service together with an integrated recycling process management.

The PLM is not like a complex equation or technology, is a platform that must include:

- the dedicated hardware;
- the core software;
- the IT management;

J. Niemann and A. Pisla, *Life-Cycle Management of Machines and Mechanisms*, Mechanisms and Machine Science 90,
https://doi.org/10.1007/978-3-030-56449-0_14

- the process modules;
- the fluxes integration;
- the management and the optimization of each stage of the product existence;
- the enabling of the inter-stages communication and their management.

The **PLM is offering a streamline for the best value creation chain** (BVC creation), regardless to particular organizations structure.

14.2 PLM—Core Features

The already presented description of the PLM characteristics and also considering the concerns of any company manager in finding the right and effective tools, for managing the product-specific process main issues, leads to the same deep instances: to keep the business quality, track the products and process parameters, reduce risk, increase the company visibility, compliance with standards and government regulations, together with a robust management of the contractual and products documentations tracking and updating.

For keeping the end-to-end management (from cradle to grave), the emitted considerations were guiding to the development of the PLM core features, as a "must-to-have" within the PLM software key features:

- Content Authoring;
- Collaboration Tools;
- Centralized Data Repository;
- Data Management Capabilities;
- Design and Visualization Capabilities.

Content Authoring. Within the development and manufacturing process, the digital technology requires an improved collaboration and communication with no place for data, documents, files, images and presentations misplacing.

The large amount of external and internal product-process documentations, including assembly and start-up instructions, internal notes, service and maintenance manuals, packaging and disposal instructions, technical documentations, or users guide must be generated and managed. The lifecycle management features must offer content authoring tools for a complete traceability and keeping the various iterations of each document.

Collaboration Tools. The PLM incorporated collaboration tools are important for the company and all the employees in different departments, not just for easy and make more efficient their work but to enable to work together as a business oriented common work. The seamless and in real time data updating, transfer and work tasks identification, is made not only within each department but also between departments, avoiding and preventing any lost in data transfer and translation that may finally lead to loss of time, resources, slowing the team activity even up to a full stop of the project. The tracking and scheduling capabilities, enables the product development management along with the upcoming milestones, constraints for further planning and resource allocation in real time.

The complete loop: managers, developers, designers, manufacturers, marketers, suppliers, distributors and vendors are kept up-to-date informed with all the available data as soon as there are generated. The larger and worldwide is the company, the more important are the collaboration tools.

The stakeholders may obtain readily available data about the manufacturer and the suppliers and notifications regarding changes on components and parts, compliances, spare parts availability and risks estimations, due to the systems that manage the components with a direct insight into the components, the process and the production data, that are all included in the collaboration tools.

Due to sensitive information regarding items, product development or manufacturing and for avoiding losing industrial benefits and working teams intellectual properties, there are promoted users security requirements through Administrative Controls that ensure the Role-Based Access so the accessed users are seeing only the data destined to their role within a Role and Group Hierarchy especially created for according different levels of access.

The added tools give users, corresponding to their role, insight into the "product activities", allowing's determined by the project phases and the assigned milestones. To keep projects change management workflow on track, the users may see all pending and implemented changes together with the resource allocation management.

Centralized Data Repository. Now is known that the beneficiary of the PLM systems are pretty large companies with subsidiaries around the world, having a lot of departments with different profiles, from engineering and manufacturing to accounting, distribution, logistic and management. Beside the products delivery, based on a certain process, all these departments are heavily producers of files, forms, documentations and machinery specifications.

The PLM must, and act, like a centralized data repository that is holding together all the evidences of disparate devices and processes. The access of at the required relevant data of each authorized individual, from every location at any time is ensured by the centralized data repository management. Beside the instant access, collaboration and the errors avoidance, the data must be structured in a convenient way for every team member and updated in real time.

The most common way of existence for the centralized data repository is the so called BOM (Bill of Materials) as a common source of information across the company. In the first instance the BOM is very important during the product design, development and manufacturing by including the list of product lifecycle management requirements and providing a "single definition" of a particular product and its components. The advanced forms of BOM are including special and specific designed features for different team and tem members in different departments, with extensions for quick list of information display, adaptable based on the professional frequent demands, helping in this way the improvement of the collaborative activity. The BOM management relays on a comprehensive management that offers a drill-down functionality, with multi-level access to the product structure to any assembly, subassembly and components, components availability, compliance documents, conformity (or unconformity) notifications, customized checking and risk assessment tools.

Following or failing to follow compliance standards and regulation can be crucial in making the difference from earning the business success or earn a business bad reputation, and therefore PLM tools and system uses relevant data and information for achieving and maintaining the compliance. Even more, depending on the product destination can be defined the required standards and the compliance system: ISO, AICAPA, DoD, JEDEC, FASB, FDA, FSA, REACH, RoHS, SEC, SOP, VSOE, WEEE, etc. or the destination of "safety", environment, medical, aerospace or nation specific.

The advanced forms are more useful within CAD activities (design and development) when updating and changes in the products configurations, adding physical attributes based on the process or technology changes, with keeping the desired functionality for all components through the entire planned lifecycle, benefit of the changes evident in the subsequent BOMs and plans.

Storing in a centralized way the data, offers a different management approach by the capability to develop and implement the so called "documentation tools". These are entering new level of collaboration including the mobile documentation, product history documentation, data audits and product lifecycle analytics (as direct text, image and video) or in connection with virtual reality (VR), mixed reality (MR) or enhanced reality (ER) modules.

The connection between the classical BOM and the new advanced forms in BOM enables to add to the already existing process quantifiable technical features new ones extending the classical ERP to the real PLM functionality.

Data Management Capabilities. Are the intrinsic value added features that offers side-to-side comparison between variants, product features and deployment methods, in creating or re-creating a product for a specific market at a certain time. For the inner side of the company the lifecycle cycles optimization, the costs reductions, labor efficiency improvement, operations optimizations or reconsidering the process or product specific functionality can be extracted from the data management capabilities.

In a way the data management capabilities made the Manufacturing Product Management function, by considering the company portfolio of products and spotlight priorities, resulting the "Best investment scenario" that generates the manufacturing planning.

The resources management will be better lead form tracking the project progress (automatically documented in the product record), determining on line the proper investment balance, maximize the innovation and the R&D investment returns.

A special attention is given to the Risk Management, feature that enables reporting the "detected" and "highlighted" risks, together with the mitigations of instabilities, as they move through product development. By RTPV (Real-Time Project Visibility) and CAPA (Corrective and Preventive Action) is made the risk analysis and the risk management.

As business intelligence, the risk modules offer reporting capability in a quick understandable form like Gantt charts. Excellent systems provide users with robust analytics tools that track the entire lifecycle of a product. You can choose a system that allows users to see patterns among successful products and also to see patterns among unsuccessful products. This enables to make better business decisions, following only the processes that are proven to work.

Design and Visualization Capabilities. The latest version of these technologies is the Digital Twin. The physical model results as the final, most advanced, "error free" solution as end product for the process validation and not anymore iterations along the optimization process.

This advanced way of work is possible due to the core features, where the industry-specific design or modelling CAD tools offers 2D and 3D graphical generation and different representations of the product.

This CAD solutions combined with CAE and PDM may go up to the fully determined and visualized product version, that interact with other products, together with specific conditions and restrictions, in a indestructible (relative low-cost) environment, before any manufacturing involvement.

The optimal accepted solution is the result of a better understanding of the physical phenomena, manufacturing accuracy, financial, contractual, technological or even social impact, as the "product requirements" specify and that before any start of the manufacturing process.

14.3 PLM—Model

When the modelling for a PLM implementation starts, usually are encountered two major opinions: exists no model as a pattern to apply; or if a PLM model pattern exists is not so useful to handle, due to the fact that each company is different.

The life shows that the reality is somewhere in between. Exists no plug and play PLM models to apply in a company, but some predefined elements grouped as a pattern can be used and then the company particularities are to match. So is to "fit in" the company profile to a success story PLM model or to "adapt" the PLM model to the company particular requirements. In both cases this is the job of the PLM Engineering Manager (or the Lifecycle Manager) implies a modelling process.

Modelling is a mathematical approach, where quantities and qualities defining batch or continuous processes are represented by all kind of numbers type (integer, real, floating, complex, binary or fuzzy) as a final job for the PLM Engineering Manager is to create the "***Design of the Processes and Products Sustainable Lifecycles***". So, the PLME-M job is the DPPSL. In order to understand the implementation of the PLM, with the modelling and management of all objects, the products and the processes lifecycle, the hierarchical collaboration between processes, people and

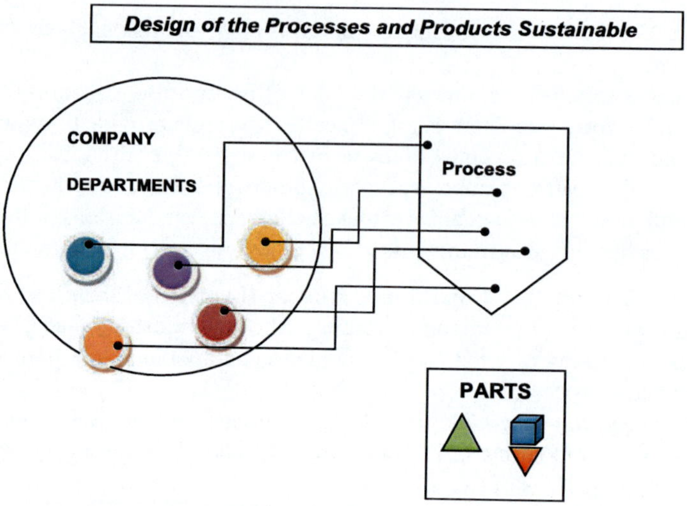

Fig. 14.1 PLM implementation modules

people tasks distribution must be considered. The simple way of their implications is briefly sketched in Fig. 14.1.

All is about interactions, starting from the tasks, subject, location, resources, people and dependences:

The task: design sustainable lifecycle.

The subject: Product(s), or Process(is), Project(s).

Location: a departmental distributed company.

Resources: devices, financial, human, IT, machine-tools, spaces, materials, robots, partnerships, patents, technologies, time, tools, royalties, spaces, etc.

People: employees, partners, training support.

Dependences: are represented by the 4 forces of change: economic, political, social and technological.

The 4 forces of change are not following the new emerging technologies or the changes in the market condition, the 4 forces are driven the development, the acquisition and the implementation of the new emerging technologies and implicitly also the changes in the market conditions at the global level.

The Product Lifecycle Management is the digital platform that support the sustainable growth based on the products and processes modelling and management. The PLM is the follower of CIM (Computer Integrated Manufacturing) the advanced manufacturing system that uses IT involving interconnection between the company engineering and management functions.

Having such good origins the PLM dynamically integrates successful product and business development strategies in order to enhance industrial competitiveness by conceptual design, production and distribution improvement.

Among this strategies are included the CE (Concurrent Engineering, '80) for a single version product with on line updating and no necessary documents management leading to shorter development times; Lean Manufacturing ('80), Agile Manufacturing ('90), continuously adaptable at the everyday change; CALS (Computer-Aided Acquisition and Logistics Support—'90); Holonic Manufacturing ('90) a distributive control on manufacturing components (cells, work stations) that are modelled as autonomous collaborative entities (Holons); Virtual Manufacturing ('90) that model, simulate and optimize the critical operations and entities in a factory plant; or KBI (Knowledge Based Intelligent Systems—'2000).

14.4 PLM—Architecture

Polices and Technologies grabbed in a methodical organized collection forms the PLM-Architecture.

The industrial business process defines the company PLM context. In theory there are some dedicated manufacturing strategies that become standards for the company behaviour. In this categories enter the ATO, BTO, CTO, DTO, ETO, FTO, GTO, LTO, MTO, PTO, RTO, VTO and MTS.

The difference between strategies relay in the moment when parts from the manufacturing process became related with the contractual forms.

ATO- Assemble to Order Is a push strategy where production is driven by medium or long-term forecast, and products are sold from existing finished goods inventory (FGI). The strategy is related with the manufacturing and logistic capacity to produce or to order in advance some of the parts that based on the marketing knowledge or statistic modelled experience (i.e. Monte Carlo Methods that principled are deterministic ones but based on repeated random sampling to obtain numerical results and using randomness to solve problems) if within the company assets a "Stocking point" can be identified at the convenient costs as the most effective spot in the production structure, as an efficient buffering activity.

The PLM BoM management decided about to types of components: the needed ones and the readily available in stock for further processing.

As central (effective) point in the production structure the Stocking point generates the existence of 2 data and 2 materials flows: an upstream for the inventory generation that determine the stock and a downstream pulling to order.

In the client-company relation, the physical disconnection between Production and Sales redirect the activity to a logical connection between Production and Sales and a physical connection between Assembly and Sales seen as a "production order" after the "sales order".

The strategy, having the disadvantage of investment in unsealed production on stock, has the advantage to soften or elevate the day to day shocks in different manufacturing demands and combined what is called the agile and the lean strategy, offering for the customers the flexibility to be swift in giving an answer and the demanded products in shorter time, and internally the possibility to reduce the waste. This is a common strategy for the Eastern companies.

BTO - Build to Order It is the oldest pull strategy that implies most assembly then manufacturing where the production starts only based on an actual customer contractual order. No product is built before a confirmed contract. The assembly capacity must be connected with the customer demands for: deadline, contractual confirmation as start line and the quantity (+quality) of the required product; especially when is a customized product.

It an approach recommended for prototypes or reduces number of highly customized products. The automotive and aeronautics are strongly following this pattern, often seen as a JIT (Just In Time) manufacturing will deliver only on demand and no previous parts stocks is ever available.

The strategy reduce waste and storing costs, increasing the efficiency but requires high performance and flexible machine-tools, expensive tools inventory and qualified staff. In this strategy PLM plays a big role in the efficient definition and optimization of the fluxes any change costing very much.

CTO – Configure to Order In classical sense CTO represents the Configure to Order strategy, close to the ATO strategy manufacturing parts to stock previous order, on a probabilistic market forecast regarding a product features, options and quantity.

The efficient element of the strategy is to have a VC (Variant Configurator), a solution applicable when a product may have variants, and based on feature options will be produced in a certain quantity. This is a complex underlying structure that enables to reach the variations but in the same time suffer from dependencies and limitations including the need to compete with manufacturing lines that may produce a totally new product that provides required options and characteristics in a more efficient way.

CTO - Chief Technical Officer The CTO that represents the Chief Technical Officer strategy is much interesting and relayed on the engineering side of a company, being based on the CTO decisions. The determining role of technology decision process and of the strategic business by understanding profitability in relation with processes, products and services implies the use of the experience, skills and broad knowledge of professional in selecting the company strategy.

Usually a CTO is a senior executive, whose occupation is focused on the scientific and technological issues within the company, with responsibilities in managing the engineering requirements, offering a successful approach where technology and innovation must be directly connected to business strategy, including mergers and acquisition aspects and "accurate translation" of the products technical details to marketing, media and VCs (Venture Capitalists).

The CTO must be always updated with the new existing technologies, monitor, evaluate, and select technologies that can be applied to future products and services, providing options on "technical bets" for the best technical infrastructure, anticipates business decisions and describe them using study cases, previsions or scenarios for clearly show the implications on the company Innovation driven overall direction.

The CTO is actually working on strategies with the company CEO (Chief Executive Officer), the highest-ranking person in a company ultimate responsible for taking managerial decisions, insuring that the CEO is completely award about the company technological capability. The CTO must provide a high risk technical vision with the highest probability of success, considered "calculated risk" in order to create a unique product containing features beyond the technical reach of their competitors leading to an extended market dominating position.

The CTO is valorizing the core company assets evolving into more efficient forms, or adding new lifeblood by developing truly novel technologies with the existing supporting technologies, and skillfully interact with legal issues (i.e. IPR, license contracts, legal counsel, regulatory), incorporate all together in a strategic planning not only for imposing the technological culture of the company but also to align it to the company goals endeavor.

In the new digital era the CTO collaboration with the company CEO must be extended to the CIO (Chief Information Officer) and the CDIO (Chief Digital Information Officer) that are every day directly working with the IT (Information Technology) and the computers systems that support the company goals, and with the CSO (Chief Security Officer) as the chief information security in charge to protect the data network infrastructure from being penetrated.

To have a clear view about the CTO position and the CTO driven strategy some examples are edifying:

- **Bill Gates**, former CTO of Microsoft;
- **Steve Jobs**, former CTO of Apple;
- **Mike Krieger**, co-founder and CTO for Instagram;
- **Werner Hans Peter Vogels**, CTO and Vice-president of Amazon.

The selected examples are showing that CTO strategy is applied mainly in automation, electronics, internet and generally computer driven technology companies [1].

Since generally CTOs possess advanced college degrees, they tend to have multiple relationships with members of academia, knowing that prominent technologists are often provide services to academia, government and professional organizations.

The studies on the CTO strategic management indicates the inclusion of CTOs in the company.

Executive Committee: 60% North America, 67% Europe, 91% Japan and in an increased number of cases the CTO actually teaches senior management about the importance of technology in their industrial activity.

As a conclusion officially the CTO strategy is recognized since'80 as technology becomes part of the company strategy and business decision strategy, separating technical facts from marketing rhetoric and adding value to the company as result of CTOs participation in by participating in academic, government or industry decisional groups, offering a sound advice on the technology involved business decisions.

DTO- Data Transfer Object In the PLM area with important reverberations within the VR and AR, the DTO is a strategy to design and use objects as data carriers between different processes.

The need for DTO increase once with the traceability higher demands and the common dialog between processes resorting different or remote interfaces with costly delays for each operation negotiation between two terminals (server-client).

In telecommunications the "costly delays" are having 2 sources: each handshake is a preceded automated process for establishing a communication by initiating the communication protocols at the start of the communication, before the full communication begins; and each call is related with a round-trip between the client–server-client. The DTO strategy reduces the communication at just one call, all the necessary data being aggregated in the object.

From the simple whole purpose to shift data in expensive remote calls, being all data related to the object, extensions to the data access object and business objects for a complete strategy are added.

The DAO (Data access objects), are patterns for specific type of data bases, providing specific data operations creating a separation between the application needs and the data access, unexposing the database to risk of damage or piracy.

The BO (Business Objects) is different categories of business processes modelled as objects that can be as large as the entire order processing system used within an information system.

ETO- Engineering to Order Is a strategy that starts from the specifications provided within a certain customer order. The difference from BTO is that each customer order is a triggering factor for a new design stage for the specified product and from here the rethinking of the entire production chain: design-engineering-manufacturing.

The need and motive to apply ETO strategy instead the BTO is the lacunar specifications, data and requirements regarding the end product.

In ETO to satisfy the customer scarce specifications and fuzzy requirements, it is heavily involved the company specialist's team from every engineering discipline: mechanics, electric, electro-technical, electronic, mechatronics, software developers, manufacturing and system engineers. The specialist team have no initial idea about the concept, materials or application until they offer the first workable solution.

The ETO address to a particular type of order from a special type of customers, customers that are requiring prototype like products and care much about the deadline for delivery dates.

About the quality risks and quality controls along with testing or quality control metrics, the customers are in a large range of acceptance.

If the company is a small size one (SME), it may not have the expertise or resources to implement quality guidelines or testing, if is a large company will use the internal quality standards or will ignore them to reduce costs.

Being a flexible and adaptive demand-driven approach to the manufacturing process, here comes the important role of the PLM, not only by the digital capacity to reintegrate and reuse parts from previous projects but also the integration of the parallel, concurrent and syncron engineering technologies, conjunctively with the integrated product team methodology and CAM.

The advantages for ETO are coming from the company tailored solutions and as a consequence increase the sale margins.

PLM is again an appropriate tool in increasing the profit margin within the EOT strategy environment providing the instruments to:

- Identify and eliminate waste;
  Considering every operation and process apply the so called DOWNTIME methodology and identify the causes and analyse possible solutions:
  - D—Defects
  - O—Overproduction
  - W—Waiting
  - N—No staff skills and talents are used
  - T—Transportation
  - I—Inventory
  - M—Motion waste
  - E—Excessed processing.
- Reduce operations expenses;
- Avoidance of any markdown;
- Improve staff tasks distributions;
- Streamline the processes operations;
- Increase the supply chain efficiency;
- Automatize the repetitive operations;
- Signalize error preventing phenomena;
- Increase AOV (Average Order Value);
- Engage JBP (Joint Business Planning);
- Put together processes and group operations for different customers;
- Improve inventory visibility by the Product Information Management;
- Implement the timing references in optimization process, "too sun" or "too late" are equal "too costly";
- Use the opportunity to create a "personal" connection with the customer implementing in the design phase a better understanding of its needs.

The new design and the lack of specifications made that ETO strategy to adopt a iterative activities in the product design and in the product manufacturing. Engineering is based on logic charts and not on the seamless of the process, especially when the customer approval is not based on testing procedures but on matching specifications. Therefore in ETO the company design its own landmarks and millstones in order to outline the border of each product development stage.

The company based on its assets and technological culture agrees upon:

- a customer order, in a contractual way;
- the products specifications.

The 3 factors: company capability, customer order and the customer specifications envisages the:

- delivery date;
- costs;
- restrictions.

The defined production landscape determines the generation of the WBS (Work Breakdown Structure) used to appoint the Management Team and the allocated resources. The management Team nominate the staff in the Working Team, stipulates the IPR and define the Milestones. In respect with the allocated resources is generated the BOM.

The design process is starting and once the product development reaches the adequate stage also the manufacturing process will start. In the end the costumer confirmation is required, that together with the design details and manufacturing specifications is included in the Report that shipped together with the product in the delivery moment. An example of the ETO lifecycle logic chart is depicted in Fig. 14.2

FTO- Freedom to Operate The FTO strategy lifecycle is related with the valorization of patents and begins with research into the IP landscape for a business potential product that may evolved through an R&D activity, considering scientific feasibility, the expected research effect on the organization is a profitable commercialization cycle.

The FTO strategy is to apply a risk—costs—competitive position balance management. The management will consider a series of correlated factors where PLM may have a major role in identifying how, when and most important whether to apply FTO analysis and strategy by recording, assisting and support the identification of:

- Financial status of the company;
- The potential product impact on the company financial status;
- Infringement risks;
- Potential rewards;
- Legal aspects impact;
- Competitors position in regards with the potential product;
- Costs of legal opinion;

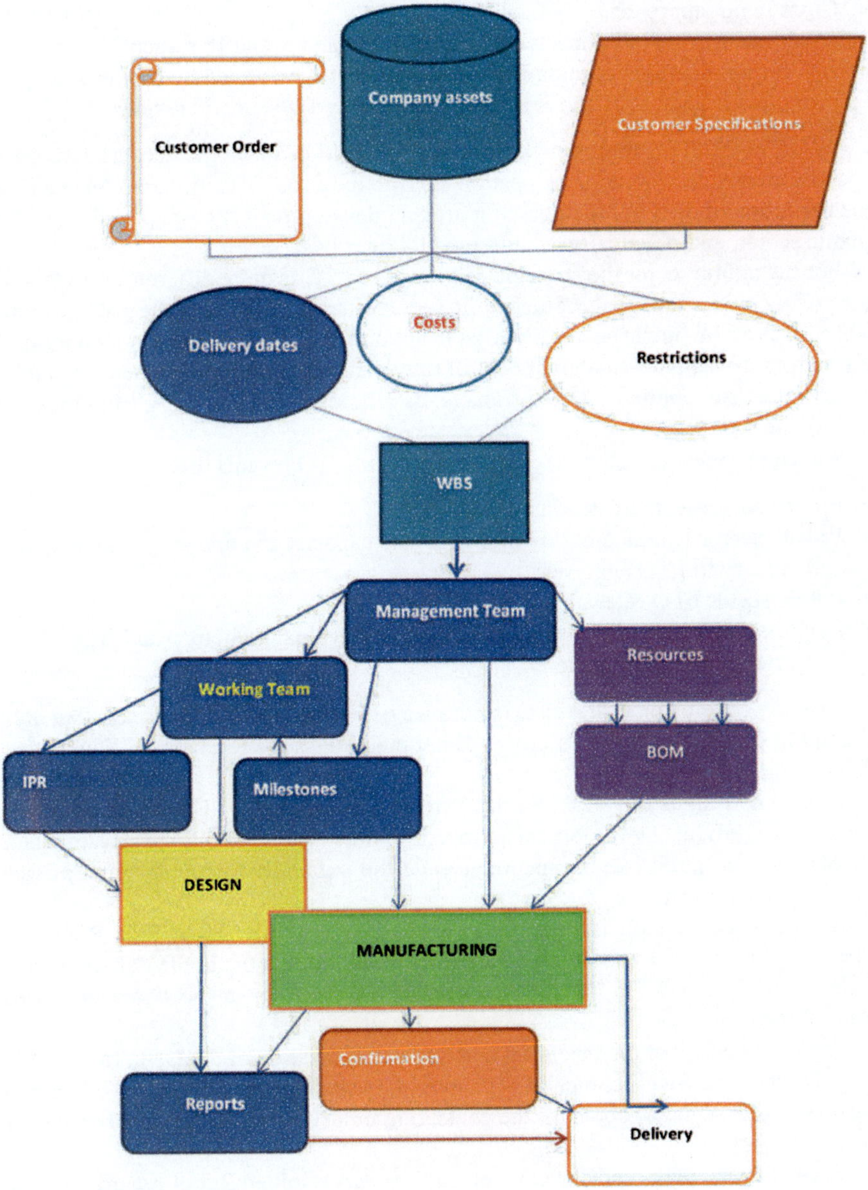

Fig. 14.2 EOT lifecycle logic chart0

- Costs in not approaching a legal opinion;
- Assessment costs of connected to the product patents and IP rights;
- Ask for a fundament opinion from a specialists team formed by business people, economists, legal staff and scientists in a recorded systematic review;

Actually the FTO represent a strategic preparation for a specific product on a specific market or geographical location, considered as the IP landscape. Considering the definition of the FTO the goal of a product development is more prevalent in the public sector, and practically is addressed to the private sector.

For the public sector the strategic relevance of FTO is quite different from that of the private sector. The goals of the public sector differ from those of the private sector since most of the public sector research is not directly intended for commercial use, a situation that applies to a great deal of university research. Even when the public sector intends to commercialize products, its mission and goals differ from those of the private sector.

A bird eye view over the theoretical aspects of FTO reveals that:

- Private sector is interested in FTO;
- Private sector is reluctant do build in-house capacity to conduct patent searches and cursory to FTO analyses;
- Public sector FTO relevance is insignificant;
- Public sector has the potential to judiciously evaluate whether and when FTO concerns should be considered.

An early conclusion is that FTO is a "rara avis" in the business landscape, but only as long as private and public sectors are not meeting as the Rockefeller Foundation demonstrates.

In this new light the PPP (Public Private Partnership) and the collaborations within PDP (Product Development partnership) made worth FTO is the private sector understands the public sector specificity and the public sectors understand private companies approach.

A considered example is the Delta Robot invented in early '80 by Prof. Reymond Clavel director of the Laboratoire de Systèmes Robotique at the École Polytechnique Fédérale de Lausanne in Switzerland one of the pioneers in the development of parallel robots.

The brief history of the patenting and commercialization of the Delta robot starts in 1987 when the Swiss company Demaurex (probably after applying FTO) started the production of delta robots for the packaging industry by purchasing a license for the delta robot.

In 1994 Prof. Reymond Clavel received the GRA (Golden Robot Award) and the company ABB Flexible Automation started selling its FlexPicker delta robot, also based on the licence acquisition, similar strategy being applied in the same year also by the company Sigpack Systems.

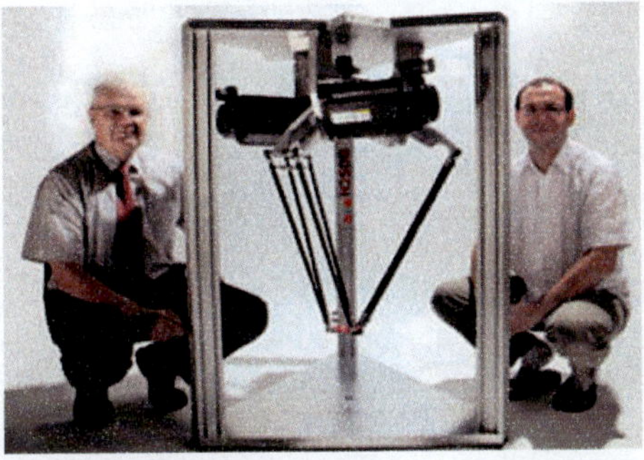

GTO - Game Theory Optimal

As simple as it looks the design of the Delta robot is covered by interconnected 36 patents like:

- WIPO (World Intellectual Property Organisation), WO 87/03,528, June 18, 1987;
- US patent, US 4,976,582, December 11, 1990;
- European patent EP 0 250 470, July 17, 1991.

Represents and exotic name for applying game theory strategies, as the Game Theory Optimal strategy representing discovering means to use an unexploited game theory in a specific situation, offering strategic, technologic and economic advantages which cannot be countered by opponents.

LTO – Limited Time Offer

As the name suggested this strategy is a niche strategy applicable occasionally in respect with a time, with a region or with a customer.

LTO is far from the idea to "cut the losses" and creates a limited time offer as clearance sales. No, LTO is much related with good quality as a "safer innovation strategy".

LTO is offering an access window to the product for connoisseur as reword or gratitude for their fidelity in using the brand, in a certain festive moment that celebrates the success of the brand or a turning point in the brand evolution to reconfirm the continuity of the good quality or simply to give access at the brand quality to a number of non-customers that base on a shift in their lifestyle paradigm will be willing to accede to the offered brand.

In the LOT strategy promotional mix, financial feasibility, guests' satisfaction, sales success is usual decision metrics. Of a great importance is the connection between the brand and the customer behaviour.

In some cases not the affordability or the taste for the brand is the triggering decisional factor, but the environment and the conjuncture where the brand is purchased or consumed.

A good knowledge of the customer, of the customer environment and of the customer behaviour (habits) made a lot of room for experiments and research, exploring configurations of the core product of the company in the comfort zone of the customer.

MTO – Made to Order

Made to Order or Make to Order is a strategy very often compared with the EOT (Engineering to Order). The core difference is coming from the tricky quality approach. In EOT exists no initial design, actually this is the imposed element in EOT, the design. Having no original design different iterations, reiterations and redesign processes may occur, but the quality matter remains at the company desired level. The PLM helps a lot in reusing previous projects design as a safe solution with intrinsic defined quality.

In MTO the "lack of defects" defines the quality of the strategy but essentially minimal level of qualitative characteristics must be specified, having from the beginning a fix design from the customer and the corresponding specifications.

Actually "Made to Order" or "Make to Order" represents a Manufacturing to Order as nosiness production strategy. The difference or better put the nuance is that "Make to Order" the represents the manufacturing in a single factory of the company and "Made to Order" represent the manufacturing within the company but in different factories of the company.

Even the manufacturing starts from a given design, the customer may ask for modifications, reconsiderations or redesign, in the end the MTO strategy is for offering a customized product to the customer. The customized product is based on an original design but also on the customized specifications that made the altering process of the original design. Anyway the real manufacturing process starts only after customer approval and a confirmed customer order is received.

Such a strategy may have the advantage of no investments in work that is not ordered but:

- Is absolutely no forecast on the products, if something is made only if is ordered, the MTO may works only in connection with a proactive management of demand, with the risk that inaccurate forecast will lead to great losses;
- If a product to be made to order is identified, for the customer is the advantage of flexible customization, fulfils the exact order, customized on their specifications but having as drawback of additional waiting times until the product is received, if the customer is willing to wait knowing that no on-time production can be really insured, and the high customization price;
- MTO is prone to obsolescence and waste, alternative activities must be designed for the no-order periods when is waiting for purchase. Alternative activities may be easier conceived for a relatively specialized company, exploiting an industrial niche and actually the orders to be practically only deviations from a core product;
- Producing "everything" from nails to airplanes a high level of customization must be provided, the company has the advantage of no finished good inventory, no stocks and no need for sale discounts strategies and management. The price for this advantages is the necessity to "preserve" skilled workers, excessive inventory or stock-out, and the capacity to manage them;

Major drawbacks may be considered: the tight conditions management; the price to fulfil the ordered; the timeline dictated by the level of customization; and the accumulated number of purchases at a certain time in respect with the manufacturing capacity. All drawbacks aspects can be easy or alleviate by using the PLM.

PTO – Paid Time Off The PTO strategy is not something really touched by the PLM. Usually in the lifecycle management the processes, technologies and fluxes are managed, but none of them is possible without people. This is a recent approach but coming from so many directions seems that the issue was here forever and looks no so new.

Recent studies show that flexible and shorter working programs seems more efficient like is nice presented by Robert Booth in The Guardian [2], remembering us that shorter day weeks start with Henry Ford in 1926 when he introduced the five-day working week. In 2014 the IZA Institute for Labor Economics, Initiated by Deutsche Post Foundation, released a study that demonstrates that long hours no gain, more than 35 working hours a week is inefficient, working shorter hours made people more relaxed and more productive.

The globalization put the necessity for more, diverse and skilled workforce at a different level. The availability of different, more and swift changing resources creates on one side increase in mobility for, smaller or widely spread families, remote work or work from home, increase number of religious interconnection within the work groups that add different number of festivities free days requirement and celebration styles.

The new trend that advocate for the four-day week, or the three days weekend, demonstrates that in large companies is a gain of about 20% in productivity and in the SMEs about 30% and is combined with the Millennials disposition for a tripartite balance: working time—family dedicated time—self-development (or leisure) time.

In this context is coming the PTO strategy for the workforce availability management, not from the point of view of scarceness of available work force and the head hunters' philosophy, but to manage the work force: absence, disability and intentness.

The fix and rigid scheduling methodologies are easier to implement, monitor and to follow, but may lead to discomfort, dissatisfaction, stress and accidents all leading in an end to losses. We may admit that historically the order and discipline (the lean management of our days) was a huge jump to reach productivity, but with the digitalization and new technologies flexibility becomes once again the key for efficiency and sustainability. So PTO strategy brings some elements to consider in the workforce efficient use not just applying a static strategy like simply replacing the five-day working week, or part-time jobs with the new trend baes on studies four-day working week (as Microsoft in Japan) or three-day working week.

Some of the PTO strategy proposed solutions refers to:

- The Floating Holidays;
- Correlation with insurance companies and social dedicated legislation (CIL);
- Sabbatical;
- Forced time off.

And this is a topic for the PLM considering the dynamic allocation of resources with timing in accordance with the contractual forms and processes lifecycle evolution.

Floating Holidays

Address directly to multicultural working group communities, where different festivities may have different meanings or no meanings at all or the same festivity has a different time distribution. A second situation may be imposed by an extreme situation like Covid19 Pandemic, where in isolation condition arrive a before planned holiday, religious festivity or marriage occasion. Staying home can be as an "every day Tuesday" or as "every day Sunday" so you may work, celebrate or rest independent of calendar or hour. The triptych: work-family-rest benefit from the multinational companies experience in working around the globe at and with different time zones. This is a suitable and perfect task for the PLM in the above mentioned condition the scheduling ticking being dictate by the person interrelation with the other members of the working group (from the inside and/or outside the company) as a knowledge of increasing the diversity and for the company to increase the productivity.

The different number of payed holidays in different countries will influence less the economic disturbances occurred due the globalization, urbanization and the logistics trends like high speed trains, regional conflict avoidance or new routes initiations (Silk Road revival).

For the moment things are probably in a very beginning as long as a study initiated and conducted by SHRM (Society for Human Resource Management) in 2017 [3], specifies that 82% of the employing companies considers that they cannot afford a floating holiday in the actual condition that 90% of the companies are closing the offices during the holiday time.

CIL

- The minimum wage;
- Paid family;
- Medical leave;
- Medical appointments for employee and employee family members;
- Extension of the employee family member definition;
- Paid-sick-leave, if also a family member is in need;
- According and tracking of the sick time;

These ere worker dedicated facilities that have different amount, interpretation and application in each country. The seamless application from the employer part (not always or everywhere are mandate measures) may be rewarded with tax and credits benefits and this could be an expansion of the PTO in the near future.

14.4.1 Sabbatical

Is a regulation that in theory is applied in Academic environment, but depending on country and University may be applied in different ways.

By definition a Sabbatical is a period of paid leave, granted to a university teacher or other worker for study or travel, traditionally one year for every seven years worked. The purpose is that have reinvigorated, refreshed, better influenced by new and exited ideas, eager to new challenges people.

In the economic environment, the Sabbatical considered for the employee the existing models there are some differences:

- The period of paid leave for a worker is less than one year, about 4 to 6 weeks, or extra hours of vacation (8–10) for each complete calendar month of work;
- The paid leave is granted can be used for travel, quality time with family and friends, or however else employees see fit.

Forced Time Off

If "Floating holidays", is something manageable by the company as an agreement between the employer and the employee and the "Sabbatical" is practical an employee decision, the "Forced time off" is an employer decision.

Resuming the idea of higher productivity by eliminating weekly long hour continuous activity, studies regarding reasons why employees are not "takin" enough time off reveal the following observations, displayed in Table 14.1

All the presented reasons in Table 15.1 are extracted in the condition that over 40% of the employees to remain with unused vacation days even if they are shifted from one year to another. The increase of the productivity temptation will lead many companies to increase their "generosity" in forcing employees in taking free time, using as coercive tool the "paid time off".

A second leverage in the companies generosity is the Millennial approach in balancing the life time activities having less financial troubles, much comfort and les responsibilities as in other generations.

The feeling of focused and productive return is strongly stimulated by the companies offering vacation bonuses, stipends or even paid forced vacations, customized on the employee profile, creating businesses for employees that must take a certain amount of time away from the office each year as an alternative of the burn out employees or an extra tool in the talents war.

As the legislation and the workforce profile evolve in a continuous change the tracking and the paid time off management becomes a task for the PLM in order to alleviate the company pain in compliance with the contracts, cash flow and payroll preparation, striving for success but avoiding the trap of overpaid wages.

The Baby boomers are going to retire and they are in big number, the "replacement" is not just fewer but with a different profile, rules, trends and expectation for the workplaces, the labor force fluctuation is high practicing the "job hop", no loyalty for the company, changing the job position in average at each 18 to 36 month. The

Table 14.1 Reasons encouraging "workaholic style" behavior

No	Amount	Reason	Obs.
1	10%	• The company management afford to impose such a schedule • The need for personal development	It is an imposed way to "encourage" the steady long hours. The result is pretty well correlated with the SHRM studies, where 90% of the companies are closing offices during holidays and 82% are imposing fix period of holidays. The result is showing the percentage of companies that keep things working during the closing times
2	10%	Guilty	An after work community and personal bonds are created among the employees. Spending much of the free time together a "guilty feeling" may appear is the free time (holiday) is spent separately knowing in what projects are involved the co-workers and what issues appears and must be solved in their absence. The "feeling" is more amplified by the use of smartphones—bringing directly the office in the holidays environment and increasing the "feel of guilty"
3	15%	Not enough resources	• The expected kind of holidays cannot afford from different reasons: direct costs; extra cost for kids, pets, relatives or elder care; credits, taxes, mortgages, investments, tuitions, etc. • The digital era technology creates also a "digital style". Instead of rerouting for alternatives the all-or-nothing (0–1) philosophy is applied
4	20%	Too much work	Assumed responsibilities, tight deadlines, time investment in future advantages, low qualification requiring more time do have the job done, bad planning, deficit in human resources, crises situations, incapacity to say NO, all are good reasons for 20% of the employees to stay working
5	45%	Saving time for later	The official, legal time for free activities is different in every country. Saving extra time for some activities is a good reason to work more, if in the end the free time is taken

PTO strategy for new hires wants not only to attract but also to stabilize the new incomes by combining the forms of the paid of time between leisure activities, body and mind building healthy activities, fitness training, professional training activities, team building activities, instructional and self-development dedicated activities, participation at events, carnivals and festivals, participation in different type of competitions not necessarily related with the company profile.

RTO – Recovery Time Objective The existence of the RTO strategy [4] starts forms the company's question: "*How to respond to unplanned incidents so the business can continue to operate?*".

Like in Pandemic 2020 year situation, oil crises, etc., the companies need to know and apply some strategies to recover after unexpected situations or disasters. In theory to approach a strategy to generate and apply a cost effective DRP (Disaster Recovery Plan) but only when it stroke you the real importance can be appreciated.

In the digital era most advocate solutions regards the digital, the data situation and recovery, but RTO aim actually or must aiming the entire company.

RTO strategy is indicate to identify and apply metrics that the company must consider as most suitable to apply in order to develop an appropriate DRP to maintain the business continuity after unexpected events. The developed metrics must not consider only the company as a standalone entity but also the situation of the company clients, the company suppliers and nevertheless the company stakeholders.

Within the DRP the metrics must identify in different scenarios the maximum of tolerable hours (days) for data and manufacturing facilities recovery within a resources frame. In generating a DRP must be modelled and calculate how quick must be the recovering process for the infrastructure and the IT services, determining another metrics to identify how long can the company survive before the operations are restored to the standard way. The modelling starts from the assumption that after the determined period for survival is considered that irreparable harm will occur.

In the modelling process there are some basic aspects to be considered:

- Cost relevance
  As RTO strategy involves the entire company business infrastructure, not just the data, the costs associated with maintaining RTO must be correlated with the recovery risk management avoiding investing excessive in the RTO assurance.
- Automation, digitalization, robotisation
  Are key factors in a company development and success. Therefore the data recovery within an RTO is essential but must be aware of the fact that is a process that must be made automatically base on the data base storage protection and data backup solution at right intervals, locations and conditions. It seems an "impossible mission" considering that involves restoring the entire IT, infrastructure and operations.
  Beforehand, put your eggs in different baskets, and consider a centralized distributed data configuration, more costly in hardware, storage space and time delays, but statistically safer in emergency situation due to a larger number of snapshots of mission-critical data.

To cut costs and keep the expenditures at a reasonable level is suggested the 3-2-1 rule, mentioning to keep at least 3 safe copies in 2 different reliable independent storage places and 1 of the two located on offsite and step-by-step procedures for restoring data and infrastructure.

- Assessments
  The RTO must reflect the entire company business and therefore consider all in all business needs. In the modelling process for the DRP determination is to be considered the company business survival time with disrupted IT infrastructure and services, the needs and the alternatives for this kind of existence.
  Also in the modelling process there are different restore basis considering different seasons, day of the week or hour of the day when a disastrous event may occurs. A similar perspective is over the time tolerances. The recovery time must be shorter than the survival time, other ways the DRP is not only useless but also harmful as long as the company and sometimes the company conformity rely on it. But even in the right situation the two values of time cannot be equal and the tolerance issue is how larger could be the survival time, logical and practically considering that the recovering time may have a minimum and also a maximum value.

Imposing a resources frame, as initial condition of recovery, the DRP is applying not only in the unexpected event conditions but all the time as a prevention plan for preserving the existence of the initial conditions in case of unexpected events defining the type, amount and location of the framed resources, how often must be updated, what backups should occur and the specifications on how should run a recovery process.

Within the RTO strategy, the management of the highlighted aspects for modelling and the entire DRP may be part pf the company PLM, having in mind the preparation for the worst-case scenarios. Remembering that in the RTO strategy also the clients and the suppliers are considered, in the PLM standard cycles can be imagine, as a result of the DRP, rerouting and backup alternatives discussed in time with clients and suppliers even in the contractual negotiations stage.

It was an old milkman in Denmark making deliveries in sharp time every morning. Once a young entrepreneur asked him: "*Dear man, it happens that you could not deliver the milk?*", the old man stayed for a while thinking, and answered: "*Yes, it happened ones, April 9th1940, when the war starts. I am sorry to admit that I had a 10 min delay in my delivery.*"

In any contractual form exists clauses regarding the major force, but just to consider disclaiming from the responsibilities. In the RTO strategy is considered a team work, to make things happens and ensure the continuity. A long term partnership is the key for the company success and also for its clients and suppliers.

It is important that the company is with the suppliers but most important with the customer on the same page regarding the long range activities. This type of strategy may surprise and bother the customers (and the suppliers) in the beginnings, in detailing aspects and requirements and tasks to fulfil if things are going wrong. But afterwards that will ensure both that the company is thinking of them and is a reliable company looking to save the business and the customers in the event of any fail. In this sort of collaboration the DRP is extended to them with disaster-recovery scenarios, helping each other to identify and eliminate the "scenarios problems" before become real problems and creates regular event recovery drills.

In connection with the company dimension and importance it will to decide over the strategy implementation expectations and the allocated budget. From here will result the dimension and complexity of infrastructure, the amount of considered variables, the need for external services and support, etc.

VTO – Vision Traction Organizer Is a strategy, applied for an organised forecast of the company, as a time management organiser for the activities or actions (Traction) that must lead to the plans achievement (Vision) as a business strategy that looks in details at every business section issues.

As a classical approach in elaborating a new business strategy happens to not seeing the forest for the trees, the VTO strategy [5] in the company new business development combine the 2 components including the vision by dressing the vision in a traction actions made coat, to make the vision a reality.

The VTO is applied by considering the company internal data, with the new business vision trying to clarify for the implied persons and in extension to the stakeholders, all the aspects that may lead or stop the vision to become a reality.

- ASSETS
  - The core values of the company must be listed to know exactly on what is rely on;
  - Identify the technological culture of the company, to position the company in the global technological landscape;
  - Staff structure analyze:
    - Staff to be reconverted or trained;
    - Staff to be hired;
    - Staff to be fired;
- GROUND
  - How old is the company;
  - On what grounds the company exist and evolved;
  - Accumulated competences and skills in the company history;
- DIRETCION
  - From now on all that in not in concordance with the core elements is to be ignored;
  - The time zone focus in 10 years;
  - Questioning the commitment, in a scale of 1 to 10 (10 the best), will be a better understanding of the attempt;
    - Humanistic role of the new business?

    - There are people outside the company that may help?
    - Risk awareness?
    - Will act regional, national or global?
    - It is well know what is targeting.
    - The new business will add new enemies or will share common enemies?
    - The mew business has a model or will be a model?
    - What chance of success is estimated?
    - Who is the idea promoter?
    - Has the promoter the energizing charisma?
    - It is the first vision or exist 'the habit" in having visions?
    - How audacious is the company on this new business?
    - Is defined a new or a clear target for the company?

- LIFECYCLE
  - Focus on market picture;
    - Frame the target market;
    - The assumed list of potential customers;
    - Identify the customer processes;
    - What assurance and promises can be offered to the customers;
    - Create a guiding Logo for the company and for the market, as a mission statement
  - Imagine the company status in a span time
    - Dimension;
    - How it looks;
    - Number of employees;
    - Profit;
    - Turnover;
    - Revenue
    - Number and type of customers;
    - Type and number of new products;
    - Investment in visibility and brand;
    - Potential awards to win.
  - Lifecycle start
    - What must be achieved in the first 100 days;
      - Specific priorities;
    - What must be achieved in the first 12 month;
      - Smart goals;
    - Faced issues;
      - Challenges;
      - Obstacles;

- Opportunities;
- Unclear ideas.

The answers will reveal how well is positioned the company, how ripe and ready to act are the traction elements in the attempt to follow the Vision with the task to made it viable.

MTS – Made to Stok The MTS is a traditional strategy based on a reliable ability to forecasting, based on the customer former consumption behavior the willingness to buy quick and cheap, from the shelf at the minimum affordable price, In the same time a prevision about the characteristics and specifications that the product must have and when the customer will come with an order. This group of prediction impose to the company that adopt the MTS strategy [6, 7] to redesign operations at specific times, instead of keeping a steady level of production, costly adjustments that are usually passed to the customer.

The MTS strategy forecast philosophy is applicable at predictable product features that can be defined and planned, specified in a catalogue form as a ready to be delivered commodity. The customers' convenience and rush with no willingness to accept long delivery time is associated the company forward-looking planning process and make profitable the production to stock and together arrive at the idea of on-time performance that offers enough satisfaction on both sides.

The strategy is decoupling the production from the sale. The production can be planed and run in a steady way based on the no events future forecasting a consumption pattern based on the customer past consumption behaviour and anticipating a customer demand. The sale is based on the incoming order using directly the availability check from the stock inventory with a no-delay offer to the customers, and offer adjustments for the planning predictions without placing any load on the production schedule.

It may happen to be no stock for selling situation. Actually is no real risk for the company, in the stock out situation the customer must just wait, but the lead time is predictable based on a steady manufacturing process. The company risk is in the case of over production and limited liquidity, and with no actual requirements for the commodities in the stock. In the fast-paced may encounter the second risk, the obsolete of the products.

As a nearby observation, the lead time is a variable dependent on the commodities stock, and therefore sometimes is defined by the customers, other times by the company based on internal constraints.

In theory, the MTS is a great strategy in buffering the variations in demand. The problem of MTS strategy is that the products are not optimally set, offering an "average product" that is accepted based on price and availability but is not a customized one, dethroning the customer as the King and creating divergent opinions and ideas between the production and sale departments about the products assignments. Therefore the products portfolio must be regularly adapted a customized part of today might be a standard part of tomorrow.

In Table 14.2 is represented a synoptic over the mentioned strategies in connections

Table 14.2 A strategies synoptic

The PLM may support and be mould on this strategies													
Management Strategy	ATO	BTO	CTO	DTO	ETO	FTO	GTO	LTO	MTO	PTO	RTO	VTO	MTS
Specificities													
MK Marketing Knowledge								■		■	■	■	■
EK Engineering Knowledge	■	■	■	■	■	■			■		■	■	
MANK Management Knowledge	■	■	■			■	■	■	■	■	■	■	■
Management Strategy	ATO	BTO	CTO	DTO	ETO	FTO	GTO	LTO	MTO	PTO	RTO	VTO	MTS
Specificities													
PCPR Previous Contract Provided Resources	■												■
LSF Logistic Storing Facilities	■								■			■	■
CPR Contract Provided Resources		■	■	■	■	■	■	■	■	■	■	■	
STD Scientific & Technology Driven			■	■					■		■	■	
Shorter SLT Supply lead time	■									■			■
SKU Stock Keeping Unit	■								■			■	■
ALD Accepted Lead Time		■	■	■	■	■		■	■		■	■	

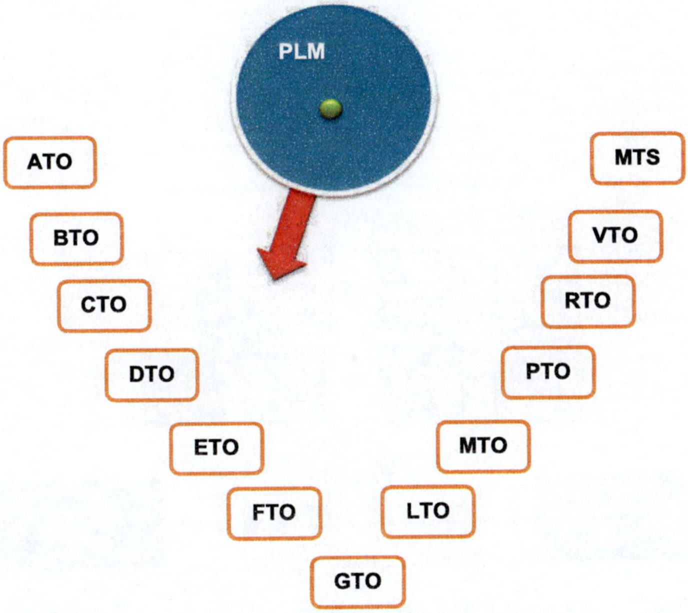

Fig. 14.3 PLM adapted to the company selected strategy model

their specificities.

The PLM molds on the company adopted strategy as suggested in Fig. 14.3.

The orientation of the PLM, toward one of the considered strategies, is made by targeting the controlling of the collaborative work within all of the strategy specificities. The PLM is enabling customization on each specificity, in relation with the strategy, the domain of activity and the company cultural technology.

In this approach PLM, as a dynamic system, is consider to be reliable if is not suffering from instability, a higher reaction speed at different strategic priorities may lead easily to that as a company is not representing anymore the activity of one team or a single business orientation.

The Multidisciplinary, multitask and multiple teams correlation is actually the job of the PLM, the common goal being represented by the envisaged product and the selected strategy. The manager of the company expect from an implemented PLM architecture to obtain a customized dashboard, that reveals and track the evolution of all the KPI (Key Performance Indicators) relevant, to part, processes or business objectives. Such a dashboard represents the upper part of the PLM architecture, displaying the organizational and technical functionality. The Technical area covers: the product design, product development, purchase, manufacturing, and quality

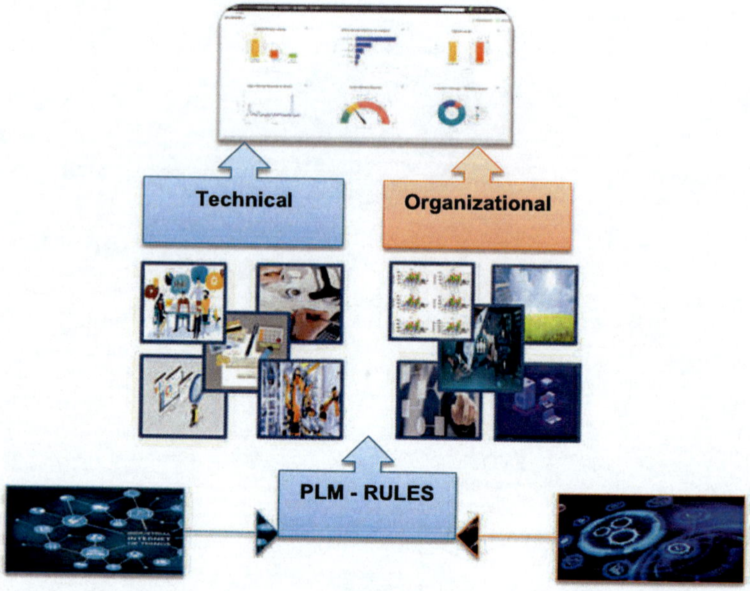

Fig. 15.4 PLM form bottom IoT to the top Dashboard

control. The Organizational area covers activities, resources, human resources, business processes and synchronization management. At bottom data is source, IoT sensors and the set of rules, Fig. 15.4 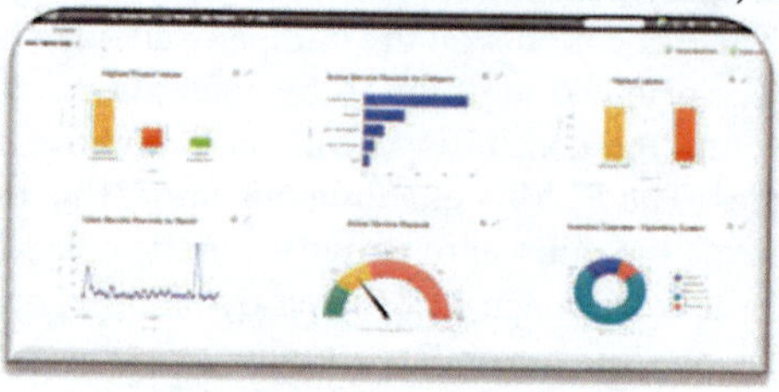 .

14.5 PLM—Benefits

To briefly resume the benefits of PLM is about on what PLM is doing and on what PLM is bringing.

Doing:

- Interaction;
- Collaboration;
- Hierarchical structuring.

Bringing:

- Costs reduction;
- Assessment of change impact;
- Assessment of changing duration.

From all this aspects the first thing first are the costs, because cost shows and gives instantly the filing about effort and the resources that must be allocated and the affordability.

The cost reduction of change is possible through the team collaboration facilities at global scale and by enabling the direct communication between all company departments. From the design phase people may have feed-back from everywhere offering in the same time not only a quick response to change but also a multicriterial right decision to change.

The impact reduction of change over the good functionality of the company is enabled to a good communication system. In this way the leadership may have an early transfer of the information and awareness messages to every member. The accessibility to the same data avoids unnecessary redundancies and errors in data definition, data updating and data perception. The quick feedback from every level (every member) avoids any tensions or misunderstandings escalation.

The awareness about the entire process, the clear tasks definition, the step by step continuous smooth habit with the elements of change and the new procedures, ensures the success of every step implementation. In short by delivering the right information to the right people at the right time is extremely important and possible by providing a single source of product and process knowledge 24/7 and for everybody.

About time, the time reduction of the change process can be impressive, up to 40–45% due to the managerial approach enabled by the engineering tools. The success is ensured by the improvement of the change cycles added by the product lifecycle management software facilities, going up to the managing of change in real-time.

The change process time reduction is not by getting the fastest ever change, but to be able to respond faster than the competition and competitors. If exist a vast unknown and uncontrolled amount of data about: the competition, the actual market, the technological challenges, the engineering problems (and solutions) you have to admit that is very easy to lose control, equal is the "unknown" exists in the concept phase, during the product development, in and after the manufacturing phase, after selling service or maintenance, or even in times when the company may grows globally or face global problems.

14.6 PLM—Implementation and Integration Techniques

About the implementation and integration of the PLM is described in more details in some of the next chapters, but there are some things to be mentioned in advance.

PLM implementation is about the knowledge management. PLM is a form of managing change but the PLM implementation is adding change in the company. That way this circular approach is about the knowledge management, how to manage the change and the management tooling at a change that you generate and try to find out how to customized it to the company profile, with no harm over the company activities, in the shorter time for starting to have the benefits of the implementation and with minimum cost, so the investments will not affect dramatically the other activities or projects of the company.

By the collaboration in the product design and the shortening of the product fabrication lifecycles the PLM is one of the major technological and managerial challenges. The PLM is leverages not inventing the core competences of a company but using PLM itself is a core competence in better addressing to the targeted markets when is dealing with the company way in running the business.

The PLM implementation starts from identifying the facts and data that exist but probably the company is not totally aware about them. The goal is to identify all the activities and the correlated data. Linearizing the activities results the lifecycle of the product within the company. The implementation of PLM will continue to ensure transition among the different activities with the goal to obtain a streamline of data over the entire lifecycle of the product lifecycle.

The integration of PLM in the company activities is starting from the ICT infrastructure: the hardware, the network, the software modules and the software engineers, briefly the implementation of the ICT structure. The product related data are gathered and visualised in a single data base. From the main 3 influencing groups of specialist: the design and development engineers, the manufacturing engineers and the supply chain engineers, are delegated the representatives that will collaborate and work with the software engineers. The goal of this collaboration is to connect the software technologies with the ICT infrastructure and the representative data and functionalities of the 3 influencing groups, to elaborate a "software strategy" that will allow the inclusive integration of all the identified product data.

The resulted technological web overlapping the company activities, connecting the design phase with the manufacturing technologies, the supply chain and the service activities. Through the web communication threats will circulate 4 types of data:

- Conceptual engineering data: design, materials and functionalities;
- Manufacturing data: technologies, tooling, transfer logistics;
- Maintenance: service activities, maintenance and repairing operations;
- Aggregated data encompassing the identified managerial patterns that will be transformed on fabrications synchronized cycles.

The PLM integration made from an extend company that may be spread around the world with many plants of different types in different locations in a collaborative enterprise that follow the same rules and manufacture with a controlled effectiveness in the end of a change management project.

The presented modalities of PLM implementation and integrations are examples of techniques that can be apply by a company in getting the PLM in the company portfolio of software tools.

14.7 PLM—Standardization

The interoperability within processes and company departments is unimaginable without specific standards. In later chapter of the book, PDP—Product Development Process, the major standards used within PLM are presented. At this moment it will be considered just an overview over the standardization within the PLM focused on the open standards.

An industrial standard is a necessary and used standard. Being a number of companies involved in an industrial standard development the standard will be not open and "democratically" managed or used by everyone.

An open standard is a recognized industrial standard, issued by the ISO (International Organization for Standardization), or OMG (Object Management Group) issuing standards since was founded in 1989, a group driven by academic institutions, end-users, government agencies and vendors, or W3C (World Wide Web Consortium) that develops international Web standards, all open standards being freely accessible and with no restrictions to every user that all must be supported by the PLM systems. With the extension of digitalization and digitalized activities is an increasing demand for the open standards. A similar trend is also in the PLM development to facilitate an accurate data flow between PLM systems and processes. The most known standards are the ISO-STEP (STandard for Exchange of Product data model), focuses on computer interpretable geometry.

Beside ISO were mentioned other standardization structures that are generating the "de facto" standards (that can be open or not) but are used because their values with the possibility to exists alternatives technologies to these standards.

In the PLM area the customers advocates for the open standards for the granted freedom to run their business but the vendors are advocating for de facto standards to control the content and price of their products.

That's why the OMG was formed to reduce the tensions, and therefore in PLM there are the STEP and GD&T standards from ISO, UML form OMG and XML from W3C.

As the PLM software starts with the CAD applications and CAD applications management, the debut of the PLM opens standards was with the geometric data definition spread after that through all the product lifecycle aspects. In the family of open standards there are two major groups, one connected with the PLM as dedicated standards and a second group consist in standards of general interest to PLM like the definition of the universal length standard, the meter. For PLM (and not only) this is an essential standard that evolve from its first definition in 1889 to the one emitted almost 100 years later, in 1983 as the distance covered by the light in vacuum in a time interval of 1/299 792.455 of a second.

For modern manufacturing capabilities and higher accuracy it can be assumed that like in the case of meter standardization also other standards may evolve, considering an improvement with each development step.

As expected the first standards were related with the parts geometry and related attributes as "engineering objects" (CAD/CAM/CAE), followed soon by the ones regarding the purchases as "business objects" (BOM-Bill of Materials, CRM-Customer Relationship Management, SCM-Supply Chain Management) and then the ones regarding industrial activities clustered around the first ones regarding the ERP (Enterprise Resource Planning) and the PDM (Product Data Management).

References

1. Smith, R.D.: *The role of the chief technology officer in strategic innovation, project execution, and mentoring*. Titan Systems Corporation, Orlando, USA (2002)
2. Booth, R: *Dozens of Firms are Trying Out Shorter Working Hours—and Finding It's Good for Workers, Customers and the Bottom Lin*. The Guardian, 13 March (2019)
3. SHRM: 2017 Human Capital Benchmarking Report (2017).
4. RTO: https://www.msp360.com/resources/blog/rto-vs-rpo-difference/
5. VTO: https://www.boldclarity.com/what-is-your-vto/
6. MTS: https://www.investopedia.com/terms/m/make-to-order.asp
7. MTS: https://www.tredence.com/blog/defining-mto-and-mts-production-strategy-and-its-implementation/

Chapter 15
Digital Product Tracking

"You Cannot Deal with Something You don't Know"

15.1 Generalities

Generally everybody is familiar with computers and uses files for different applications. In the beginnings everything is simple and easy but in time the data and files of all kind accumulates.

In the same time with the increase of the data complexity and quantity, also the data management activity is increasing. The common activities of storing, scanning, deleting, purging or updating the data is going along with an increasing demand for swift data search, certainty in data identification and data recovering, creating a complex self-standing but costly process that may lead to success or failure and data stuck.

Imagine now that within industrial activities raise almost the same problems. Especially in the case of complex products is needed to manage the following:

- technical drawings;
  - new generated;
  - retrieved from archives;
  - or combined with technological specifications;
- the resources availability;
- the resources allocation, on processes or within the production lines;
- external components characteristics monitoring;
- the amount and the quality of the accumulated products and components;
- identification and record of the production results, to know at any moment;

J. Niemann and A. Pisla, *Life-Cycle Management of Machines and Mechanisms*, Mechanisms and Machine Science 90,
https://doi.org/10.1007/978-3-030-56449-0_15

  - what and in what quantity is ready;
  - and what is processed at that time;
- parallel monitoring of conflicts or malfunctioning (errors);
  - on the manufacturing lines;
  - on resources allocation;
  - on suppliers;
  - on contractual requirements;
- and so on…

Now you have a better image of how complex and important is the data management and what great challenges rise the I4.0 industrial revolution.

One cannot control the unknown things; therefore the control activity starts from **MONITORING**.

This activity is relative simple if a small amount of parts of the same type are considered, but the problem became much bigger in the **case of large pallet and quantity production** especially in the **FMS** (Flexible Manufacturing Systems).

The main issue is the identification and the traceability of the product characteristics and the product characteristics modification, together with the responsibility for that changes everything requiring a high accuracy and data availability in real time.

Traceability is not only to localize parts in faulty situations and the responsible for that but also describes the stakeholder's ability to understand and follow relationships between the components that play some role in process lifecycle, including technology-independence in defining semantically rich traceability relationship, and creation of a generic implementation template reflecting generic problems, common to traceability applications(e.g. OWL—and TGraph ontologies).

The Web Ontology Language (OWL) is specially designed to represent rich and complex knowledge about things, groups of things, and relations between things.

The unreliable product quality is one of the most difficult problems in manufacturing companies. In case of frequent faulty products, the profitability may greatly weakened, especially when the products arrive to the customer followed by costly claims for repairs or replacement, due to nonconformity or malfunctioning within the warrantee period. This situation is decreasing a lot and very fast the value of trademark.

Internally, the faulty products are directly reflected in material waste together with useless manufacturing and assembly work that actually represent extra unnecessary costs and loss of time. This is the result of other problems that exists within the manufacturing companies: the insufficient knowledge of product material, the estimation of the product value, the originality level of the product and the difficulties in the manufacturing lines maintenance.

Therefore to identify the errors, locate the faults, prevent faults continuity, avoid by forecasting possible errors and minimize the waste is an absolute necessary process starting with the trace of the product throughout the entire lifetime of the manufacturing process and to keep the levels of problem under control.

A clear formulation, for understanding the meaning of product traceability is given by [1].

> Generally, **product traceability** is the user ability to trace a product through its processing procedures.

The user may be the manufacturer, the supplier, the vendor, etc., concretely the product traceability deals with **information's records maintenance** of all materials and parts along a defined lifecycle (e.g. from raw material purchasing up to finished good selling) **using a coding identification**.

The product process traceability starts from planning, over the product generation process, considering every moment the actual process in progress. The tracking creates the foundation for the advanced information management, necessary for the quality assurance improvement [2] and the traceability of the order-delivery-process contains the tracking of an individual product and the delivery to the customer [3, p. 252].

> Product traceability is by definition a PLM topic, since it is related to a product centric approach, where product data and information might be retrieved and managed along the whole lifecycle [3, p. 58].

By improving the traceability of products it is possible to prevent those faulty products at an early stage and to minimize the costs. The importance of the early stage came from the fact that the cost of changes begins to raise dramatically in later stages of the lifecycle. The whole product lifecycle contains the entire list of phases, from the mechanical design, material selection and acquisition (including the complementary or the equivalent suppliers' logistics data and characteristics), the tooling and the direct manufacturing, up to the withdrawal point over the dismantling and/or recycling stages.

Considering this approach, and the technological capability to solve things, it seems that all the problems are the result of a single reason: ***the lack of information or/and knowledge***.

The key to solve the lack of knowledge over the system or process is to design and integrate a system which allows efficient and constant access to information, during the change process of the product status throughout its lifecycle.

Within the system the **backbone** must be considered the **capability to trace the product** and the order-delivery-process, in order to close information's gaps between different phases and processes of the product lifecycle to any individual product or assembly.

From this point on may start the optimization of the holistic product management?

Yes, it is the necessary starting point, the holistic product management contains not only the own view of the company but also a **knowledge network** that together with the external lifecycle partner access available additional product knowledge [4–7].

To create and implement such a network it is necessary to use , data capture and processing technology, the Auto-ID.

Table 15.1 Sensors family

Optical	Electronical
1D Barcode	Radio-frequency-identification RFID
2D Codes (QR and Data Matrix)	Global positioning system GPS
3D scanning	

Where to Look Now?

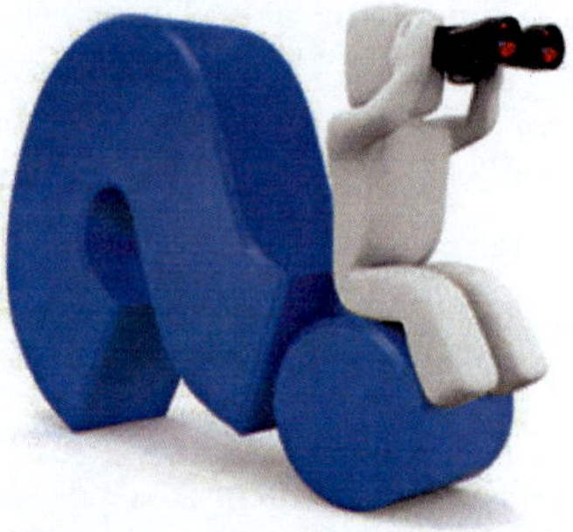

[Source: www.me-mo-tec.de]

When we start to measure an "entity" it is to consider first the nature of the measuring "dimension" and thinking about the adequate readers. The result of the first conclusion is about the type of readers, Table 15.1.

Basically there are two families of identifiers (sensors): the optical and the electronically one. For the optic sensors it is to specify the number of dimension to measure: 1, 2 or 3 dimensions. For the electronically sensors case, dimensions don't count, important is the data transferring technology and the components characteristics data storing capacity and coding modality.

A barcode is a discrete self-checking symbology that encodes different characters for data with additional characters for start/stop. The barcodes are used by professionals in many fields were entities must be tagged and monitored (tracked): administration, aerospace, banks, education, healthcare, libraries, logistics, manufacturing, military, police, post, research laboratories transportation, etc., being easy to print the barcode on any impact style printer also mentioning that internally (having as disadvantage the miss of portability) the barcode can be created by anyone that may define a rule for consecutive numbers.

15.2 Tracking Labelling Overview

It is a connection between and the type of identification code, the sensors and the code reader. To know what is suitable we have to know first the possibilities and to understand the processes were is applied the tracking system.

Each tracking system is bring unique opportunities, as qualitative benefices or imposed restrictions and therefore looks like a real challenge to select "the best one" from a wide range of solutions. The compromise follows to apply an adequate solution that provides the highest business efficiency, with the lowest investment, covering the entire business and facilitate the most secure communication with all the partners.

It will be always looked for the simplest solution with the minimum number of sensors and readers. The journey to find the right solution for the envisaged process starts with the decision regarding the number of dimensions in which the tracking is necessary (1D, 2D, 3D…) [8].

The 1D (one dimension) barcodes is materialized by parallel lines of the same length. The data are coded by varying the width of the lines and the space between lines. It is a very common representation that can be found out on most of the domestic products (consumer goods and retail industry) and therefore it is well recognized in any representation.

15.2.1 The Linear Barcodes

The linear barcodes is in a way the classical technology that almost everybody is familiar form the shopping activities. The linear barcode reader is used for the product identification in order to extract the part price and keep the logistics evidence in the stocks databases.

Being the oldest model, exists many types of barcode classes. In Table 15.2 are summarized depicted the most used versions.

With a limited number of digits (from 6 to 13) is only possible to describe a limited data and this is the major restriction of the linear barcodes. A second disadvantage comes from linear barcode scanning technology. The scanner has a short range to read the code and have also a low reading speed that may lead to many reading errors. The result is that if a company looks for a safer and much rapid reading and with more registered data it is required another Auto-ID technology.

15.2.2 The Two Dimensional Barcodes

The two dimensional barcodes (called also matrix code, bi dimensional or 2D codes) are representing data using two dimensional symbols or shapes that may represent

Table 15.2 1D Barcode classes

New class name	Former class	Obs.
GTIN-6	UPC	UPC-E is a light, smaller variation to encode on 6 numerical digits, most common to be used in USA, UK, Australia, New Zeeland and associative countries
GTIN-8	EAN-8	Is a small version, with encoding on 8 digits applied where the space is limited (like on candies)
GTIN-12	UCC-12, UPC	Label and scan consumer goods around the world UPC-A encodes on 12 numerical digits 234567 899992
GTIN-13	EAN-13, CIP	I mostly used in Europe, is the standard EAN form, comprising 13 digits. Used also in the ISBN and ISSN coding and JAN-13 (Japanese Article Numbering) as barcode Symbology 4 191841 707909 or 1 234567 899992 This Barcode enables to code the data on 13 digits (numbers), based on the following rules: First two digits, 1 and 2, nominates the origin country of the product (i.e. Germany 40–44 and Romania 59) The following five digits (3–7) contain information about all the separate companies coexisting within a concern. With this information every user will know exactly in which company was made the product The next five digits (8–12) are forming the article number The last digit is actually the code check digit of the barcode

(continued)

Table 15.2 (continued)

New class name	Former class	Obs.
GTIN-14	DUN-14, ITF	Interleaved 2 of 5 (ITF) is a 14 numerical digits barcode, using the full ASCII set Is a code with high printing tolerance and therefore applicable on corrugated cardboard and therefore used to label packaging materials across the globe 0 1 2 3 4 5 6 5
	CODE 39	Is a code using 3 from 9 digits, with most of applications in Automotive industry and Defense (especially US Department of defense). It is an alphanumeric coding, enabling both characters and digits, having the full ASCII support. In the initial form is enabled only 39 digits set, but in the newer form 43 digits set is considered 'ABC 123'
	CODE 93	Compliment to CODE 39, was originally developed in 1982 by Intermec, using a combinatory of 2 characters Like CODE 39, CODE 93 is having the full ASCII support, but is more compact and affords higher data density as CODE 39, being with about 25% shorter than CODE 39 for the same coded data. Enabling additional security, the barcode is very suitable for logistic, manufacturing or retail The symbology derives its name; every character is constructed from 9 modules arranged in 3 bars with the corresponding adjacent spaces. The new barcodes reader supports camera based scanning on IoS and Android through smartphones, tablets and wearable devices [11] CODE93

(continued)

Table 15.2 (continued)

New class name	Former class	Obs.
	CODE 128	Is a more compact and high data density code used in ordering, distribution, transportation and logistics industries. Supporting the entire number of ASCII characters (128), capable to store a very high diversified data like supply chain label units with serial shipping containers code (SSCC)
GS1 DataBar	Reduced space symbology	Was introduced in 2001, in omnidirectional, truncated, stacked, expanded or combinations o this versions, mandatary for the retail coupons in US, but generally used for retail (retail outlets, perishable products, etc. or small products in healthcare industry [12]
MSI Plessey	Modified Plessey	For the inventory management in retail environments

more data per unit area than the linear barcodes, Fig. 15.1 [9]. The 2D barcodes includes new large spread in applications representatives like the QR or the PDF417 codes [13].

From industry, retail, logistic, healthcare or government, practically any social, politic or economic structure can be the beneficiary of the data matrix codes. This codes can be labelled on any item large or small goods, buildings, documents, wearable items with a tiny (actual flexible in size) footprint [14].

Except the higher quantity of data on the same are the 2D offers the following advantages:

Fig. 15.1 2D barcode versus 1D barcode

- Easier to read: no special scanners, a smartphone or a digital camera will fit, in some cases maybe an free App is required to install;
- High Speed in reading;
- built-in error checking systems;
- Redundancy in data representation, the same information is hold in different ways, enabling the reading of the partial damaged code;
- Easy to transmit, supporting the SMS text message communication;
- Higher security by encrypting facilities

The first organization that uses the data matrix was NASA in 1980. The major differences between the data matrix and the QR code is that in data matrix codes may "only" afford 2335 characters for coding data but the QR codes may store 4296 characters; and QR codes are using more recognitions area when data matrix is using only the perimeter. But actually what data are represented on a QR code?

1. Alignment pattern
   Ensures that the code can be read at high speed even when the coded surface is viewed at an angle, or is printed on a curved or distorted surface;
2. Data cells
   Each nonstandard feature individual black or white square contains some data in the code;
3. Finder patterns
   Large black and white squares in three of the corners to confirm that this is a QR code and not a different one;
4. Quiet zone
   An empty white border that makes possible to isolate the code from among other printed information.
5. Timing pattern: This runs horizontally and vertically between the three finder patterns and consists of alternate black and white squares. The timing pattern makes it easy to identify the individual data cells within a QR code and is especially useful when the code is damaged or distorted;
6. Standard features
   The timing, alignment, etc.
7. Version information.

There are different versions of standard QR code. The version identifies which one of the standard QR versions is being used. The version information is positioned near two of the finder patterns.

In Table 15.3 are briefly summarized the characteristics of some of most popular 2D barcodes.

If one is looking for a proper selection of barcode for a certain application, the decision is based on a list of considered factors:

- Firstly depends to who is dedicated the coding: to a new or to a growing business, to an organization or just for personal use;
- The dimension or the time horizon for the application activity is next question and finally;

Table 15.3 Popular 2D codes characteristics

Character set	Length	Check digit	Size, module width X, print ratio
AZTEC code			
	5 bit/character, encode up to 3832 numeric digits 3067 alphabetic characters; or 1914 bytes data codes 0–127 are according to ANSI X3.4 (ASCII), and 128–255 are according to ISO 8859-1	Check words are appended to fill out the available space	Quiet Zone left/right/top/bottom: 0X
Applications	The Aztec codes are frequently used in people's transportations, for tickets and airline boarding passes, where poor printing or poor images on the smartphones may causes difficulties or problems by using some other codes		
Notes	Aztec Code can encode small or large amounts of data with user-selected percentages of error correction. That means that even with a bad resolution the barcodes still be decoded The symbol size adjusts automatically depending on the amount of input data, not requiring a "quiet zone" like some other coding systems, (the surrounding blank)		
Data Matrix code			
ASCII	1556 ASCII (8 Bit), 2335 alphanumerical or 3116 numeric 2300 characters	Error correction	Dynamically variable physical size from .001 inch square to 14.0 inches square
Applications	Ideal for marking very small containers in manufacturing or pharmaceutical industry, for unit dose and product marking		
Notes	The Data Matrix is a two-dimensional matrix code, with visual representation of binary code (2D code optical solution), used for encoding up to 2300 data characters (2300), originally developed for SSP (Space Shuttle Program), where millions of parts must be tracked for years		

(continued)

Table 15.3 (continued)

Character set	Length	Check digit	Size, module width X, print ratio
	The data matrix consist of black and white modules that can be created (the number of rows and columns), the more information the company wants to save in a little area		
QR code			Quick Response system
ASCII (0-127) + Extended ASCII	3067 alphanumeric 3832 numeric 1914 Bytes 4200 characters	Error correction	4 different symbol sizes (M1-M4) or Maximum symbol size: Version 25 (117 × 117), Minimum module width: 0.4 mm [12]
Applications	The technology was used in the production logistic. Now companies used it to share data with the customers (i.e. QR tag placed on the company homepage) that means in many situations the technological solution used for advertising and then for fast detailed tracking process		
	QR was developed for Japanese automotive industry (1994). Fast readability and large storage capacity made it popular also for healthcare applications: patient identification, patient-safety applications, blood products and specimens, mixtures, wristbands and labels for medication units		
	QR code is used also in augmented reality (AR) for items position or characteristics determination, delivering AR client experience in any type of application, simulating the functionality of different items in different environments from luxury cars and airplanes, Lego toys or IKEA items		
Notes	QR is a public domain free to use barcode, the follower of the Data Matrix. As optical machine readable code, consists of black squares arranged in a square grid on a white background. Can be read by a camera imaging device, storing and pointing data about the labeled item to an identifier, locator or tracker, then transferred to a website or that may include item identification, product tracking and time tracking together with the document management		
	Nowadays is the largely applicable barcode as an easy to use application contains fast readability large amount of recorded data, flexible in size, with a high fault tolerance, and beneficiary of a large variety of opens source or payed QR code software generator. Most applications are in advertising, business cards, healthcare, marketing and tracking.		

(continued)

Table 15.3 (continued)

Character set	Length	Check digit	Size, module width X, print ratio
	As difference form Data matrix, the QR code offers not 2300 but 4200 alphanumeric characters storage, including automatic error correction. Up to 30% of destroyed data code can be automatic repaired by the code himself 		
	There are two possibilities to read the QR-Code: by scanning and by using a dedicated App on the Smartphone. The scanning is similar with the GTIN 13 using similar type of mobile or fix scanners like the popular ones used in POS  Mobile scanner  Fix Scanner		

(continued)

Table 15.3 (continued)

Character set	Length	Check digit	Size, module width X, print ratio
	QR can be read directly with a snapshot on Smartphone App  Try it!		
	As an example of quick application within the University, could be an "*Information Tag*" on the office door similar with an automatic e-mail feedback.		
PDF 417			
CP437	1850 alphanumeric characters, 2710 digits or 1108 bytes	Built-in error correction method based on Reed-Solomon algorithms	Takes about four times the area of a DataMatrix or QR Code
Applications	Beneficiaries like aerospace, logistics and robotics that requiring storing over 1.1 Kb of data (graphical recognition)		
Notes	A machine readable code, much powerful as any other 2D barcode, and is a public domain free to use		

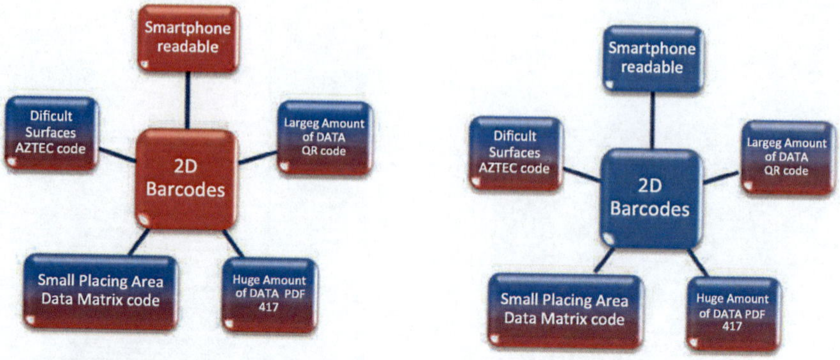

Fig. 15.2 Suggestion of decisional block diagram in a barcode selection

- It is advisable to use a decisional block diagram in the digital code selection like the proposed one in Fig. 15.2, guided after a set of questions:
    1. Destination market (EU, USA)
    2. Type of Data (Alphanumeric, Numerical)
    3. Amount of data (required no of digits)
    4. Application support
    5. Imposed Reading devices.

Specific graphical modules are created for each form the 5 decisional questions. First will be selected the destination area, for an easy continuity EU was colored in blue and the USA influence are in maroon.

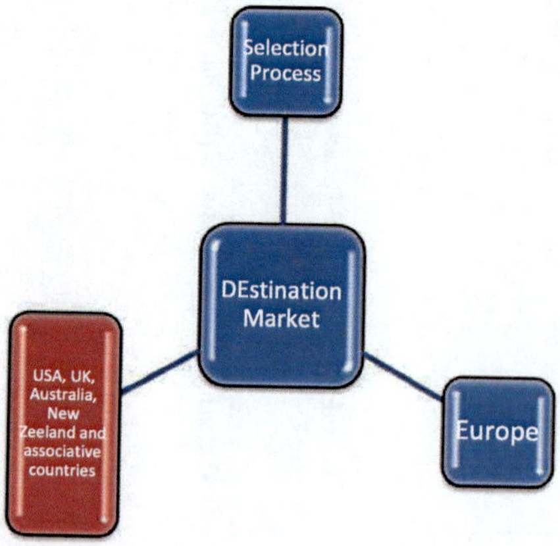

Once selected the area we may select IF there are numerical data or alphanumerical data and one of the adequate modules is selected and within the module adequate branch will guide to a suggested solution.

A double colored module indicates a valid solution for both geo-political areas.

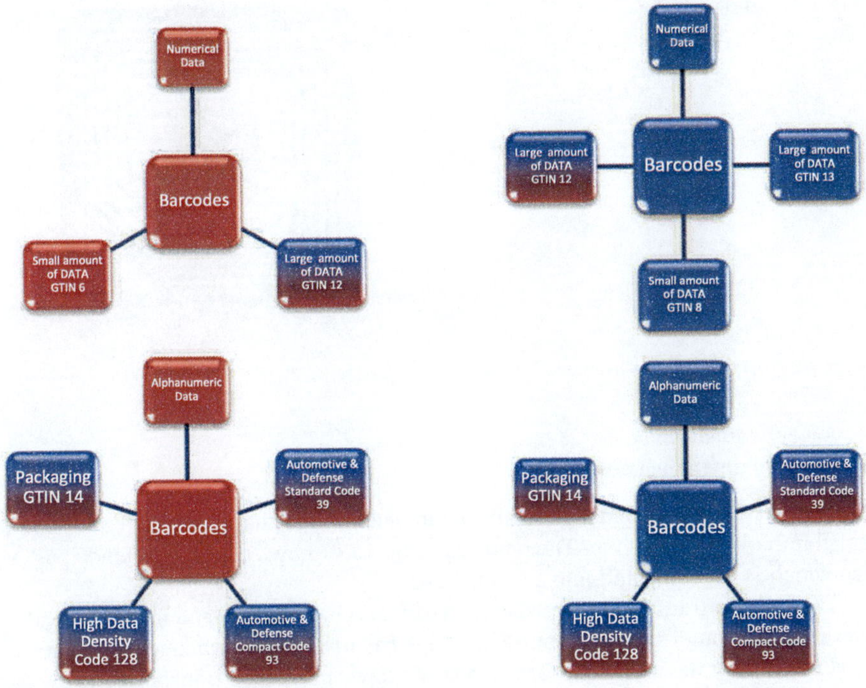

If special reader, in this case the smartphone is required the following modules will guide to one of the desired solutions.

15.3 RFID Tags

This type of tracking system works with radio waves. It is a non-optical tracking technology based on the radio-frequency-identification (RFID). The data can be received and transmitted wireless to any business activity.

An example of a RFID device is depicted in Fig. 15.3. The RFID solution combines a lot of benefits like: not necessary optical connection to the product (direct view), minimal lightning facilities and easier possibilities to integrate a lot of sensors into the holistic supply chain tracking and management.

Basically there are two different types of RFID transponder: the passive one and the active one. The passive transponder needs no external power supply to transmit the data, and all the benefits resulting from here:

- small dimension;
- reduced weight;
- continuous functioning without the danger to lose the transmission power;
- no need to change batteries;

Fig. 15.3 RFID tags for tracking

- lower acquisition costs;
- lower maintenance costs.

The passive transponder is totally maintenance-free, has a high lifetime and is smaller than the active one. The picture in Fig. 15.4 shows a high frequency passive tag which is implemented into a screw head.

The passive transponder consists in a silicon chip and the antenna. The system works only through inductive coupling, a high frequency electromagnetic field, which is generated by the aerial coil of the reading device that also transports the needed energy to the RFID transponder.

If the transponder comes close to the magnetic field, the antenna can absorb the energy to move the microchip throughout the communication process. This now activated chip can decode the sent commands of the reading device. In the next step the tag send is own serial number and coded information by using field weakening.

Fig. 15.4 A passive RFID Tag

Table. 15.4 Communication characteristics connected with the frequency

Frequency	Maximum range (m)	Type of tag
Low frequency	0–5	Passive
High frequency	0–5	Passive
Ultra-high frequency	3–6	Passive
Microwaves	~10	Active

The RFID tag does not create a new own electromagnetic field to send information, but it takes influence of the sending field. The transmission range that is able to reach depends on the inductive solution and may be from the contact situation (0 m) up to 5meters.

In Table 15.4 are detailed some of the characteristics that depend on the electromagnetic field frequency.

For using the ultra-high frequency it is necessary to change the coupling to an electromagnetic dipole field. A possible disadvantage of this technology is the reduction of the real time work capacity. The transmission needs to use the combination of many components for the technology implementation, and each component introducing a specific delay. In the end the total delay is equivalent with the sum of all cumulated delays. The minimal required components for a RFID communication are:

1. Tag;
2. Data carrier;
3. Reader;
4. Antenna;
5. Controller;
6. Sensor;
7. Indicator
8. Actuator;
9. Computer
10. Specialized software.

Only having all this components together may handle the information. Generally the RFDI tags are more expensive then the optical ones, depending on the quality and the characteristics. The usual price falls in a range between 50 cents and up to 100€ per Tag. In theoretical similar conditions a linear barcode or a QR-Code currently requires at the most 1 cent per Tag.

The RFDI Tag prize generates also more benefits:

- the RFID tags lifetime is longer;
- the RFID tags can be covered with protective material;
- the RFID tags can be placed in different locations, on or into the parts.

These mentioned advantages cannot be offered as upgrades for the optical tags, furthermore the optical tags requires a direct and clear view to the data, an adequate lightning level and being less protected can be easily destroyed within careless handling, process, or accidently masked by dirt.

15.4 The RFID Tracking in Supply-Chain-Management

Any tracking process starts with the goods reception and ends with their delivery to the customer. If in the logistic chain as the parts, the trucks or other transportation devices (electric tug, AGV, etc.) are equipped with a RFID transponder, they can be automatic register at the factory or the workshop gate and directly connected with the required input that must be confirmed during the unload activity.

This process is not only a registration of the truck and the delivered products in the company; it can be used to transfer the delivery ticket data in real time. The company can match the ordering data with the transferred ticket at the point of the reception of goods, the RFID system sending the acknowledgement data of receipt simultaneously to the internal database and to the delivering company.

The next step in the tracking process is to verify the goods quantity and quality to achieve the "merchandise control". The entire checking process is a time consuming one and for the receiving company also an expensive one, adding into consideration also the possible faulty flow of goods.

A well done tracking system implemented in the logistic chain is shortening the time to market, increase the importance of logistics, creates a constant components flow and finally is leading to an economical efficient manufacturing. From the stock reception point to the components control and the supplier delivery confirmation the entire investment will be justified.

The wireless communication capability offers swift and complete data about the transport (including i.e. temperature, pressure, acceleration, air moisture, etc., if is required), simply by combining the RFID system with the adequate type of sensors. The best examples came from the food industry that uses different specialized sensors and parameters detectors for checking the entire set of data continuity during the supply and logistic chain.

Similar set of sensors can be useful considered within the manufacturing process. The aerospace, automotive, health and robotic sectors are the most interested beneficiaries. The transport conditions and the contamination possibilities may produce important damages of the entire process or processing system. The data may be used locally, for immediate actions and by buffering centralized for the management of the entire cycle: production planning, reception, optimization and strategic management.

RFID transponder ensures the components identification, components location and the status of the stored components, making de the connection with the delivery schedule.

The storage management includes the duration of stay for a component, the storage costs for every storage location being.

Scenario: if a component in stock, must be physically located in production like:

- plant 0003 storage location 79,000

and the stocked component is located in the warehouse as:

- storage location 0101

The goods are moved physically into production, and posting that a change takes place from the warehouse to be managed in plant 003 storage location 92,092.

In this case the RFID technology is used like an electronically Kanban-system. The demand of components at different processing points automatically starts the orders for components by checking the internal reserves and the warehouse reserve and the suppliers' capabilities profile, everything being based on the customer order.

Following this procedure a transport request is sent to the logistic. Therewith it is possible to implement a pull manufacturing process which controls itself. In this case is no need to control every process step by step, the downstream process ordered the needed items and launch the production of new items. The company benefit is the need for a low control task. The process is called the pacemaker process.

Beside the parts flow, other fields of RFID application are the production control and the manufacturing system maintenance. With the correct sensors network it is possible to check the manufacturing data and the quality control in real time. That enables to start counter-measures if the set point value of the product is broken. Therewith the company can ensure a stable product quality and can also observe the machine-tools status. Every part has his own identification number; so that all sets of data (color, next assembly, part versions, etc.) can be send in a direct way to the processing place. In this way every part is collected together to the final product, the RFID tags make possible to create a holistic documentation of every process in a tree structure.

The result: the company may identify bottlenecks and optimize the production process, in faulty parts cases being able to remediate the failure with minimal costs.

15.5 Trends in 3D Scanning

The 3D scanning is applied for the 3D barcodes materialized by engrave, or apply 3 dimensional bars. The reading is not by sensing the difference on reflection between black and white but by the difference in height from a reference level, as depicted in Fig. 15.5.

The trend in 3D scanning is connected with the field of application and its specificities. As it becomes an affordable technology for many industrial and commercial applications the 3D scanning has its own development lifecycle evolving in coding algorithms and data cryptography, design and attributes of the scanning devices, development of reading apps dedicated to smartphones and nevertheless extending the type of applications and their integration in larger data driven applications like the PLM.

Fig. 15.5 3D barcodes

The 1 or 2D barcodes scanning are not only needed to track the purchase but also to store product information, process information and new attributes added to the product within a process, combining the quantitative data with the qualitative data.

The qualitative jump to the 3D barcodes derives from the part storing, processing and functionality. The lightning, moisture, temperature, chemical corrosion may lead to a different type and material support for the barcode landing to the 3D barcodes (engraving solutions), as reliable and long lasting solution but with the need to read also the depth of the barcode, the differences in the height of each line. For this reason in the 3D barcodes reading laser scanners are used, by measuring the time of laser beam travel form the source and reflected back to the sensor the distance between the scanner and the posted code is determined. The small wavelength of the laser radiation enables the "sense" of small differences in high of the coded lines the scanning device being named DPM (Direct Part Marking).

As the DPM is measuring the difference in height is no need to consider any colour or to print in black and white, the colour being meaningless.

The 3D barcode enter in the area of optic sensing but the small engraving dimension, an electron beam may engrave microscopic dimension coding that are not visible with the necked eye especially that they are colorless.

Like the other optic coding using optic sensors to read the 3D coding and scanning probably will follow the trend of 2D codes to be used for vehicle and products tracking in logistics, manufacturing or in economic and selling area for customization and marketing. The advantage is the incredible small dimension, reading accuracy, capacity to be integrated in data driven processes and connected with other type of optical or radio wave sensors.

The 3D barcodes incorporate ultrahigh encoding capacities at microscopic scale in range of 2–6 μm. As an engraving not printing process, the 3D "writing" may be integrated in the parts manufacturing process used as an inventory system but also connected with a manufacturing process where each stage can be verified and marked on the part by an engraving extension, or in the assembly procedures to connect the parts that must be assembled together with the assembly process, tools and staff.

Clearer evidences are on the part manufacturing average time, delays in processing or nonconformities, in the end all influencing the product pricing, the quality control, planning, logistic and delivery process, Fig. 15.6 (Free 200 + Images, 2020).

Going back to the trends issues in the 3D scanning, we may consider the cyclical evolutions pattern. In this context as 3D scanning start from the 1D and 2D development and applications, also in the trend is following the two predecessors.

Fig. 15.6 Clever use of the 3D barcodes, from assembly, through delivery, to recycling

So the trend is for reaching the UII (Unique Item Identifier). The idea, the advantages and the solutions for UII is coming from the DoD (USA, Department of Defense) [15]. That may be a keystone in the business transformation process in every company, using also the 3D scanning.

The UII is destined to identify an item attributes with a set of data in a globally and unambiguous way, data that are marked on the product. The dedicated symbology is the one use to mark the 2D symbols, is not the QR but a specifically Data Matrix with the Error code 200.

The Data Matrix technology may encode a large amount of data on a small area, having the capability to increase the number of modules (rows and columns) ones that the amount of data to be encoded increase. Usually is a square shape and can store up to 2335 alphanumeric characters, each code is unique the symbol sizes vary from 9×9 to 49×49 in the version ECC000-140 to the 10×10 to 144×144 in the new version ECC 200. Some of the strong elements of ECC 200 version are:

- the shape may be square or rectangular (8×18 to 16×48);
- The numbers of rows and columns are always in even number;
- Error rate of less than 1/10,000,000 characters scanned (idautomation, 2019);
- All symbols using the ECC 200 error correction can be recognized by the upper-right corner module being the same as the background color (binary 0);
- The routine reconstruction capability, assuming the matrix can still be accurately located when the symbol has suffered up to 30% damage and the entire code can be rebuild;
- Capabilities that differentiate ECC 200 symbols from the earlier standards include:
  - Rectangular symbols
  - Inverse reading symbols (light images on a dark background)
  - Structured append (linking of up to 16 symbols to encode larger amounts of data)
  - Specification of the character set ECI (Extended Channel Interpretation), containing embedded references for the data encoding, used automatically by the barcode reader.

By applying the Data Matrix the tag is consisting at least the manufacturer ID, the part number and a unique serial number, identifying sensitive components as the Military standard 130 or the ATA (Air Transport Association) impose, as generally considered "the best choice" for assets identification and tracking within data-driven applications, when a lot of data needed to be encoded.

As expected from the previous description the QR Code and its standard enabling infrastructure does not support IUID (Item Unique Identification) compliance and therefore cannot currently be used as the ECC 200 Data Matrix. The ISO standards allow the ECC 200 encoding structure to be placed in a QR Code, but then each mobile device or scanner have to be customized and configured, a totally not typical approach for QR Code and smartphones, as Nicole Pontius from camcode, specialist in Total asset lifecycle tracking affirmed already in December 2012.

The difference in data coding in the QR code and Data Matrix code are presented in Fig. 15.7, where in Fig. 15.7a is represented what RQ code may store, in Fig. 15.7b how is represented in Data Matrix code what may be stored in QR code and in Fig. 15.7c the entire amount of data stored in Data Matrix code. For the same amount of data the Data Matrix code requires less space, on the same area as a barcode may be registered data for 100 products. In the same time the Data Matrix code is covering extended specifications and attributes available as standard for IUID only with Data Matrix code.

From the above description, within the trend analysis, besides the UII (Unique Item Identifier) arise the issue of the IUID (Item Unique Identification). The IUID enables the lifecycle traceability distinguish one item from another, preserving the data quality and integrity, facilitating a fast and automatic capture of involved data across multi-operational at globally interoperable networks.

The IUID is actually a system establishing the UII by assigning a number of characters string applicable on a discrete specific item (product) as a machine–readable inscription.

The number, the characters string or both is selected based on the UID (Unique IDentification) another system that is establishing globally unique and unambiguous identifiers serving to distinguish a discrete entity or, attention, relationship from other like and unlike entities or relationships. Initially was defined, designed and dedicated to the DoD, but in time was extended to be used as an enterprise or commercial identifier issued by combining the a company EID (Enterprise IDentifier) with the agency as IAC (Issuing Agency Code). Where the EID is a code uniquely assigned to an enterprise by a registered issuing agency and the AIC is the code representing the

Fig. 15.7 Clever use of the 3D barcodes, from assembly, through delivery, to recycling

a) b) c)

agency that issued the enterprise identifier. The IAC can be derive from the data qualifier for the enterprise identifier and does not need to be marked on the item being assigned by the Registration Authority for ISO/IEC 15,459-2, Registration Procedures. The current Registration Authority of ISO/IEC 15,459-2 is NEN–Nederlands Normalisatie-instituut.

The UID generates the UI (Unique Identifier), the number, the characters string, or just a sequence of bits that will be associated with an item or its discrete attributes as the unique distinguish entities for any others.

Going further other two issues appears: the RPUID (Real Property Unique Identification) and the APUID (Acquisition Program Unique Identification).

In the classical restricted form the RPUID is related with land, buildings and structures as facilities. In the extended form the land may be connected with plants as sites, and the plants with the assets, may having the entire inventory in single identification. This is facilitating a total assets accountability and visibility and with a real time production of reliable and timely information for decision-making and reporting.

The APUID is related with a contractual approved need extended to the suppliers, supply chain and logistic system, with the departments' acquisition portfolio and the considered project. Within the data lifecycle there are different relationships. The data related to the actual APUID with the preceded or retired APUID(s), or even concurrent APUID(s).

Before considering the contribution of the IUID within and for the PLM, it must be highlighted the components within the IUID lifecycle.

1. The UIDI lifecycle starts by aiming the business systems by provision of accurate, complete and reliable data for the company management, engineering and accountability purposes;
2. The IUID "mandatory role" for tangible Items: procurement of new Equipment, re-procurement of consumables, equipment, tools, and spare parts;
3. The IUID "mandatory role" for personnel allocation and contracting;
4. Definition and the extension of the IUID management;
    a. Implementation and usage milestones;
    b. procedures and processes of legacy items;
    c. procedures and processes in inventory;
    d. procedures and processes in operational use;
    e. procedures and processes for private as well as government furnished property.
5. Key points in the IUID management:
    a. Identification of IUID International standards;
    b. IUID Quality Assurance Standards;
    c. Planning guidelines for IUID implementation;
    d. Guidance for preparation of Program Plans/company plants;
    e. Establishment of Depot capabilities;
    f. Rules for data capture of legacy items;
    g. Rules and policy on IUID data for embedded items.

With the increase of the digitalization 3 important elements must be reconsidered in the IUID lifecycle:

- the transition from the paper-based form in the digital based form for the other two elements;
- the PICC (Property In the Custody of Contractors) and
- the PIPC (Property In the Possession of Contractors).

From these concerns the next points in the IUID lifecycle are dealing with:

6. Key points related with the property
   a. Logistics policy to support IUID
   b. Milestones for PIPC compliance with IUID requirements;
   c. Principles definition for the electronic property management;
   d. Clarification procedures associated with Part Number changes;
   e. Acquisition milestone reviews address to the IUID Implementation;
7. The Lifecycle ends with the part end of life.

As observed the IUID have a life for itself but implying the registration and the visibility of property, personnel and personnel roles, together with processes, means that is playing a big role in the PLM in the registration and evaluation of:

- the property;
- the connection between the property items with plants inventory, staff and staff roles;
- the management of all historical data;
- the assets management;
- the assets status in connection with staff and staff allocated equipment;
- the plants, equipment, operating materials and supplies;
- the inter plants relationships;
- the contractual relationships with suppliers;
- the contractual relationships with customers;
- the departments initiatives (Projects and Innovation);
- regional legislative and regulatory frame;
- the new business implementation;
- the actual business;
- the value adding activities within the value chain;
- the non-value adding activities.

References

1. Terzi, S., Panetto, H., Morel, G., Garetti, M.: A holonic metamodel for product lifecycle management. Int. J. Product Lifecycle Manage. Intersci. **2**(3), 253–289 (2007)

2. Wuest, T.: Identifying Product and Process State Drivers in Manufacturing Systems Using Supervised Machine Learning. Springer, Cham (Springer theses). Online verfügbar unter https://search.ebscohost.com/login.aspx?direct=true&scope=site&db=nlebk&AN=984670 (2015)
3. Kapoor, B.M.: Life management emerging paradigm. Global India Publications, New Delhi (2009)
4. Niemann, J., Westkamper, E., Tichkiewitch, S.: Design of Sustainable Product Life Cycles. Springer, Berlin. ISBN: 978-3-540-79081-5 (2009)
5. Niemann, J.: Eine Methodik zum dynamischen Lifecycle Controlling von Produktionssystemen. Dissertation Uni Stuttgart, Germany (2007)
6. Niemann, J., Stierle, T., Westkämper, E.: Kooperative Fertigungsstrukturen in Umfeld des Werkzeugmaschinenbaus: Ergenbnise einer empirischen Studie. Wt Werksttstechnik **94**(10), 537–543 (2004)
7. Niemann, J.: Lifecycle Mangement, Neue Organisationsformen im Unternehmen – Ein Handbuch für das mderne Management, 2 edn. Springer, Berlin (2003)
8. Barcodes: https://www.scandit.com/2015/01/27/types-barcodes-choosing-right-barcode/
9. QR code: https://www.qrcodeshowto.com/what-is-a-qr-code/2d-barcode-versus-1d-barcode-with-pictures/ (2020)
10. QR: Wikipedia, https://en.wikipedia.org/wiki/QR_code (2020)
11. SCANDIT: https://www.scandit.com/blog/exploring-code-93-barcode-type/ (2019)
12. SCANDIT: https://www.scandit.com/products/barcode-scanning/symbologies/gs1-databar/ (2020)
13. Woodford, C: https://www.explainthatstuff.com/how-data-matrix-codes-work.html (2019)
14. TEC-IT: https://www.tec-it.com/en/support/knowbase/barcode-overview/2d-barcodes/Default.aspx (2020)
15. DoD Regulation (2010)

Chapter 16
Boosting Performance

Creativity is what keeps organizations ahead in the marketplace.
Tim Brown

16.1 Introduction

It is no news that nowadays industrial context is considering of strong competitiveness, not only within Europe but also Europe in the international market. The globalization is coming with greater challenges than ever before, complex products and systems are developed with shorter lead times and more cost effectiveness. Therefore, the reduction of costs and the time to market has become a strategic task for any manufacturing company.

Nowadays a company may have subsidiaries on every continent or evolve towards large-scale partnerships with thousands of people that are working together and in parallel, on complex products and systems in a distributed synchronized manner regardless from what discipline are coming or what software platforms and modelling languages are using. At this level not only engineers, programmers and scientist are involved but also economists, lawyer, managers, or staff representatives. In that context the "engineering data" is to be processed in a most consistent way, available to be used by all the partners in the entire pallet of different activities 24/7.

The digital integration aims to offer all the tools and methodological approaches to support communication channels within a collaborative digital platform, as much as possible neutral for each industry. From the product development processes, to the implementation, exploitation, service and maintenance and nevertheless recycling phase integration, the digital system must provide a robust design optimization and collaborative engineering with enhanced simulation capabilities for the entire product lifecycle. In short: PLM.

Digitally the products are described by ***features***. The ***features*** of your product are totally defined by the ***technological culture*** you belong. The ***benefits*** of your company are defined by the short term advantages and are the result of your product

J. Niemann and A. Pisla, *Life-Cycle Management of Machines and Mechanisms*, Mechanisms and Machine Science 90,
https://doi.org/10.1007/978-3-030-56449-0_16

features. The solving of your challenges and meat your goals by stabilizing your position on the market and your targeted clients are the ***outcomes*** that results from the obtained ***benefits***.

The concept and marketing phases are identifying the actually criteria and the product features that must cope together in order to survive in the market.

The performance boosting is coming in many ways, from the design phase to the predictive maintenance and recycling solutions. After the product (process, project, etc.) concept is defined, the importance of the design phase is crucial, every mistake in the design phase is reflected not only on the time and resources costs of the phase but in all the other phases that follows and finally on the product success.

The design phase deals with the geometrical features, combined with the material capabilities, from where result the product performances in: mechanical strength, thermal resistance, available colours, surface smoothness or engineering possibilities for machining and volumes malleability. Beside the "deductible" presented performances, in the design phase are also defined the derivate characteristics like parts critical mass, the functionality and manufacturing safety, the resources flows, the conformity with standards but never the less the manufacturing, functionality, maintenance, repairing and recycling costs.

The design activity leaves no room for speculations, is a very precise and accurate one, and therefore needs to be effective. And here is the role of PLM, the market putting on companies a high pressure in creating new products, bring them fast to the market, to be better than competition, optimize the development, manufacturing and delivery operations, identify customers and new niches, optimize the logistics and the supply chain, and always reduce costs of any kind.

The faster product development starts with the product computer aided design (CAD), a common tool for performant industry in our days, which make engineers more productive in achieving a larger variety of products, in a modular or/and multi versions of the same product, more complex but reliable in the same time, ready to be manufactured in a shorter time and with fewer errors.

CAD is impossible without an adequate IT platform, which means that all the data are capture in a central database, enables concurrent engineering (engineers from everywhere in the world may work on the same product in the same time at any hour—24/7). The errors generators detection, the engineering changes management together with the costly updating are tremendous reduced and the conflicts between different subassemblies design are solved faster or completely eliminated semiautomatic.

The price that usually was determined after the first well tested prototype, is estimated now. CAD can be considered as a structured and manageable instrument in achieving a proper design, accepted by all the departments within a company, defined also as a "way of the communication system" needed in the decisional process.

On the other end the spare parts stocks costs, the accidental interventions cost, the programmed or not maintenance activities costs, rise many problems and resources allocation in any type of industry. Therefore the performance boosting considers the continuous monitoring of the mechanical components condition, in hundreds of points including bearings, compressors, gearboxes, linear and rotational motors, levers and wheels, etc. creating the basis of the early detection faults and predictive maintenance, a relative new trend with multiple opportunities and instruments in various applications.

Using the signals processing technology, each physical condition is accurate and virtual continuously recorded offering in detail the knowledge about the design and the functionality of the systems.

But the customers of the manufacturer plant also benefit from the CM (Condition Monitoring) and the ongoing condition diagnosis: higher availability and increased production output bring great advantages, since unplanned downtimes are avoided as best as possible. You can also see at which points faults can occur. At the same time, they have a basis for deciding whether a bearing or gearbox replacement can wait until the next planned shutdown. Since many spare parts are delivered just-in-time, plant operators can use this information to implement relatively accurate order management.

An efficient predictive maintenance offers reliability that means advantages, like a better position in the client's preferences. In the same time offers an almost zero spare parts stock, a non-disruption normal operation with all downtime effects and cost. The classical routine-based predictive maintenance, gain new technologies like IIoT (Industrial Internet of Things), Machine learning for signal processing, decision making algorithms, co-robots and digital-twins, satellite communication for weather control and prediction, communication, environmental conformity, etc.

Speaking about IIoT we may consider that it already completely changed the game and implicitly the innovation competition landscape, especially for manufacturers but also for the services and IT providers. In every way the business process changed the relations between parts, humans, resources and decision makers, becoming blunt with every day and every more neutral decision and very controlled responsibility. That is imposing a new way of thinking firstly for designers, manufacturers, and maintenance specialists but also for salesman, representative and assets managers and nevertheless the business drivers.

One of the actual most performant solution named by the American market research company Forrester as "leader", is the Siemens cloud-based, open IoT operating system MindSphere, that may connect and integrate an entire manufacturing process.

Such a solution is possible to apply by starting from a product and/or process digital twin. Considering the digital twin in a close loop, via integrated IIoT, it enhances operational efficiencies starting from the customer experiences. We cannot escape form our technological culture but we may change it by adding or bring new innovative technologies.

Every company is very fond on its technology intimacy. But what can add the IT technology in business by AI (artificial intelligence), AR&VR (Augmented and Virtual Reality), Robotics (build, bayed or rent) for dull or repetitive, dangerous or complex tasks as self-standing or in collaboration with humans (the cobots), blockchain and digital outsourcing, where an already common practice is the RPA (Robotics Process Automation)?

1. The business driving, from the supply chain to every aspect;

2. The digital insights facilitate deep analysis, process transparency, entire connections from machines and processing systems to complete factories all over the world 24/7.

The digital age has a particularity that every industrial revolution brings: even the company is not investing in the new devices the technological wave will hit it. The new digital business bring IQ tests to provide scores that gauge employees cognitive abilities and helps in the training process; and RQ tests, scoring the company and the individuals ability to work with robots.

In any type of activity or application exits a critical infrastructure that must be monitored and maintained. A wider view over these problems shows the growing importance of the predictive maintenance, enormous potential and the huge risk in ignoring it.

The altogether influences, change the maintenance costs, the productions stability, the products cost, the human safety, the human resources, the human training process, services externalization and human factor stability.

Where the CNC machine-tools, robots, manufacturing lines or system downtimes lead to enormous downtime costs, manufacturers, managers and users have to think about condition monitoring (CM) a system that justify the investment because is fitting very well in previous automation and control technologies.

Within the company, the monitoring can be considered as a common tool for managerial and predictive control activities as the predictive maintenance can be assimilated as a strategic business function within the company managerial activity. The early fault detection, diagnosis and remedy is playing an important role with a better resources allocation possibilities.

These aspects are not reflecting only within the manufacturing sector but also on logistics, mobile components and infrastructure, together with intervention scenarios; company energy supply equipment inspections, energy entries and distribution.

Why from all performant system we select MindSphere, firstly due to the features performances and secondly because is an open internet of things cloud-based operating system that real ensure a worldwide connections of a company triad: systems—machines—management.

16.2 The Contribution of Industry and for Industry

All big players from industry, banking, healthcare or energy are BIG because they always look for efficiency and effectiveness. In the digital era is the same but this is not anymore only about skills connected with costs and time cuts, niche identifications and innovation. The big data analyse, the AI applications, Robotisation, and generally everything that is dealing with data, communication and decision is the "privilege" of the ones that still wants to be efficient and effective for being BIG. So let's go BIG (Build-Innovate-Goad)!

1. **Build**. Without the critical mass of resources and without having the right instruments, we as Homo Faber we are stick to the ground, we cannot do much.
2. **Innovate**. But having the tools, that cannot just stay there. We need a way to use them, a technology, so we may (must) invent technologies for our tools and tools for our wishes, and new technologies for our new tools and ...innovation will continue.
3. **Goad**. The quitting, the self-satisfaction, the comfort zone, NEVER leads to performance or success. Every person that you "envy" for its success was and it still is a warrior that suffers and is still suffering, even if is in a different way. And these are the **leaders**, the people that inspire. In Latin "to inspire" (Inspíra) means to blow into, to inflame. Therefore the Olympic game inspire people not only to practice sports but also the patriotism. In leadership means to breathe life, to encourage, to excite, to give the example like blowing air over a low flame make it grow. And very often that must be followed by "pushing" resources and goad the people.

If things could be so simple not so many discussions, debates, research and investments would be necessary. Why? Because seems that this digital story is in a way still fragile. In this sense the industry is going for, looking for and investing in, trying to implement the best available solutions. But, they still need help. So is needed to defend the digital village.

The cybernetic attacks and threats that are much visible in bank attack, personal data, political or celebrities' data leeks, or the fake news, are also increasingly present in industrial and commercial environment, and against them are built firewalls, upgraded defense algorithms and defense strategies. In this struggle are in front line companies like Siemens (Germany) and Bitdefender (Romania) that offers adapted solutions to the changing threats.

In this battle a lot of people and smart ideas are involved. Different layers of hardware and software protection are developed and used. From biometric recognition, virus detection, passwords protection, firewalls, private networks, AI in depth malware search algorithms, back up servers, cloud secured applications, the conformity with the IEC 62443 international standard for automated systems IT security levels, cryptography and blockchain that uses cryptography and a time line during transactions, all together are IT instruments for the Industry.

To be functional the industry must be aware and conscious about the need to introduce a "Password policy" (stronger than the complex "123456"), ensure the continuous security updates, promote the monitoring of the cyber security system, and invest in the AI adaptive or self-learning security systems research to face the escalade of cyber security war. From the beginning of the industrial existence also the blackmail, the industrial espionage or sabotage also exists and will exists. Digital factories are using an increasing number of servers, computers, sensors, increasingly networked (IIoT, WANS and LANS).

In January 2020, both the UK and the USA California state, introduced new laws that will significantly tighten security requirements for IoT devices and networks, going below the requirements that IoT devices be provided with unique passwords.

This example will be followed by other countries in order to introduce new concepts in connection with 5G, machine learning, trusted environments, and "security by design" with bigger role as the spending on IoT security projected reach around USD 830 billion in 2019,

16.3 Responsibility—A Must

Nowadays a lot of struggle is in finding the balance between the profession, family and free time.

What has the digitalization to do with that? Almost everything.

The gaming industry with all the virtualized and digital systems offers a multi billions market. The family ambient with "smart home" with all kind of gadgets in monitoring children, training, lecturing and leisure devices, automation of different activities, service and maintenance assistance, computers for "work at or from home", internet and wireless applications, security and heating systems, fire protection, home remote video surveillance, all are digital.

What about the professional life? And here came probably the most of the quantifiable responsibility. The key function is not about productivity, reliability, efficiency, or the profit but about the sustainability.

In the digital company the sustainability embrace many faces. In a digital company we may reduce waste. Waste of time, waste of valuable people, and waste of materials, manage the energy consumption; reduce the water consumption and the company emission.

Monitoring and managing almost everything, a larger comprehensive "Environmental program" can be built from the beginning or developed in time. The "Smart factory" and the "Digital city" start to merge and the "Environmental program" covers much more from the company location, the company building structures and their energy free behavior, to a climate neutral behavior where the employees location and logistics to get to the company and back home is combined with their behavioral habits, cultural, educational and recreation habits (and needs).

The increasing comfort and progressive desire "worth living" is balanced with the regional sustainability, dictated by the resources, economic concentration and development distribution.

Everything is framed in a socio-political structure. The key words in a digital managed company start with the idea of sustainable profitable business, based on sustainable products and processes, practical achieved by flexible productivity with enhanced competiveness and with renewable solutions for water, air end energy consumption.

Technically this are not "just words" real measures can happened using all the 4 changing driving forces into action: managerial, political, social, and technological. This is not a random order is the order determined by the latest time, money and consuming criteria, but each entity may select the driving forces order or the combination of them that better suits, or are just available.

Managerial, is a way of administrate better the processes, by adopting different criteria, thinking out of the box and being innovative.

Political, is to adopt a right decision and adopted it in standards, regulations legislations that are imposed in an equal mode to every participant at the process.

Social, is to adopt consensus, necessarily and sustainable ideas, and to implement them in the general behaviour of the people as a second habit, an accepted and a learned one.

Technological, is to identify and accept a cultural technological change in doing things. To invest in devices, software and processes, to restructure the organization accordingly, to fight with the people resistance, to convince people, to inform people, to train people, to specialise people and keep the things on running with an appropriate maintenance process and appropriate change management methodologies.

Managerial, sustainability can be.

The managerial sustainability is based not only on managerial concepts and methodologies but is driving the process relay on the instruments offered by the other 3 driving forces.

For example, in the managerial "policy" may be introduced a concept from ISO 50001, to gradually reduce with a certain % yearly the energy consumption. That can be done by technological meanings (acquisition of new performant devices, implement a new technology, etc.) or by manage in a different way the energy management system (EMS) where the power generation, buildings, overall lighting, motors and other type of resistive consumers are individually monitored and integrated in a single managerial system within a single data base and a unitary approach.

Such an energy management system may link the energy data with the wheatear parameters, employees schedules, process information, work areas climate, legislation, contractual data from the supply chain and clients, energy stock market, new technologies and investment strategies.

Create an innovative environment that opens potentials for new business models. For example the products and productivity improvement can be done by creating a transparent, secure, safe and reliable data communication and data management through the entire life cycle.

In this chapter enters also the IT as a crucial important thing that the new technologies must arrive with the **systems commissioning** and a **training** program for the involved peoples.

An interesting point is the so called "virtual commissioning" useful for both the technology supplier and the technology beneficiary. The virtual commissioning is more than a virtual reality is actually the digital twin of the delivered machinery, systems and technology that validate in a virtual environment the functionality and the interactions between mechanical parts, mechatronic components, actuating systems, sensors and controlling algorithms.

The digital twin enables the functional simulation, the process testing and also the functionality optimization due to faults identification scenarios.

In this way is avoiding any real accidents, downtimes and cost due the unknown behavior of the systems. All mentioned characteristics made from the virtual commissioning systems an ideal training platform for the operators and not only. Practical for every stakeholder that may intervenes trough different scenarios, is a way to test or improve his skills and learn new way of application or/and development of the new technology in a safe and risk-free environment.

Political sustainability.

- Considering **planting trees** in degradable areas, offers;
  - o An environment rehabilitation;
  - o Redesign the biotope;
  - o Ensure the biocenosis;
  - o Creates work places for locals;
  - o Ensure better education facilities for locals;
  - o Creates a sustainable resource;
  - o And clean the air;
  - o Filter the water;
  - o Some rebalance of the climate change.
- **ISO 50001**
  - o Impose the modern energy management starting for meeting the ISO 50001: 2011: 2018 standard, the international standard for Energy Management Systems, created by the International Organization for Standardization.
  - o Is a standard that follows the ISO 9001 Quality Management System and the ISO 14001 Environmental Management System (EnMS), creating a modular integrated unity [1, 2];
  - o Once imposed the ISO 50001 standard, several problems are solved like the valorization of the untapped energy efficiency potential, cost savings and reduction of CO_2 emissions. The standard specify about the energy management system and a systematic approach [3] about the:

    implementation;
    maintenance;
    and achieving a continual energy performance improving by following specific goals:

      the energy use;
      the energy security;
      the energy efficiency;
      the energy consumption.

Consistent energy management helps organizations to realize they will benefit from and make a significant contribution to environmental and climate protection. The standard should alert employees and in particular the management level to the immediate and long-term energy management gains that can be made. The organization can discover potential savings and competitive advantages. Furthermore, a huge image boost for the organization can be created [4].

Social sustainability

Push the digital to digital processes that change the continuous, stochastic or batch manufacturing systems to the more customized ones for the personal and industrial acquisition zone. More important for the manufacturing or service zones in offering digits = drops of material, like in 3D printing the additive materials or in agricultural applications for fertilizers, pesticides and the more important the water.

Running toward accountable persons about the environmental impact, education and training systems, company sustainability and innovation capacity.

Technological sustainability

- A direction of technological sustainability is to create or make conversion from conventional power plants (CPP) to combined heat and power plants (CHP, or COgeneration plants).

The week technological development in atomic plants and especially in radioactive residual materials treatment made that the energy hunger to be heavily based on coal and oil and natural gases (fossil fuels) in spite of the fact that are very polluting and inefficient technologies.

The switch to renewable energies, in spite of the diversity: biomass, geothermal, hydro, oceans waves, solar or wind is not continuous and not so affordable like the one from fossil sources. Therefore a primary step is to continue to use fossil fuels within another technology that reduce the consumption and the environmental impact.

Where is the catch? By producing electric energy in a fossil fuel based power plant, burning the fuel that release heat makes a fluid to boil (usually water) transforming the water in hot steam under pressure that drives a steam turbine that drives an electric generator. The electricity that we may use as final consumer is about 20–25% I the best scenario, depending on the distance from the power plant, the distribution hardware and distribution management efficiency. All the rest (75–80%) is heat waste that together with the burning gases (mostly CO_2) contributes to pollution and global warming.

The CHP power plant beside electrical energy is delivered also thermal energy that from the 75–80% heat waste is reduced at 40–60%, and by using ORC (Organic Rankine Cycle), another technology) we may increase the electrical energy amount with 35–40% and decrease the heat waste at 30–45%.

Finally investing in technology offers work places, comfort, safety, and sustainability that are good for the planet, for the people and for the pocket.

- **3D printing**, a totally digitalized technology used from the milimetrical implants to big houses made from a large variety of materials. The 3D printing derives from the classical CNC machining, and if also the sintering process is added together with the casting and metal forming technologies all are the ancestors of the 3D printing technology. While the classical technology is dealing with the removing and reshaping the material the 3D printing is dealing with adding material.

The material efficiency is very important, if in the first case the material efficiency is somehow between 20 and 40% in the cutting process and 60–70% in the casting process, within the 3D printing the material efficacy is minimum 95%. At this advantageous characteristic is added the capacity to materialize very complex shapes, very difficult or impossible to be made by other technologies, the energy efficiency in producing the materials and shaping the parts, together with a shorter time for design and manufacturing of the single parts.

The drawbacks are the difficulty in materialize series or mass production, the lower resistivity of the part, the porosity and the surface accuracy.

- A next important, new and very innovative technology, after 2010, is related with the "Precision agriculture" in different forms like "Precision Farming", "Precision Dairy Farming", etc. This is a really complex zone why?
  - o The world population is increasing;
  - o The urban population is increasing;
  - o The demand for food products will increase with about 70% in the next 30 years [Aniţei]
  - o A side effect will be an increase in energy demand with a similar 60% to 70%;
  - o The energy availability may decrease, the energy price may increase;
  - o Degradation of the planet climate;
  - o Reduction in the availability of the agricultural areas;
  - o Considering the PPP (Purchasing Power Parity) a theory that measures prices in different areas using a common good or goods to contrast the real purchasing power between different currencies, shows an increase capacity of the emergent countries (especially Asia), that will put them in a dominant position;
  - o Lowering of the agricultural labor force;
  - o Increase the plants disease (anthracnose) and animal disease (zoonosis) due to the higher density of exemplars and globalised logistics;

An answer at this challenges could be the Precision Farming, which is very connected with the industrial digital applications like robotisation, not only for handling and manipulation but also for different processing tasks in cooperation with humans or as standalone units; 3D printing for fertilizers, pesticides and water plants precise dozing; drones for fields and animals monitoring, and precise farming of large area with added robots and payloads up to 2000 kg, satellite that provides global observations and analysis regarding of wind profiles, along with information on aerosols and clouds, atmospheric dynamics, and weather forecasts and climate research.

- **Blockchain**, for industrial applications offers at least 3 important elements. On the business side is creating a process transformation enabling the cutting of any intermediary link, account the two links ends and in this way rationalize the process. On data management side, enables the process automation, reduce the redundant data volume, simplify the data management and therefore reduce the unnecessary costs. On the communication side, increase the data transfer security, ensure the data transfer (transaction) transparency all the value chain stakeholders having access to see the transaction details.

The blockchain is working like a computer file that records and totalling a transaction data in terms of accounting units by accounting type with initiated values (beginning of a balance) and ending units (ending of a balance) for each account, as distributed, decentralized Account Ledger. The blockchain can be programmed to record, but not to manipulate, every digital transaction and if connected to a subject every item of information.

A digital transaction is named "block". Beside a time stamp, every block contains a link to a function that converts a data value into another with the purpose to create a checksum generation for the data security and data compatibility test, to compress the data for lowering the memory storage necessity and speed the transmission, to cryptograph the data for the content security, indexing and mapping the data, called hash.

The function associate a matrix structure (tables) for data storage and retrieval, accessing the data in small, and nearly constant time for any retrieval, requiring just a small fractional space for storing the data as the data itself requires.

- **5G** It is a new and strong trend around the 5G networked communications. The smartphones manufacturers, mobile networks operators, together with national and international and regulatory bodies (like international mobile network committee 3GPP) are already active from some time. Is expected that half million of devices per square kilometer may became active using the new elaborated standards. The industry is following heavily these trends with the ambition that industrial 5G to be available for automated manufacturing systems, robots and virtual reality applications by 2023, in leading countries like Germany, UK and USA.

Once the frequency spectrum is allocated, millions of applications will booming for administration, agriculture, health care, industry, research centers, urban management, etc. So, proof of concepts TRL 1 and TRL2 solutions are already developed within project worldwide. The 5G technology not only is consuming only 0.001 of the actual energy per bit transfer but is also 10–20 times faster than the actual 4G communication standard for wireless broadband communication for mobile devices and data terminals (LTE), but is perfectly adapted to the internal private communication networks, with a higher data security level, real time products and locally processes tracking, the so called "edge applications" with no need for the on cloud uploading.

16.4 Smart Manufacturing

Server based computing or cloud computing, blockchain and 5G are new born of the digital edge but the core element of the Industry 4.0 and the real challenge is not to increase the memory capacity (clouded or not), not to increase the communication speed, or the amount of data generation points, not even the data protection and security, but the countless benefits to business and society by the decisional capability: The smartness.

This is what helps to improve the results based on high data computational and processing capacities.

The digital smartness is "downloaded" from top to the EDGE. Moving to the "edge" we got already a lot of edges:

SolidEdge, a 3D CAD, parametric feature and synchronous technology solid modelling software that runs under Microsoft Windows operating system but also under Apple OS and Linux, providing parts solid modelling, assembly modelling and 2D orthographic view functionality, drafting and analysis for mechanical designers.

Solid Edge and CAMPro bundles combine professional 3D CAD software with computer aided manufacturing (CAM) software to significantly enhance the product design and manufacturing performance. Solid Edge manufacturing solutions help manufacturers to define and execute a wide range of traditional and new manufacturing processes as additive manufacturing (3D printing), assembling, bending, moulding, nesting, robotized welding and precision CNC machining. The previous in house knowledge and experience is valorised in the process of maintaining a competitive edge;

EDGECAM, is a CAD CAM software for 3D milling, mill turn, multi axis machining, a market leading computer aided manufacturing (CAM) system for NC part programming, with a complex toolpath generation, with tools collision avoidance and providing a safe operation environment.

The geometrical inputs can be formed by point's coordination, part wireframe or the solid geometry, valid for a variety of multi-axis, multi-turret CNC machine-tools configurations. That includes the "Waveform roughing" a high speed machining technique that maintains a constant tool cutting load by ensuring the tool engagement into the material. That is consistent, moving in a smooth path to avoid sharp changes in direction and maintaining the CNC machine-tool velocity. Additionally the modules EdgeCAM Designer and EdgeCAM Workexplore, fill the gap between the CAD and the CAM applications by considering also the part fixture design enabling also repairs by modifications and adding's, in combination with application of metrology tools and knowhow.

Edge Device is a dedicated device that provides an entry point into enterprise or service provider core networks. Examples include routers, routing switches, integrated access devices, multiplexers, metropolitan area network or separate wide area of network access devices.

Edge Gateway is a device like system that includes a variety of wired and wireless connections to communicate (Wi-Fi, WWAN and Ethernet, including serial connections). The I/O on the intelligent device makes it easy to connect industrial systems and new mesh networks.

Edge server is basically an edge device, providing an entry point into a network. Within this edge device are include other types of edge devices like routers and routing switches, usually placed inside Internet exchange PointS (IxPs) to allow different networks to connect and share transit.

Edge TPU is a dedicated hardware-software solution that combines the custom hardware, open software, and state-of-the-art AI algorithms, with the aim to provide high-quality, easy to deploy AI solutions for the edge.

Microsoft Edge is a web browser developed by Microsoft, first released for Windows 10 and Xbox One (2015), then for Android and iOS (2017) and then for macOS (2019).

Microsoft Edge includes integration with Cortana, a voice-controlled virtual assistant for Microsoft Windows Phone 8.1., comparable to Siri, the intelligent assistant enabled on Apple devices. It is mentioned that Microsoft's Cortana will use the Bing search engine and data stored on the user's smartphone, for personalized recommendations [5], with extensions hosted on the Microsoft Store a digital distribution platform owned by Microsoft.

Microsoft Edge includes the information in the User-Agent HTTP header whenever it makes a request to a site. A user agent (UA) string is able to detect what version of a specific browser is being used on a certain operating system.

Internet Explorer 11 and Firefox will continue to run Java on Windows 10, but the Microsoft Edge browser does not support plug-ins and therefore will not run Java. The new Microsoft Edge is based on Chromium, released on January 15, 2020, compatible with all supported versions of Windows, and macOS. Downloading the browser, will replace the legacy version of Microsoft Edge on Windows 10 PCs.

For industrial applications the Chromium open source in the new Microsoft Edge creates better web compatibility for the customers with less fragmentation of the web for all web developers.

The GitHub, the home for over 40 million developers "working together to host and review code, manage projects, and build software together" [6], contain the commitment to be an active contributor to the Chromium project, as Chromium is a free and open-source software project from Google.

The source code can be compiled into a web browser, however, Google does not release an official Chromium browser instead, and Google uses the code to make its Chrome browser, which has more features than Chromium.

EDGE stands for Enhanced Data Rates for Global Evolution network, a third generation mobile data technology, according to AT&T. It is used to provide fast Internet service to cell phones, and can be used to fill in the gaps of coverage networks from the cell phone providers.

Edge 5G applications; require a new approach to edge computing that is a distributed cloud, a solution which expands the computing possibilities.

Edge computing represents decentralized data processing offering advantages to the manufacturer by avoiding the challenges in dealing with huge amount of accumulated data that must be stored and transmitted accurate and safe at long distances.

The edge computing is a distributed computing paradigm which brings computation and data storage closer to the location where it is needed (to the edge "just edge"), to save bandwidth and improve response times, implying no need to be sent on cloud or other server centralized data processing systems. Data transport distance and time compressing boost, improve the edge devices and edge applications.

Fog computing is a standard, defining how edge computing should work facilitating the compute operations, storage and networking services between end edge devices and the cloud computing data centres, as a jumping-off point for edge computing.

All the 4 driving forces of change are leading us to the edge but each company is different and therefore each situation must be in detailed analysed to decide which industrial centralized-edge computing architecture make sense and how the business processes might call for a more or less centralized computing model for an optimal techno-managerial mix.

References

1. Ramesh, L., Sayyad, N., Kranti, D.: ISO 50001:2018 Energy Management System Requirements & Implementation (2019). ISBN 9781090748300
2. Charles, E., Frederic, M., Timothy, C.: Inside Energy: Developing and Managing an ISO 50001 Energy Management System. CRC Press (2011). ISBN 9781439876701
3. Deutsche Gesellschaft zur Zertifizierung von Managementsystemen (DQS): ISO 50001 Energiekosten sparen—Klima schützen—Verantwortlich handeln (Produktblatt). Frankfurt, (2012).
4. Sergio, T., Hervé, P., Gérard, M., Marco, G.: A holonic metamodel for product lifecycle management. Int. J. Prod. Lifecycle Manage. Inderscience 253–289 (2007)
5. Microsoft official home page, www.microsoft.com
6. https://github.com/microsoftedge/msedgeexplainers

Chapter 17
Industrial Digitally Prototypes

"I've seen a lot of basic vendor comparison guides, but none of them come close to the technical depth, real-life experience and hard-hitting critiques"
Alexander T. Deligtisch

17.1 Holistic View of Product and Process Design

The whole product lifetime can be firstly defined into 5 major phases:

- No 1 Concept-Design;
- No 2 Tooling—Acquisitions;
- No 3 Manufacturing Quality control;
- No 4 Logistic—Costumer handover;
- No 5 Recycling—Circular Economy management phase.

Considering the digital business management, the considered 5 phases are coming to be overlapped over the classical 3 phases of the PLM: the start, the growth and the dead.

The new integration paradigm comes from the fact [1]:

> Digital manufacturing is a key point of integration between PLM and various shop floor applications and equipment, enabling the exchange of product-related information between design and manufacturing groups.

Basically is a matter of decision based on engineering data and managerial data. In many cases the problem is solved based on heuristically or biases methods, instead of a Holonic approach that our day technology enables. For a deeper understanding the actual decision process was pretty described in the book "Thinking Fast and Slow "of the Nobel laureate Daniel KAHNEMAN [2].

The paradigm can be sustained by using the concept of "generative design" Fig. 17.1, that is considering the previous engineering experience and documents databases (the captured knowledge), the actual technological culture (the available

J. Niemann and A. Pisla, *Life-Cycle Management of Machines and Mechanisms*, Mechanisms and Machine Science 90,
https://doi.org/10.1007/978-3-030-56449-0_17

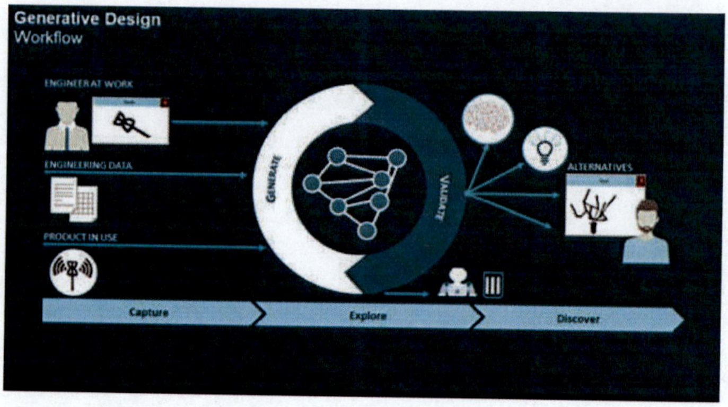

Fig. 17.1 Generative design concept and workflow. [Courtesy, Siemens blog]

knowledge) the considered products, tools, resources and technologies, all together in the “Capture phase”.

That is followed by the “Explore phase” where millions of algorithms signals, components and simulations are trade-off in order to synthesis in a parallel engineering a valuable and valid solution.

In the last phase, the “Discovery phase”, the innovative novelty put the company in the marketing position. This is practically the way in which the entire investment is justify, being the added value element of the company that integrates the internal capability with the external requirements and opportunities to promote and validate every mechatronic component.

The potential of digital manufacturing is in creating eFactory models faster and ensure that the models are working in an optimal layout and appropriate material flow, before the production ramp-up.

With this digitally networking direction the companies are able to have a better cost control, can realize cost saving and have not so many problems to achieve their time-to-market.

The connections of the whole eFactory make possible to create a complete manufacturing process in a virtual environment. On one hand the production process can be optimized before any product is manufactured and on the other hand it allows a real time feedback of the process. The information that the companies get, in fact of real time feedback, allows taking advantages during the planning stage, because the acquired data can be integrated into the product design process.

Example of a Digital Manufacturing. The Companies can design the entire manufacturing process in a digitally way. So they are able to do the:

- Tooling
- Machining
- Factory layout
- Assembly sequencing

At the same time the designer can create the new following production program with the opportunity of the collected information integration by the digital manufacturing process. Therefore they create a holistic view of product and process design.

Within the holistic view, primary there are two essential elements the "manufacturing process" and the "set-up time optimization by scanning the company environmental positioning". The two components are integrating the part lifecycle and impose the part lifecycle management (the local PLM) and the company evolution life cycle and its management (the company PLM).

The overlapping of the two managerial systems is not necessarily hierarchic as expected, but represents the digitally networked PLM, where everybody (every manufacturing line or department) is engineering its own score but the entire symphony must be played together under the conductor management.

[Courtesy, industrie.de and Siemens PLM]

The life cycle of each development process is based on the initial conditions and expectations but is changed by any new considered design or implementation condition.

17.2 Training and Commissioning

The Training necessity was imposed in time by the complexity of the new digitalized hardware and the afferent software packages and operation systems.

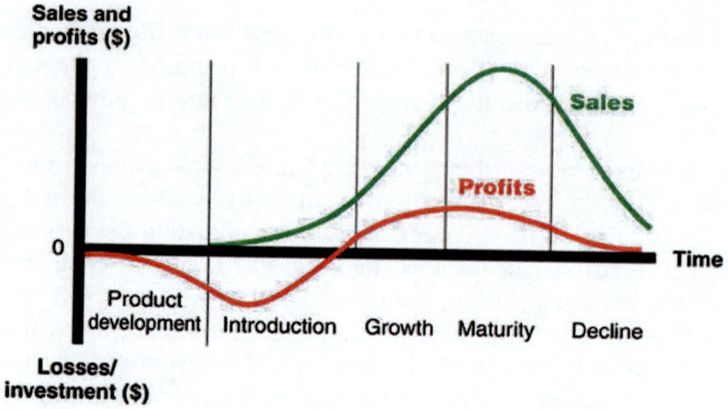

Fig. 17.2 Classical lifecycle diagram. [Curtesy: Marketing Insider]

A real necessity based on an increasing demand leads to the resulting classical diagram for the product lifecycle, Fig. 17.2, where things look very linear, safe and successful.

But people are optimistic, they consider that after maturity stage (when is obtain the saturation) a rebirth could happened, Fig. 17.3.

What was happened? In the real life, we cannot always win, so how can we prepare to reduce the loosing chance? In the beginning of the new digital approach, the training costs were part of the beneficiary. In most cases both, the beneficiary and the supplier were not happy, why? The new acquisition is not working properly or not at all, due to fact that the beneficiary is not investing in training and relaying on how smart are the employees as self-teaching subjects, or by choosing the wrong training package.

On the other hand the suppliers face a lot of errors and repairing intervention during the installation, or warranty period after commissioning, basically being the result of the miss operation made by the new customer.

Fig. 17.3 The lifecycle diagram with extension of the lifecycle. [Curtesy: economicshelp.org]

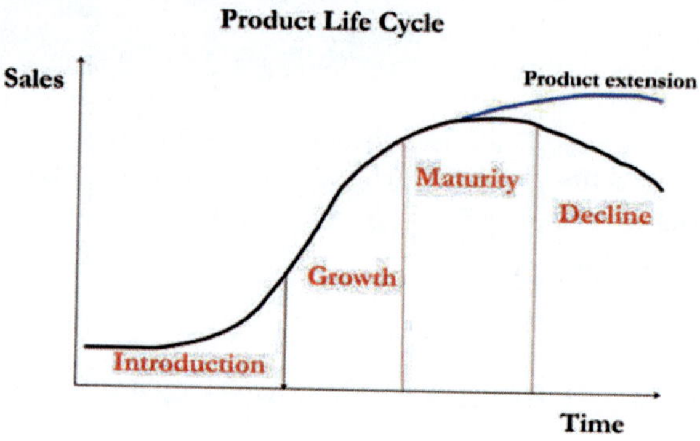

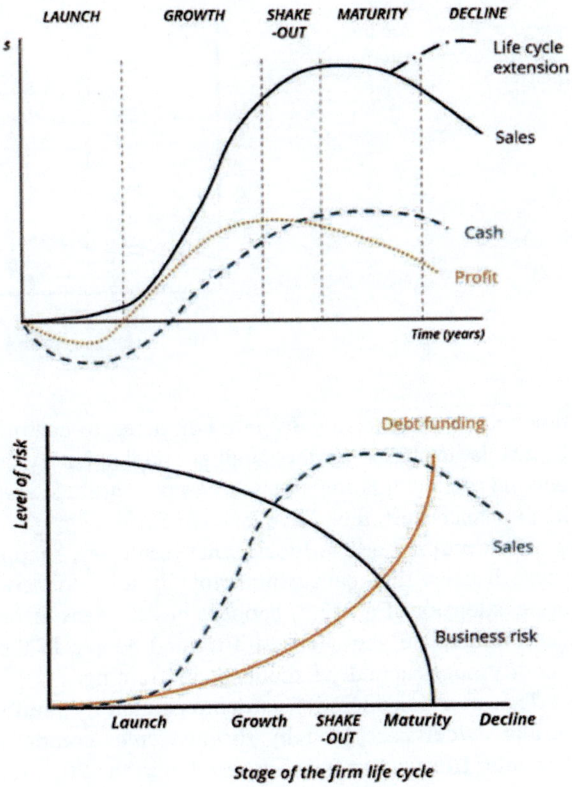

Fig. 17.4 The lifecycle diagram when the acquisition is not working properly. [Curtesy: CFI]

Finally, everybody is losing. So how looks from the PLM perspective when things go wrong, Fig. 17.4. The lifecycle expectation is an illusion, sales decreasing more rapidly than expected (right from the start of the maturity phase) the cash flow is low, the profit disappear, debts are rising and the business risk is growing mostly because the business disappear at all.

No business—No errors, but also nothing happened. That means that "no errors" is not leading necessarily to a happy situation.

And then the entrepreneurs, as main business suppliers, developed in time a new business model. The novelty comes by including automatically the training within the delivery and installation process. It was not only an economic decision but also a safety measure, trying to ensure the best operational way and the customer satisfaction. Even if the beneficiary could be less happy in the beginning by increasing the investment costs, in little cases the training was ignored, so on long run everybody wins including the desired new product lifecycle type and this time like an end product.

In parallel with this process was discovered that are developed other two business strategies regarding the quality of the product and the after sealing service. That led to a new theory: The Service Paradox, Fig. 17.5. Customers with product break-

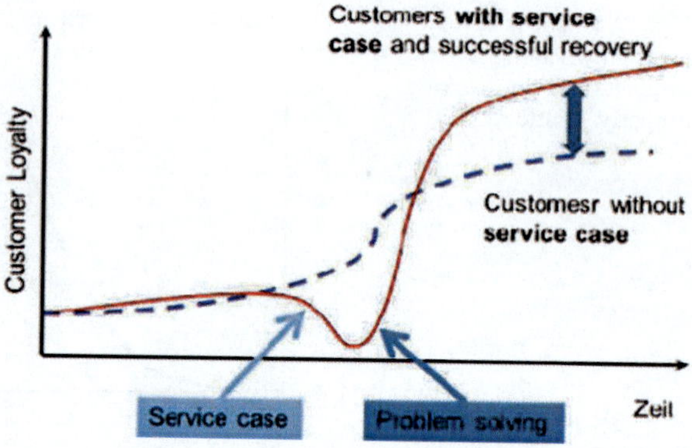

Fig. 17.5 The service paradox: Excellence in Service fosters customer loyalty

down have a higher loyalty rate compared to customers without breakdown. The loyalty is not based on the supplier supplementary competences, costs and efforts into the product quality but on the supplier attitude and investment in the service and maintenance staff, the "Excellence in Service!" [3].

The caution made to look at that result as a "happy situation" not as a defining cause, because that real customer loyalty, the sale increase and in conclusion the life-cycle extension (Fig. 17.3) happens based on the affordable better customer service due to adequate and successful trained people in the company. So Excellence in Service means actually Excellence in Training.

This new consideration is demonstrated by putting together different research results, widely accepted by the academic community. Firstly we consider the customer lifecycle framework resulted in the 2017 research "Taking an Outside-In Perspective of the Customer Lifecycle", developed by Nick BONFIGLIO, Mickey ALON, and Myk PONO in a book chapter of the book "*Mastering Product Experience: How to Deliver Personalized Product Experiences with Product-led Go-to-Market Strategy*" [4], Fig. 17.6. From this framework we consider 2 important stages: Retention and Expansion.

We added the research presented in 2012 by Irina MAKAROVA, Rifat KHABIBULLIN, Artur BELYAEV and Eduard BELYAEV, "*Dealer-service center competitiveness increase using modern management methods*" [5], Fig. 17.7, and finally express the idea of "Qualify the customer".

From Makarova and others, it is to be considered the table with "The cause-and-effect diagram of the firm service system".

From this table can be extract the entire interconnection system, where the technological area is concerning with the diversity and severity of the defects at the customer location, together with the control of spare parts and maintenance delivery. The maintenance aria is what early called "Excellence in Service!".

Considering the digital technologies the "technological area" deals also with the data collection through the IIoT smart devices and sensors bot never the less the IIoT management and so we arrive to the Software area.

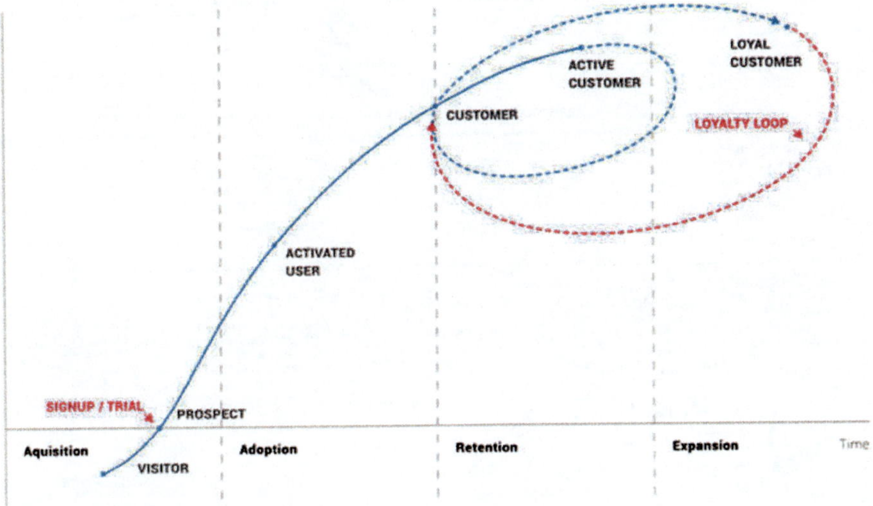

Fig. 17.6 Outside-In Perspective of the Customer Lifecycle framework [4]

In the middle of the Software area stays the data logistics the connection with the data sources, the DSN (Data Source Name) and the security elements like DMSS (Digital Mobile Security Systems), where actually all the real disturbances of the implemented systems (the errors) are gathered in a safe way.

In the same "location" are coming also the data from the external suppliers and customers through dedicated interface portals and the internal data based on the company adopted digital technology.

No doubt that for both levels special trained personnel is required and in this way we arrive at the "personnel area", were the data and information from the external sources (remember: customers and suppliers) are defining the quantity of work and the work/time distribution, and the internal digital platform defines the required training activity.

And so is arrived to the SERVICE area, where the data logistics and the connections plays a great roll in defining the quantity, quality, visibility and service opportunity, as business factor lifecycle extension. In this context is required a service activities forecast. That may leads only to the customers adaption "reconversion" due to the big role of the ***defects reduction***, in the potential new implementations.

But, if the quantity and required quality of the service is coming from the information gathered from the external sources (that every company likes to apply), the service management and the direct interaction with the customer data, represented as the actual key of success, is offered by the TRAINED skills of the involved personnel that is making the bond through the ***cost reduction***, of the actual customers and leading them to continue to be also the new customers, results: Creates loyalty!

In this way to forecast the PLM in the new digitally strategy we arrive at the idea express also by the Lucidchart Team to "Qualify the customers" by performing

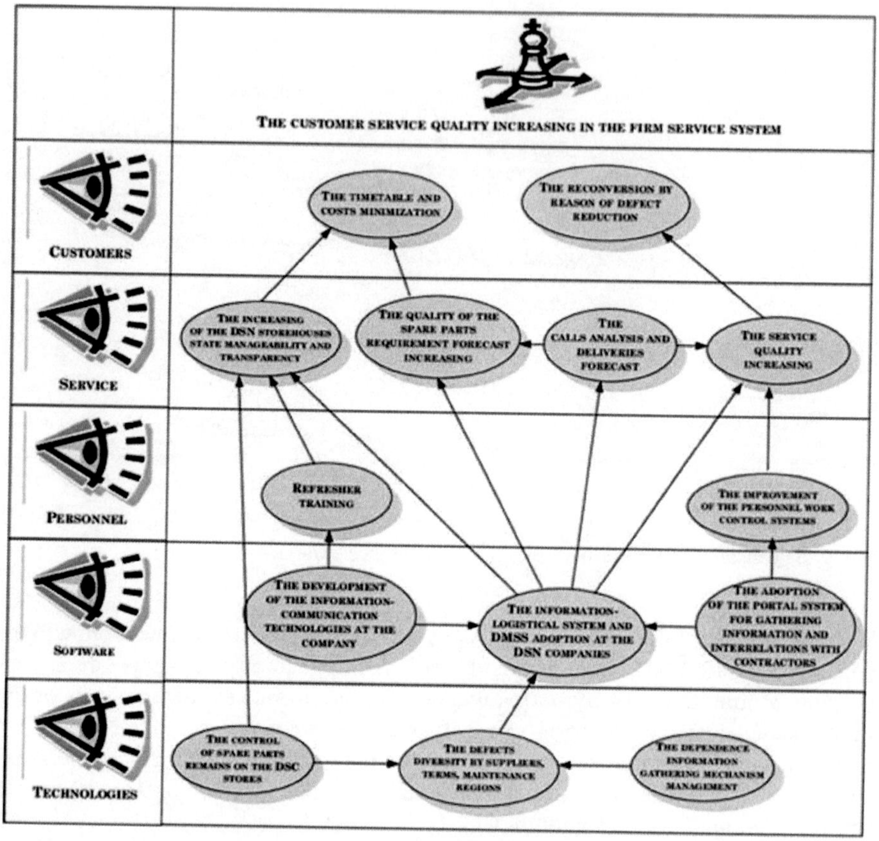

Fig. 17.7 The cause-and-effect diagram of the firm service system [5]

a background check, fact-checking, to determine and to convince them about their usefulness as much as possible from the first meeting or appointment.

As expected result the customer will be formed by qualified leads, capable to understand the product and the service and to make profitable decisions.

17.3 Knowledge Engineering and Management

The developers of technical products are facing the customer requests for a specific product, in the form of product adaptation or an invitation-to-tender (ITT) [6]. In the case of digitally networked PLM prototypes, the amount and the quality of data (details) increase in significance for the success of the company. As long as the

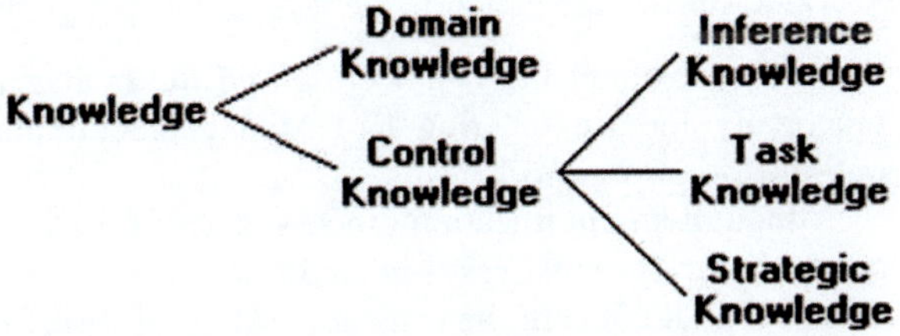

Fig. 17.8 Structured levels of knowledge engineering by KADS

amount of time and money increases substantially for complex, prototypes (one-of-a-kind) or customized products (build-to-order) for preparing a bid will grow exponentially and therefore the demand for digital support is higher and higher.

Ambiguity, incompleteness and inconsistencies in the requirements data are influencing negatively the entire chain: acquisitions, installation, training, commissioning and maintenance, regardless of the products specificity (transportation systems, healthcare or military).

Besides the cost of a bid itself, the cost for later rectification of inconsistencies and incompleteness of the initial specification will lead to significant higher costs in spite of the fact that a proper bid, as an attempt and effort to obtain something, is often considered as a time-consuming process and a major cost factor with so far little automation support.

In the domain of knowledge engineering and management, based on the KADS (Knowledge Analysis and Design Support), the knowledge engineering models can generally be categorised into 3 major structured levels under the control of a knowledge management and knowledge engineering tool (Sch 1998): Domain, Inference and Task knowledge (Fig. 17.8).

KADS represents the name for one of the knowledge structured methodology developed in Europe [7, 8]. The KADS is a systematic, theoretical analysis of the methods applied to a field of study, as an implementation independent methodology for developing KBS—Knowledge-Based Systems (other name for the expert systems) that is in practical use in many places in Europe and beyond.

The KADS comprises the theoretical analysis of the body of methods and principles associated with a branch of knowledge. As modelling methodology KADS is still difficult for domain experts in using a non-graphical nature modelling languages, even if is a successfully used one.

- The **Domain knowledge** contains the structural information of the domain and describes the main static information and knowledge objects in an application domain, differentiated in:
  - the requirements engineering domain knowledge that is needed to represent requirements;
  - the enterprise domain knowledge that is used to support knowledge functions operating on requirements engineering domain knowledge.
  - **Inference knowledge** describes how the static structures of the domain knowledge can be used to carry out a reasoning process;

- **Task knowledge** supports the way to achieve goals by applying knowledge.

Putting together the engineering and the management tools and methodology, practically the Engineering and Management Knowledge (EMK), leads to an optimization framework.

Although an optimization process is about finding a local or a global minimum or maximum (in most cases related to a function) it is far away of being mundane or trivial as it concern the critical problem of decision as long as is unknown how "global" is a solution and on how many domains is extended based on the considered restrictions.

An example may be the one from the parallel kinematics robotics in defining the hyperparameters. Hyperparameters are usually used in the machine learning model (a decisional system) but starting from their definition:

> A parameter whose value is set before (the learning process).

What is happened and how can be used the hyperparameters in the serial kinematics and parallel kinematics robots models?

Basically the motion control of a robot starts from positioning a point, the TCP (Tool Center Point). In order to control the robot (the robot kinematics and the robot dynamics) we need to know the motion equations, but there are actually to models the IKP (Inverse Kinematic Problem) and DKP (Direct Kinematic Problem) that in parallel kinematics are the real problems.

If you have an industrial SERIAL kinematics robot (the most known ones), Fig. 17.9a, by rotating each actuating motor result the position of the TCP. This is actually the DKP: where is positioned the TCP if we have the motion equations for each actuating motor.

The IKP is more difficult to solve because for each TCP spatial position we may have several configurations of the robotic structure, so is not anymore a biunique

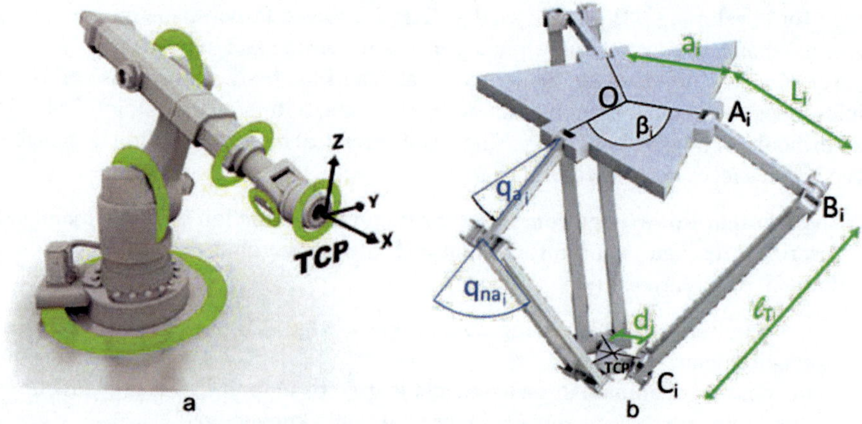

Fig. 17.9 Structural differences between serial (**a**) and parallel (Sch) kinematics robots [Courtesy: Heidenhein and ResearchGate]

correspondence. This is the moment when the hyperparameters are necessary to define an acceptable or a start position (necessary for the motion learning process).

In the case of PARALLEL kinematics industrial robots (about very few people knows about) Fig. 17.9b, the situation is almost vice versa as it is in serial kinematics. If the TCP trajectory is known the close chains of the parallel kinematics determine a unique robotic structure position between the TCP and the actuating motors positioning equations being a biunique correspondence.

This time the DKP is more difficult to solve because for each TCP spatial position we may start from several configurations of the robotic structure, so is not anymore a biunique correspondence. This is the moment when the hyperparameters are necessary to define an acceptable well defined start position, necessary for the motion validation process.

In spite of the "problematic" looks of the optimisation process it is still duable as long we are dealing with deterministic variables, but the hyperparameters are also dealing with the statistical situations. In this way came in discussion the probabilistic models like the Sequential Model-Based Optimization (SMBO) in finding the minimum of any function, in the surrogate or response surface that maps input values to a probability of a loss, by the capability in returning a real-value in a standard of measurement (a metric).

From a simple function like $y = x^2$, to a many-dimensional complex function that validates the robot starting (or admitted positions) as a hyper-parameter choices, with respect to tens of robotic structural models and hundreds of structural working environment conditions, the probability model being easier to optimize than an actual objective function.

It seems that finding the minimum of any function is not the single advantage of the method. It seems to be more efficient like a manual method, if we deal with a low amount of input variables:

- A grid search, were all the "gridded values" are considered; or
- Random search.

The advantage of the SMBO is based on two main elements [9]:

- the number of solved problems, offering better overall performance on the test set;
- less time required for optimization.

But real good news is that is not necessarily hard to use the method. It all depends on the problem complexity, considering that there are a number of libraries such as "Hyperopt", "Spearmint or SMAC that for simple applications allow using Bayesian optimization, not "on line", but in "one line"!

In any situation the problem must be described and the formulation of the problem optimization must be defined. The POF (Problem Optimization Formulation) consist in four parts:

- Objective Function definition—that returns the search minimum;

- Domain space: the range of input values to evaluate for leading to the Objective function;
- Optimization Algorithm:
  - o Defining the start "surrogate function" to evaluate the outcome;
  - o Selection method for the next values to evaluate;
- Results: the algorithm selected value pairs, used to build the model

Once the optimisation problem leads us to the model it is time to deal with the prototypes.

17.4 Model-Based Optimization

The industrial prototypes once digitally networked benefits from the advantages of all specialists involved in the process, but also suffers from all the restriction imposed by each product lifecycle component. Therefore the lifecycle management once digitally networked follows the necessity to consider every rule and every restriction, leading at what is known as optimization process in obtaining an industrial prototype.

What is optimization?

In real life optimization is actually ***a compromise*** between all contradictory factors, end to be accepted by all the implied players, the stakeholders.

In early time the specialists basing on their expertise, deduction, educated and trained guess or intuition, may develop new products and solve new problems adding impressive technical, economical and/or managerial solutions.

Now days the complexity of the problems imposed the use in modelling and decision process a large amount of restrictions and product characteristics in a digital form, namely the data. In this way the modelling and simulation process becomes indispensable in finding the optimal (the compromise) design and functional parameters of a product.

If a combination of effects are considered like: mechanical, thermal, chemical or radiation stress, together with non-linear, discontinue or non-convex process mix with the stochastic and non-deterministic processes, working with different types of variables from Boolean and real ones up to complex or the fuzzy sets, the low abstraction is not working anymore so more complex approach is necessary.

In conclusion without modelling and simulation process is impossible any new performant and complex product.

But the problem is not simple and not ready; beside the necessary algorithms that manage the components characteristics data also other digital algorithms are necessary to estimate the computing and data processing power, the computing time and the computing costs. A balance between all that is another huge problem of itself, tens of thousands of programmers working permanently to solve and improve the solutions.

17.5 PDP-Product Development Process and the Impact on PLM

A model base has the aim to ensure an adequate collaborative environment for the product design. Basically is having the right data in the right context and all the persons involved in the design can generate, share and exchange data along the entire development process.

The PLM is a concept innovated and used within the industrial context. A product is functionally a complex entity where the real part is representing the fix characteristics (practical given as immovable by the decision or by natural properties) and the imaginary part is what we may change or alter based on the freewill (freewill, that in many occasions is just following some more or less arbitrary decisions). Each domain is represented by its stakeholders and by its challenges.

The lifecycle of each product is determined and follows the effect of the two types of components, real and imaginary. Being always subject to risks, uncertainties, difficulties in the correlation of interdisciplinary design and technologies the lifecycle as a process not as a result must be managed.

The different tighten between the interacting phenomena defines the great challenge to find not a local but a global optimum for the entire system and therefore is needed to adopt an integrated simulation-based design approach and a FBS (Function-Behavior-Structure) sort of framework, where the product data is corresponding to an appropriate behavioral structure. The FBS is tributary to the simulation context, the field of application of the product.

Actually the technical, staff and managerial data are blended in the PLM process as finality within the digital management. We are facing with 3 major situations: the data consistency, interoperability and an integrated reference.

Even in a highly standardized situation the data consistency may be loos within the transfer process between different digital tools. Like FBS the digital tools are tributary to the simulation context.

If different platforms are common used, as the ones that are used to manipulate technical data (CAX) or the ones for engineering data management (EDM) or globally for the product data management (PDM) that may present a lack of efficiency within the data exchange process, reflected finally in the loose of efficiency in the entire PLM process.

The integrated reference, refers at the use of a product characteristics design and not at the product draws so the design data with analyse data are totally integrated aiming a prototype design with a full traceability of the generation steps, timing and responsibilities, considering all the new list of constrains or release regarding the scope and the objectives of the project.

Regardless all the specified difficulties we have to look at the bright side of the problem by looking back where we were 20 years ago?

Without past exists no future, so we have to be thankfully and appreciate all what we have imagine how it is that we are the people from 20 years ago and what effort implies to obtain the actual results.

On other hand we do not have to let the past blocking the future. The identified gaps and miss alignments must be eliminated in the most professional way by using all the available instruments and by that we are back in the sphere defined by our technological culture.

As a practical conclusion the DATA is the King and the product data management (PDM) the key element, the star, in the data exchange environment between the 3D model-base product definition, the digital mock-up creation (DMU) enhancing the 2D-3D simulation model analysis.

In this process the PLM is a relative new entry. With SLM (Simulation Lifecycle Management) functionalities that ads and integrates into the DMU behavioral simulations data. By enhancing the list of characteristics increase the level of details, the representations fidelity, and the abstraction of the lifecycle phases and opens the application for different disciplinary domains.

The actual practice still suffers from the lack of flexibility of the different software platforms and applications interoperability to link the product characteristics to the process activity.

This desire is similar with what is happened within the PLM platform: each module is practical biunique connected and finally every module is connected with each module around the considered part (project). The goal is to have a biunique bi-directional interface between the virtual part (the totally of the product characteristics) and the considered process solved within the SLM kernel.

The envisaged solution is to consider the DMU as a complex value that offers a flexible product definition and an appropriate PDM as a PLM/SLM interoperating environments.

The practical result is the definition of a conceptual and logical multi-domain data model enriched in attributes implemented in a relational data base, with a specified data structure that supports collaborative data exchange, with a dissociation-integration property regarding the system configuration definition data, management data for the topological data, structural data, functional data, and behavioral data with the necessary data import/export functionality.

Finally the digitally AI is bringing the capacity of the web semantics for communication and data exchange and the semantic data mapping for the decisional process in connection with the engineering dimension of reconfigurable BOM and Tooling.

Most of these functionalities can be already found in commercial software platforms like the Siemens TEAMCENTER FOR SIMULATION or ENOVIA V6 from CATIA.

For the working teams that starts to create industrial prototypes based on the digitally networked PLM a powerful tool (as solution) is the so called SysML, the Systems Modelling Language (SysML) an Open Source Project dedicated to the systems engineering applications that Software Development Times awarded with SD Times 100 for industry leadership in Modelling category (2007). Even a genuine open source specifications projects, the SysML is a de facto standard for the MBSE applications (Model-Based Systems Engineering).

Conceptual the SysML is a dialect of UML profile that customizes the language via three mechanisms: Stereotypes, Tagged Values, and Constraints.

The UML (Unified Modelling Language) adequate for Agile Architecture & Design, that after was standardized in 1997 has been widespread among the software engineers and developers community as a standard architecture modelling language for specifying software-intensive systems across diverse industries, ranging from finance and aerospace-defense to healthcare and automotive, including model simulation and code generation.

In this way using SysML allows benefiting from a hierarchical multi-level, multi-physics and multi-domain system and afferent interfaces to improve the lead time for setting-up and simulating integrated products by acquiring, analysing, structure and validate the data.

Within the definition of an integrated system (product) model the modelling language is essential in explicitly enabling to share the data (information) in different perspectives (views).

For identifying the modelling impact on the PLM must be clear from the beginning that PLM is not a perfectly defined TOOL but STRATEGII for dynamic complex systems. Actually is a system that integrates different systems of interest: for designing, assembly logistic, manufacturing, operating, supporting, workflows and their specific lifecycles in order to ensure the adequacy between the industrial process, resources and the company capabilities. All these elements and the included systems are evolving during the considered lifecycle so we face beside the Product life cycle management (PLM) also with the Process lifecycle management and the company lifecycle management and practically landing on the application (the Project) lifecycle management (ALM) were also the organization structure and the software configurations are considered creating an complex to handle extensive configuration management. This approach shows clearly the ERP or the Systems Engineering cannot face anymore in the face of complex activity and engineering and management approach through the PLM is necessary, enhancing the technical interactions and automated reconfiguration with data exchange, data sharing and multi-disciplinary interoperability between different specialized, organized and located teams.

Dealing with previous projects, documents and implementations actually the Lifecycle Management and the simulation of the Lifecycle Management is dealing with the capitalisation of the intellectual property (data, methods and processes) tacit capturing the knowledge of designers, engineers and experts in order to form a knowledge-based engineering platform capable to perform a decisional process in justifying modelling or solutions generation specifications, capable to reproduce a cognitive mechanism and automatic reuse of drawings, models or procedures in creating evolution scenarios in the area of PLM/SLM interoperability and standards.

Reducing the conclusion to the main functionalities the simulation models needs standardized simulation interfaces and standardized object-oriented modelling language for managing the structural data, the behavioral architecture, reduce redundant functionalities and functionalities units for a better integration of the geometric and specifications data and for the data exchange multi-level, multi-physics and multi-domain design teams. The PLM interoperability eco-system embrace the wiliness of the industrial association stakeholders in aerospace, automotive, defense, healthcare, robotics and trade, form the 11 of Europe's leading industry sector associations and (Business Europe), together with the quality checkers, research institutes, standardization organisms, universities and software vendors.

About some of these standards we are discussing in the next pages, in order to consider some PLM/SLM collaborative hubs for the next developing applications with a model-driven approach (MDA).

In spite of the fact that seems no way of assessing the quality/conformity of standards implementations in PLM solutions, due to the standards limitations regarding the implementation and usage vis—à—vis of the specific business needs, there are some examples of standards and norms to consider from ISO [10], ANSI [11], Purdue University standards database [12], Airbus [13], Boeing [14]:

First About STEP

STEP—STandard for the Exchange of Product model data and the actual designation of the STEP standard is ISO 10303 Industrial Automation Systems—Product Data Representation and Exchange.

STEP—provides a mechanism capable of describing the product data throughout the life cycle of a product, being an independent description from any particular system, suitable for neutral file exchange, and an implementation basis for sharing product databases and archiving.

STEP standards are developed for specific application domains and referred to as Application Protocols (APs). The AP is a standardized representation of product data in a specific application context, including:

AAM (Application Activity Model), a description of the component functionality;
ARM (Application Reference Model), the user's point of view application-oriented reference model;
AIM (Application Interpreted Model), a representation of the reference model through objects from the Integrated Resources as implementation view;
The Implementation guidelines and conformance conditions for implementations

STEP got also a modelling language, the EXPRESS, a standardized (ISO 10303–11:1994) lexical, object flavored information modelling language and the EXPRESS-G is an iconic language that provides a subset of the lexical modelling capabilities.

As illustrated in Fig. 20.1, to cover the entire life cycle of product data across all functions throughout the supply network a STEP series of standards has been designed (White Paper AP 239 PLCS ed3—V1.0) [15] (Fig. 17.10).

The original concept of monolithic STEP APs led to incompatible information models which did not interoperate, leading to a demand for "Recommended Practices" which facilitated common use. This was largely addressed by the introduction of the modular approach, but not all STEP APs have yet been converted to modular form, as AP235 (Materials properties) and AP238 (STEP-NC).

For modular APs, interoperability issues can arise when the SMRL (Semantic Markup Rule Language) containing all the component parts is updated to reflect new APs. Changes to modules and their architecture through subdivision and aggregation can demand the regeneration of the AP based on the updated SMRL.

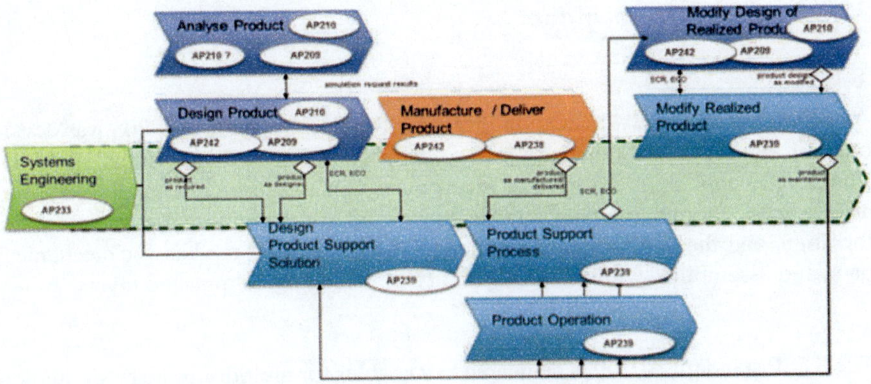

Fig. 17.10 The data flow, covered by STEP standards through Application Protocols

If an AP is not updated due to lack of resources, then there is a risk that it will be rendered non-interoperable with those generated from the new SMRL version.

17.5.1 Selected STEP

AP202 Associative Draughting;
AP203 Configuration Controlled 3D Designsof Mechanical Piece Parts and Assemblies
AP207 Sheet Metal Die Planning and Design
AP209 Composite and Metallic Structural Analysis and Related Design
AP210 Electronic Assembly, Interconnect, and Packaging Design
AP212 Electrotechnical Design and Installation
AP213 Numerical Control (NC) Process Plans for Machined Parts
AP214 Core Data For Automotive Design Processes
AP218 Ship Structures
AP221 Functional Data and their Schematic Representation for Process Plant
AP224 Mechanical Product Definition for Process Planning Using Machining Features
AP225 Building Elements Using Explicit Shape Representation
AP227 Plant Spatial Configuration
AP232 Technical Data Packaging—Core Information and Exchange
AP233 Systems Engineering
AP235 Engineering properties and materials information
AP238 STEP-NC
AP239 Application Protocol for Product Life Cycle Support (PLCS).
AP242 Managed model based 3D engineering

17.5.2 STEP Description

AP203

AP is coming from the Application Protocol. The STEP standard AP203 was developed for the aeronautics industry and is one of the most used ones in the 3D CAD. The AP203 applies in the general mechanical CAD design, parts and assemblies and for the Configuration Control Design. The AP203 is defining the geometry, the topology, and the configuration management data in solid models for mechanical parts and assemblies, without the capability of managing colours and layers.

AP 209

The STEP standard AP209 is dedicated key enabler for multidisciplinary simulation interoperability, with the ability to exchange, share, visualize for sharing and long term archiving of the engineering design and the multi-disciplinary simulation data as the "Engineering Analysis and Simulation information" (EAS).

As international standard, is having also an ISO equivalent the ISO 10303–209—"Multidisciplinary Analysis and Design", and can be purchased through ISO website: https://www.iso.org/standard/59780.html?browse=tc.

Within the simulation-driven context, doing the Concurrent Engineering and Extended Enterprise design and analysis information, in a continuous growing size of data exchange and management with strict certification processes and traceability requirements, is a strong need for multidisciplinary interoperability between analysis environments. This is the factor that allows decreasing the time to market for new products, and supports the long term retention of knowledge, and traceability of new virtual simulation processes, as described in Fig. 17.11 [Courtesy, AP209 Org].

Multidisciplinary Simulation and Interoperability is de developed for individual disciplines based on discipline-specific simulation workbenches. The interoperability is not always very strong and therefore special modules are developed to cover individual analysis disciplines such as aerodynamics, electromagnetics, flow, mechanical, noise, optics, thermal, vibrations, etc. and the links between disciplines.

STEP AP209 including simulation metadata and processes, creating a co-simulation system from the two, the SPDM–Simulation and Process Data Management with the following functionalities:

- Simulation of data quality;
- Master multi-physics effects;
- Model exchange and co-simulation.
- Long term data archiving and retrieval;
- Manage larger heterogeneity of analysis formats;
- Communication worldwide with the extended enterprise;
- Handle simulation data representation formats and management information;
- Partial validation of product physical behaviour in the early product development phase;
- CAD-CAE-PDM applications, integrating both the downstream and the upstream of data with heterogeneous authoring

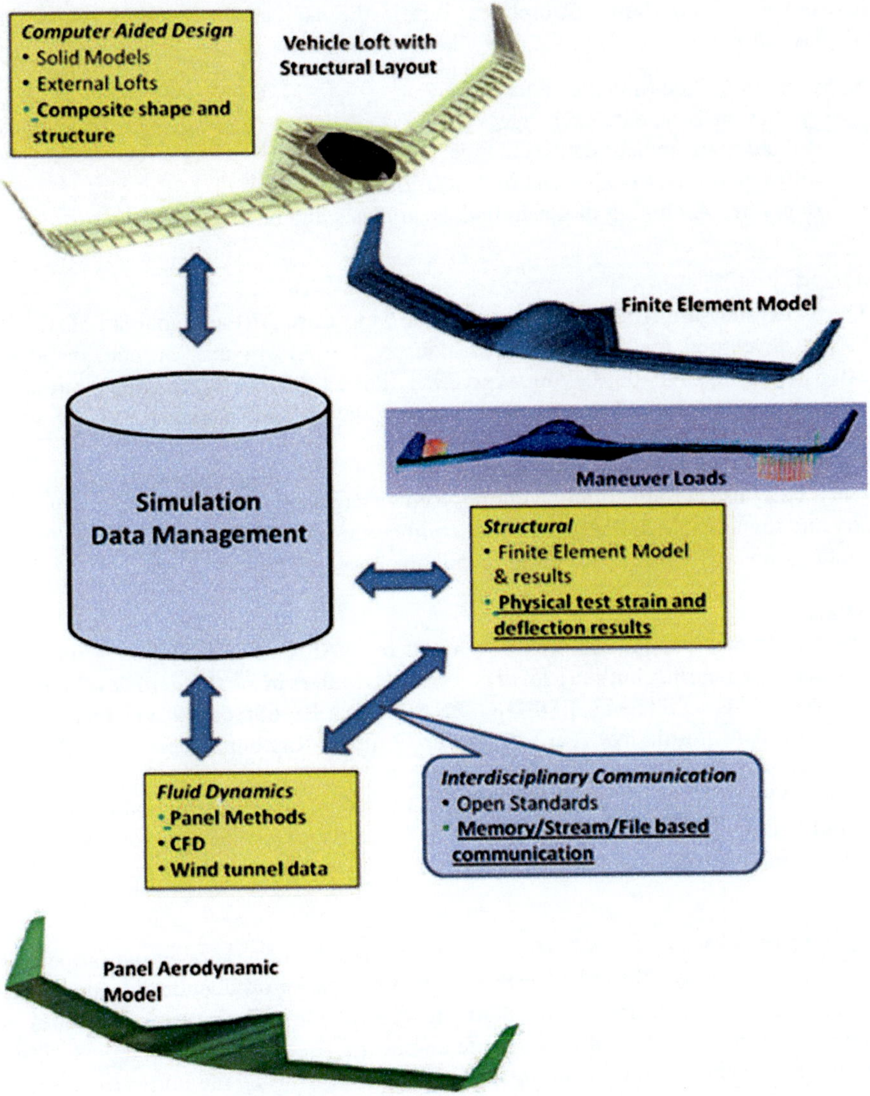

Fig. 17.11 Example of interoperability between analysis environments

With applications in a multitude of use cases the systems is aiming the added business value (ABV) by remaining solver-independent (Not being tied to a vendor proprietary format); confidence to archive valuable engineering simulation and analysis assets in a stable format, ensuring future retrieval and reuse; providing an alive the link between simulation and design; full traceability of the context in which a product was designed through capturing the metadata attached to it; sharing data across various simulation disciplines. Within the use cases a higher frequency is displayed by the.

- Simulation Data Management;
- Supplier analysis data exchange;
- Demand management data exchange;
- Multi-disciplinary design and optimization;
- Long Term Archiving of mesh models, analysis and results.

AP214

The STEP standard AP214 is another standard most used in the domain of 3D CAD. Firstly developed for the automotive industry, the AP214 may be considered an extension of the AP203, having everything that an AP203 file includes, but adds colours, design intent, geometric dimensioning, layers and tolerance.

AP227

The AP227 use certain common entities known as generic resources, such as geometry and topology, to define solid models and pertains to spatial plant technology and contains several entities pertaining to that area, such as pipe or elbows.

AP233

The STEP standard AP233 was developed by NASA as Information Model for Systems Engineering, but sun a lot of company rise inters in the standard development like: ATI, BAE SYSTEMS, BOEING, ESA, EADS, EuroStep, Georgia Tech, IBM, John Deere, Motorola, NAVSEA, Northrup Gruman, Raython, Rockwell Collins or Volvo Aero.

The AP233 is having an approach from the STEP side, information model side and the implied companies (organisations). A sample of implementations is offered in Fig. 20.3 [Courtesy: NASA] (Fig. 17.12).

AP235

The STEP standard 235, ISO 10,303–235, is related with "Engineering properties and materials information". The standard is dedicated for product design and verification, create specifications for the digital representation of engineering properties to support interoperability and through-life engineering services that can represent the measurement of any property of any product measured by any method in a computer processable form that is independent of any proprietary software. This enables the information to be more easily exchanged between different software systems and archived for long-term data preservation.

AP238

The STEP standard AP238 is related with NC (Numerical Control—of machine-tools), the now days CNC. The AP238 is replacing the RS274D standard (ISO 6983) used for general language (G-code) numerical control machine-tools language specifications for the M and G instructions that by this standard are replaced by an associative language. The new language connects the CAD design data, input attributes

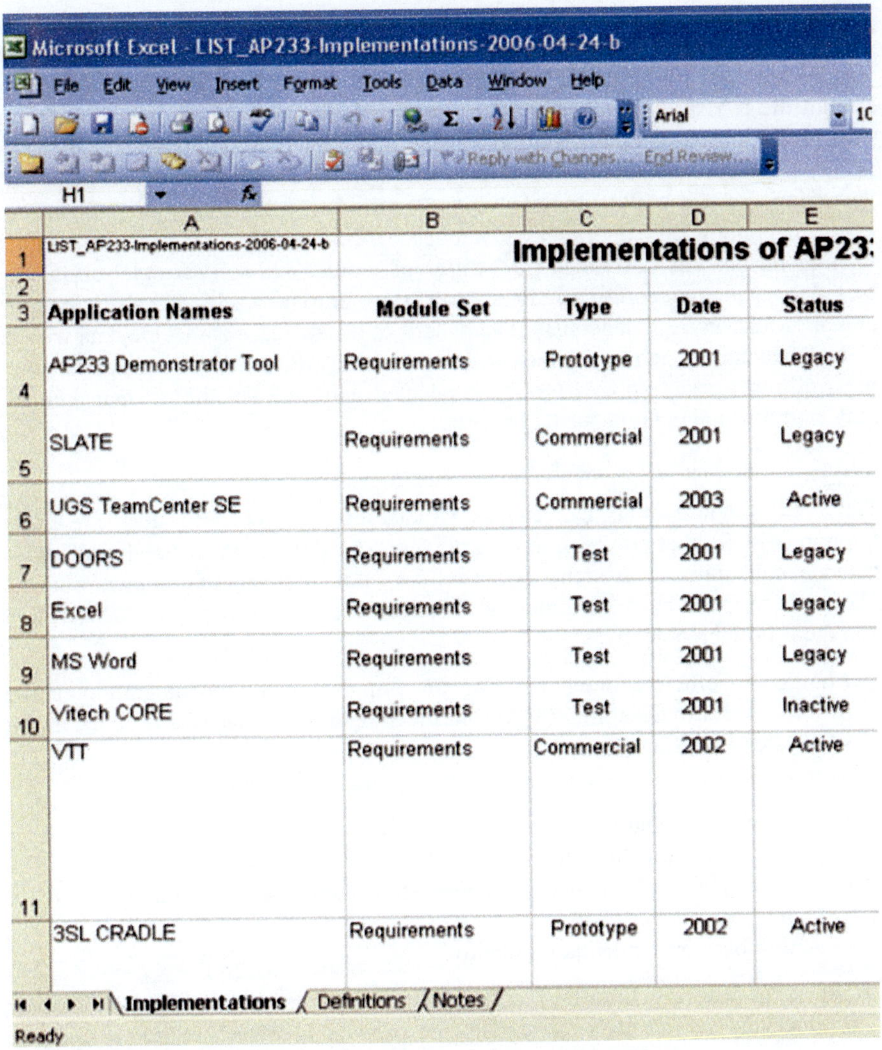

Microsoft Excel - LIST_AP233-Implementations-2006-04-24-b

LIST_AP233-Implementations-2006-04-24-b — **Implementations of AP23**

Application Names	Module Set	Type	Date	Status
AP233 Demonstrator Tool	Requirements	Prototype	2001	Legacy
SLATE	Requirements	Commercial	2001	Legacy
UGS TeamCenter SE	Requirements	Commercial	2003	Active
DOORS	Requirements	Test	2001	Legacy
Excel	Requirements	Test	2001	Legacy
MS Word	Requirements	Test	2001	Legacy
Vitech CORE	Requirements	Test	2001	Inactive
VTT	Requirements	Commercial	2002	Active
3SL CRADLE	Requirements	Prototype	2002	Active

Implementations / Definitions / Notes

Fig. 17.12 Example of interoperability between analysis environments

for the machining requirements, with the CAM process. The AP238 is a neutral data standard for CAD data, using modern geometric constructs to define device independent Tool paths and CAM independent volume removal features.

This new characteristic enables manufacturer to share seamlessly machining and measurement data between machine-tools over an internet type connection. The compliance check measure performance shows between 15 and 30% reduction in machining time and a real improvement of the yielding process. The automated measurement and compensation features offer higher accuracy at a lower cost.

The integrated simulation and checking procedures offers a correct and safer way of machining avoiding the downtimes due to wrong cuts, away from the design requirements, enabling the so called "**e-Manufacturing for Mechanical Parts**". Within the e-Manufacturing the CAD are redefined as "***working steps***", where a post processor that is embedded the specific machine-tools characteristics with the library of specific operations that might be performed on a CNC machine tool "breaking down in valid semantic steps, required to perform the manufacturing, every machining operation. The advantage of this new approach is a streamline manufacturing. The CAD data from different sources and software platforms may be sent over internet connection to any world corner to be machined by sharing reliable information instantaneously. The benefit of the standard is that the data will be instantly identified and recognized with a direct targeting to the adequate machine-tool (milling, turning etc.), the tool-set is selected, and once with the tool management may start also the manufacturing process.

AP239

The STEP standard 239 is dedicated as Application Protocol for Product Life Cycle Support (PLCS), that proves in its over 50 years of existence the capability of maintenance and support data exchange across the whole lifecycle of complex products like those in focus by A&D Industry. From the business point of view the AP239 aim the following objectives:

Manage an efficient Integrated Logistic Support (ILS), by integration of the different logistic disciplines, covering the supportability aspects over the entire life cycle of a product;

Minimize cost and the development delay including interfaces by

- Clarify the standards;
- Business specifications mapping facilitation by a template;
- Simpler and performance implementation methods, based on main stream technologies;
- A web base implementation method.

The manufacturing Industries and adjacent customers to manage the design, product and service data throughout the product lifecycle, including rigorous configuration management and the long term retention of information, where the data is 'created once and used many times', by ensuring the interoperability of AP 239 ed3 with other STEP APs (incl. AP 242);

Reliable and harmonized information model through the product life cycle and across domains thanks to the harmonization between AP 239 and AP 242;

Improve interoperability within domains and across domains thanks to the management at international level of a sharing semantics mechanism for the common information (e.g. classification and reference values defined in common Reference Data Libraries).

As can be seen the PLCS (Product LifeCycle Support) is an evolving concept, aiming on the product support lifecycle, ensure the product support from the design phase to the operation exchanging data with the other parties: buyers, certification authorities, customers, design domains, seller, suppliers etc., over the lifetime regardless if is a product or a service. That is the so called philosophy "From cradle to grave", where the entire lifecycle of a product, system or process shall be prepared, monitored and managed through a set of key indicators:

- The DMC (Direct Maintenance Cost);
- The LCC (Life Cycle Cost);
- The TaT (Turnaround Time);
- The TOC (Total Ownership Cost)
- Downtime.

Therefore the PLCS is closely related to the integrated supply chain of physical products and services, ensuring the timely delivery and availability of proper data, to the right place, at the right time.

For the PLCS success the key problem is not (as expected) the software challenge but to identify the appropriate business processes that may operate effectively assuring the right data for the internal and external processes and that data must be exchanged between the parties. A survey conducted by "The Economist Intelligence Unit" on 198 senior executives' shows that the major concern (about 40%) reveals the necessity of the right data better shared between departments for a better decision making process.

As described in the [15] and a list with the challenges (tasks) for the market players:

1. Enrich the customer satisfaction;
2. Provide Innovation and novelty in products & services.
3. The product continuous quality and reliability improvement;
4. Drastically cost reduction in all stages (design, manufacturing, service);
5. Flexibility and rapidity in creating alliances to meet the market demands;

In conclusion the AP239 standard offers the power of PLM by offering the interoperability and in this sense the approach to a new implementation technology named OSLC (Open Services for Lifecycle Collaboration), supported by STEP standards, with the objective to simplify the key integration scenarios across heterogeneous tools as an emerging approach to the lifecycle integration based on Web and Linked Data technologies, to ensure cross-disciplinary relationships.

Corroborating the standards landscape for the manufacturing industry with the major challenges that global marketplace put in face of the competitors results the standards impact vision, Fig. 17.13.

The OSLC core specification is published and available (issue 2.0 at open-services.net).

The benefits and business drivers for AP 239 are related to the STEP use. The report, "Economic Impact Assessment of the International Standard for the Exchange of Product Model Data (STEP) in Transportation Equipment Industries" offers good examples of the benefits from implementing international standards [13, page 41/111 3].

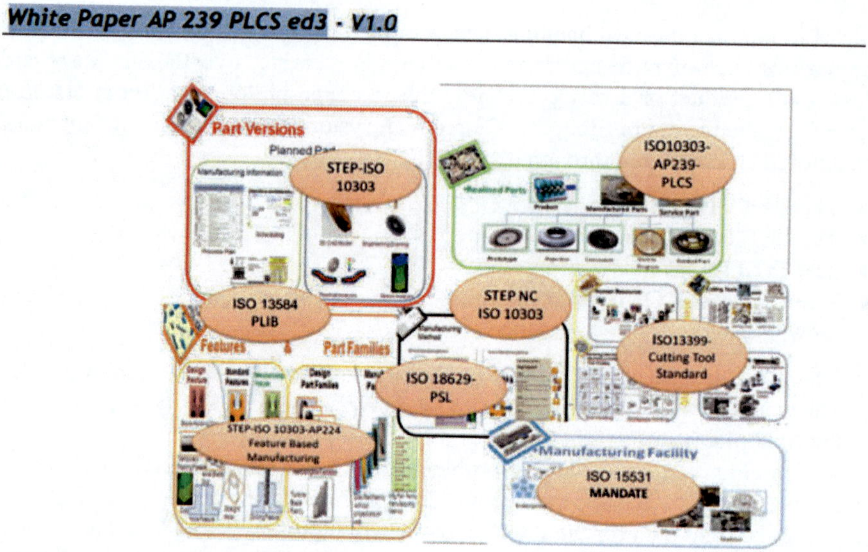

Fig. 17.13 Standardisation landscape for manufacturing standards as described in the With Paper AP239 PLCS ed3

The given examples outcome from transport connected industries data collected surveys: automotive, aerospace, shipbuilding, robotics and specialty tool and die industries, targeting an economic impact assessment of STEP's by quantifying the current realized benefits and estimating the full potential benefits of STEP existing capabilities. The estimated STEP saving s potential is at about 1 billion USD/year, only by reducing the interoperability problems.

Once with a larger penetration of digitalization and effectiveness of the data interoperability, data management and CAx applications, the costs benefits will be segmented and the savings will come firstly from:

- Decrease in IT hardware investments and maintenance;
- A united database "a single source of truth" for product and support information, eliminating the useless software service, repairing and maintenance, having as errors source the "different type of data";
- General and transparent data access across the entire company;
- Create and consolidate the necessary link between different sources of data, from different platforms (CAD, CAM, ERP, MRP, CAE, FEM, etc.);
- Decrease the costs in the supply chain;
- decrease avoidance costs, or imputed costs for preventing environmental deterioration by alternative production and consumption processes, or by the reduction of or abstention from economic activities, expenditure in order to reduce emissions and more generally pressure on the environment;
- decrease mitigation costs, that includes a range of changes, such as behavioural changes and the switchover to alternative technologies. Estimates of aggregate

economic costs considering these changes stem from different integrated models based on different assumptions;

- decrease delay costs, usually applied being a way of communicating the impact of time on the outcomes we hope to achieve. A partial derivative of the total expected value with respect to time. The "cost of delay" combines an understanding of value with how that value leaks away over time, "What would it cost us if this was delayed by 1 month?" So next type when you like to postpone a task think about applying the advantages of PLM and STEP standards.

The development STEP's standard implies costs and expenditures from governments, agencies, software vendors and industry users, but shows that all together are about 15–17% form the benefits. This is a truly quantifiable support for the digital business vision, together with improvements in product quality and process efficiency over typical practice across different life cycle phases within the company.

AP242

The STEP standard AP214 is having the ISO equivalent in ISO 10,303–242, aiming "Managed model based 3D engineering" by merging other already mentioned international standards, the STEP AP203 and AP214:

STEP AP203 aerospace, "Configuration controlled 3D design",

STEP AP2014 automotive, "*Core data for automotive mechanical design processes*".

The primary goal is to ensure the communication related to the common development and the primary use by the automotive and aerospace industries of the AP 242 standard, especially for end users (Governmental agencies, OEMs, Small and Medium Enterprises).

"Managed model based 3D engineering" is targeting the overall maturities of the associated COTS STEP interoperability solutions of the PLM vendors. COTS—in the context of the U.S. government, the Federal Acquisition Regulation (FAR)—is defined as "Commercial iTems including Services".

COTS are available in the commercial marketplace that can be bought and used under government contract, for example Microsoft is a COTS software provider.

The STEP AP242 integrates a COTS interfaces list of PLM vendors. The Business Object Model (BO Model) was introduced in AP242 (and AP209), to have information definitions at a level which was easily understood by business in their own terminology.

The BO Model allows an information model in the language of the business discipline experts. The scale and level of detail in a set of data is more suited for the communication with and understandable by domains experts (e.g. Aerospace & Defense, Automotive, Healthcare).

Using the XML markup language (Extensible Markup Language) that defines a set of rules for encoding documents in a format that is both human-readable and machine-readable, is useful to generate an XML schema (a description of a type of XML document, typically expressed in terms of constraints on the structure and content of documents of that type), may derive from the BO Model, formalized with EXPRESS, in order to support the implementation of data exchange and sharing.

The ISO 10303 standard is critical for small business participation in manufacturing (SME), allowing them to use low-end CAD systems that are in some cases one tenth (or lower) of the cost of a high-end CAD system that large manufacturers are using. Using STEP, small manufacturers need only maintain a single design system rather than multiple systems when working with multiple original equipment manufacturers.

The core standards for PLM interoperability are sharing the common data and requirements, Fig. 17.14 [16, 17].

The following modules must be considered and harmonized for evidences, conformities and acceptance as suppliers:

- Security;
- Classifications;
- Product Structure;
- Project Management
- Change Management;
- Document Management;
- Persons & Organization;
- Requirements Management;
- Configuration Management;

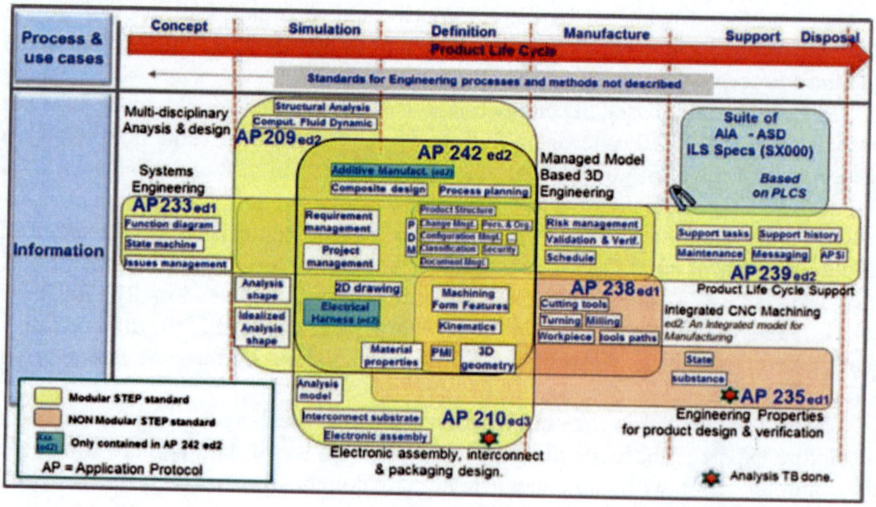

Fig. 17.14 STEP core standard AP242 for PLM interoperability [Courtesy; AP242 org]

ISO/IEC 15288

The ISO/IEC 15288:2002 is a systems-engineering standard covering processes and lifecycle stages. ISO/IEC, represents International Organization for Standardization/International Electrotechnical Commission, that may have common agreements or joint technical committees (JTC), with the purpose to develop, maintain and promote standards in the fields of Information Technology (IT) and Information and Communications Technology (ICT).

The ISO/IEC 15288 standard started in 1994, when the need for a common systems engineering process framework was recognized. The standard establishes a common framework of process descriptions, for the describing the life cycle of systems created by humans. It defines a set of processes and associated terminology from an engineering viewpoint.

- These processes can be applied at any level in the hierarchy of a system's structure;
- Selected sets of these processes can be applied throughout the life cycle for managing and performing the stages of a system's life cycle. This is accomplished through the involvement of all stakeholders, with the ultimate goal of achieving customer satisfaction [ISO];
- provides processes that support the definition, control and improvement of the system life cycle processes used within an organization or a project;
- organizations and projects can use these processes when acquiring and supplying systems;
- concerns those systems that are man-made and may be configured with one or more of the following system elements:
  - o data;
  - o humans;
  - o software;
  - o facilities;
  - o hardware;
  - o materials;
  - o occurring entities;
  - o procedures (e.g., operator instructions);
  - o processes (e.g., processes for providing service to users).

The ISO/IEC/IEEE 15289:2019, international standard for—Systems and software engineering—"Content of life-cycle information items". The standard specify the purpose and content of all identified systems and software life cycle and service management documentation. The document identifies records and information items based on analysis of references, which in some cases provide partial or complete outlines for the content of specific documents. Information items may be combined or subdivided as needed for project or organizational purposes. The management documentation IIC (Information Item Contents) is defined according to the specific purpose of the document and the generic document types:

- description;
- plan;

- policy;
- procedure;
- report;
- request
- specification.

17.6 The Compromise

Knowledge engineering was introduced in this chapter Sect. 17.3 and represents the main tool for optimizing a product or process characteristics. The optimization process usually is based to maximize or minimize a parameter that offers competitively in respect with similar products. But is just a parameter, a criterion. When multicriterial considerations are defining the optimization process thinks to get complicated.

A parameter increase (or decrease) is not an independent matter but in correlation with other parameters, the one increase may lead to another decrease or to another also increase but on an unwilled direction.

As a life guiding experience is advisable to consider the so called "Iron triangle", or "Triple constrain". The triangle expresses the "Wishes!", regardless if they are yours, your customer decision or is just prioritizing the process based on the actual conditions.

There are three main factors in all decisions processes that incorporate all the other auxiliary factors. Some auxiliary factors seem sometimes to become very important but finally they subscribe to one of the three main factors and this is the basic premise in considering the Iron triangle.

Optimization is a decision process and in the Iron triangle case only two of three factors can be controlled. The first important decision is not how to control the two but which one must be expelled from our control?

Every situation can be covered by the Iron triangle so we arrive at the idea that in any situation we have to make a compromise. When you do it, do not forget that any compromise will bring in the equation another compromise!

Very often the Iron triangle is used in the project management helping also for prioritizing tasks not only for decision making, and therefore is also known as the "Management triangle" or the "Project Management Triangle" (PMT).

I Want It All (give It All I Want It All)

I Want It All

I Want It All (yeah)

I Want It All and I Want It Now

In this context the Freddie Mercury wishes [16], seems pretty feasible if you are ready to pay the cost.

Regardless of the name, finally the "triangle" is just a tool and became a powerful tool only in the hands of a competent and professional leader.

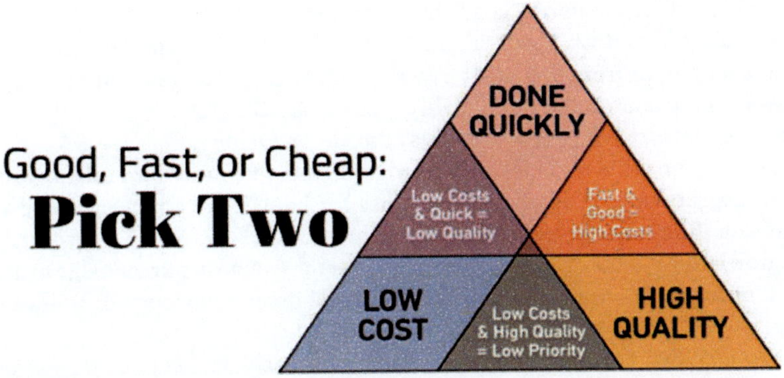

Fig. 17.15 PMT—the Project Management Triangle

But how looks this triangle? Figure 17.15, [Courtesy; OfficerReports.com].

The boss wants it all, of course. Bad news for the boss, he cannot and you knew it. So having two options:

- convince him;
- manipulate him.

YOU discovered that YOU have to apply now the PMT for YOU:

- How much time do you like to spend with the boss decision?
- How much energy do you like to spend, for the boss decision?
- How important is not to solve the matter of the problem solving?

The answer my friend is blowing in the wind.

Joan BAEZ.

Why?

Because it will be any way a subjective decision and nobody have a clue about the implications 50 years from now.

The triangle decisional model is taught in many courses, named in many ways, only here you got 3 of it. The question is what they don't tell you?

In a project one of the first things (even you know about topics and tasks) is the budget. Almost everybody calls "money" resources and are the bedrock of any projects and in conclusion of any business decision.

As long as everything is seen not like an investment but like a cost, even if sometimes an "extra budget to spend" is available, the budget will be a so fix element, a CONSTANT.

The conclusion is that in any situation we have the **"CHEAP"** implicitly as "the chosen one".

Now think about any of your lately projects. You always had a deadline for START and a deadline for END? Yes, of course, you have to get in time about everything. We are working now with another CONSTANT.

So the "FAST", implicitly, is a "chosen factor".

The reality is that we speak about the Project Management Triangle, were we have to choose two from there, but exactly for Projects this assumption is not valid. We always may control only the quality factor, "the GOOD".

This a COMPROMISE that we always made, in Quality.

Resuming the problem in most cases to the Quality Control this is always possible only in respect with the company capability for knowledge engineering, so always depends on the company technological culture (CTC).

"**Knowledge workers**" is a term typically for those who use knowledge to analyze, design, make decisions, or do some action based upon knowledge they have or can gather.

Knowledge workers are mostly considered the designers, engineers, researchers, program coders, scientists, etc., but it is applicable to anyone needing to seek or use knowledge for their tasks like: inspectors, mechanics, planners, etc., think about prototyping, single parts like space vehicles a great amount of knowledge and agility is required from everybody.

Even some prototypes transferred to the series production are not based on mundane, routine, or mindless task, but of ingenious and innovative solutions that is not the appanage of any one, as position or type of job.

We started from the question: **what they don't tell you?**

and we must consider also the question: **why are we telling you?**

Because we are looking from a different perspective in "Industrial Prototypes of Digitally Networked PLM", we need clear and powerful standards but most of all is the well trained and awarded people.

Now that the people matter, is clear that the need for Standards must be explained. By digitalization any manufacturing process becomes High-Tech.

High-Tech needs always effective and efficient technical and non-technical data, and so we may land on the problem of hidden costs in the information work.

Data and information are used in a process if is digitalized or not, but if not about:

- 25% of time is used for "data search", average 9–10 h/week;
- another 25% of time is used for "discovered data analyse", average 9–10 h/week;
- 7% of time is used for "unproductive", data that "I know that exists but I cannot find anymore!", average 2–3 h/week;
- As a result of the previous 7% is about 9% of "recreating contents that already exists:, average 3–4 h/week;

And these are the sources for hidden costs in the information work, about 66% of the knowledge workers time.

And a Knowledge worker is not really cheap, that leads that in a specialized company with a totally of 1000 employees (the knowledge workers activities is reflected in any other workers activities), the annual estimated loss is about 6,000,000 Euro.

If you have a different scaled company you can make now the math.

References

1. https://www.plm.automation.siemens.com/de_de/plm/digital-manufacturing.shtml (2020)
2. Daniel, K.: Thinking, Fast and Slow. Penguin (2012). ISBN: 9780141033570
3. Jörg, N., Claudia, F., Martin, S: Measuring the economic impact of life cycle management and service performance. Int. Confe. Compet. Manuf (2016)
4. Nick, B., Mickey, A., Myk, P: Taking an outside-in perspective of the customer lifecycle. In: Mastering Product Experience: How to Deliver Personalized Product Experiences with Product-led Go-to-Market Strategy, in SaaS (2017)
5. Irina, M., Rifat, K., Artur, B., Eduard, B: *Dealer-service center competitiveness increase using modern management methods*. Trans. Probl. Int. Sci. J (2012) ISSN 2300–861X
6. Heimannsfeld, K., Müller, D.: Requirements Engineering Knowledge Management based on STEP AP233 IMW. Institutsmitteilung Nr. 25, pp. 73–78 (2000)
7. Wielinga, B., Schreiber, ATh., Breuker, J.A.: KADS: A Modeling Approach to Knowledge Engineering. In: The KADS Approach to Knowledge Engineering Special Issue, Knowledge Acquisition, 4(1), Academic Press. UK, London (1992)
8. Schreiber, ATh., et al.: Knowledge Engineering and Management: the Common KADS methodology. The MIT Press Cambridge, Massachusetts, London, England (2000)
9. Bergstra, J., Yamins, D., Cox, D.D.: Making a science of model search: hyperparameter optimization in hundreds of dimensions for vision architectures. In: Proceedings of the 30th International Conference on Machine Learning, vol. 28. Atlanta, Georgia, USA, 2013. JMLR: W&CP (2013)
10. ISO https://www.iso.org/standard/63711.html
11. ANSI https://www.ansi.org/
12. https://guides.lib.purdue.edu/standards/MajorStandardsIssuingBodies
13. https://services.airbus.com/en/aircraft-availability/material-management/material-seminars/standard-and-supplementary-provisioning-documents.html
14. https://www.aviationsourcingsolutions.com/bac-standard.aspx
15. https://www.asd-ssg.org/c/document_library/get_file?uuid=3e01%20b539%E2%80%937d77%E2%80%9348dc-a7a5%E2%80%93076511664054&groupId=11317
16. AP 242 https://www.ap242.org/
17. Queen Song https://www.lyricfind.com/

Chapter 18
Siemens Plm Platform Structure

Embrace what you don't know, especially in the beginning,
because what you don't know can became your greatest asset.
Thomas Alva Edison

The most common representation of the PLM platform structure starts from the one described in Fig. 12.1, renamed as 18.1.

A Platform is actually a foundation upon which functional capabilities, data, and processes are enabled and executed. The PLM platform may be seen as a Business and Innovation platform:

- As a business platform because enables heterogeneous, configurable functional capabilities to support standardized end to end business processes and the related data;
- As an innovation platform due to the capability to assist in the management and reuse of the existing projects, enabling products and processes improvement and comparative functional simulation across generations of products over the entire lifecycle.

18.1 The Kernel

The central "point" the sphere is the spirit of the platform, the kernel, containing the ontology of the PLM platform. All-encompassing approach for innovation, product development and product data management are exiting in the kernel.

In big companies, like Siemens, 8–9000 programmers are working in parallel for the kernel and corresponding software development, because the Kernel is a piece of software that refers in the same time to computing, mathematical modelling, manipulation of the software objects and functions. Functional is a routine always resident in the memory, separate but used by a main program as the core of a computer operating system (COS) with complete control over everything in the system, facilitating the interactions between the hardware and the software components.

The Kernel handles memory and the peripherals, translating the I/O requests into data. In parallel the kernel methods contain class of algorithms for pattern analysis

J. Niemann and A. Pisla, *Life-Cycle Management of Machines and Mechanisms*, Mechanisms and Machine Science 90,
https://doi.org/10.1007/978-3-030-56449-0_18

Fig. 18.1 The PLM Platform (*Source* ADA Computers srl)

using numerical approximation for engineering and physical sciences, having as the best known member the SVM (Support Vector Machine).

As geometric modelling kernel, in the international market are two dominant competitors, the Spatial Corporation of Dassault Systèmes with ACIS and the Siemens with Parasolid.

ACIS stands as an acronym (Alan, Charles, Ian's System, coming from Alan Grayer, Charles Lang and Ian Braid as part of Three-Space Ltd.). The ACIS Kernel is used by many industries and software developers for CAD, CAM, CAE or AEC (Aided Engineering in Construction), having 3D modelling functionality from the ships building and coordinate-measuring machines up to animation. With object oriented architecture ACIS enables 3D modelling with hybrid modelling features integrating wireframe, surface and modelling functionality with a large set of geometric operations.

The Parasolid is named by joining the noun "solid" that represents a non-fluidic substance that is keeping a firm and stable shape, with a very interesting prefix "para" form Greek loanwords with many meanings like "side by side" or, "beside", and "to one side", or "at side" and the most interesting "beyond" (as paradox) that by extension is designating auxiliary objects or activities going up to naming occupations in English.

But for us is a great opportunity and sense of what the Parasolid Kernel is performing. From model creation and editing by having Boolean modelling operators to support advance surfacing, blending and filleting, hollowing, feature modelling, sheets modelling or thickening Parasolid, with over 900 functions is used in hundreds of software applications, integrated by more than 130 software vendors.

The Parasolid strong communication and migration of 3D Solids and 3D surfaces is mitigated by the incapacity to communicate and migrate the 2D data, like lines and curves (circles and circle arcs).

But the powerful tools of Parasolid are capable of geometric model direct editing with functionalities like: geometry removal, feature details replacement, offsetting and tapering with automatic regeneration of the surrounding data, with hiding lines, details regeneration, rendering and tessellation.

Furthermore Parsolid is enhanced with the capability to export form the parent software package the files in binary format more computing machine independent as other kernel solutions and without being tributary to the errors due to the binary-to-text conversions.

The Parasolid key elements are not dictated only by the innovation requirements but mainly are based on the demands form the actual and potential PLM platforms clients.

18.2 The Hardware

The hardware–software escalade of development, defines different implications at different stages. Without a solid hardware platform the software development is not possible and once developed requires some effort from the end-user to adapt its investments in acquisitions not only for hardware that enables the running of some software version but also in the applications.

Powerful software may run on so called "light platforms" that benefit on a single strong hardware component like motherboard, processor or memory capacity. The light platforms will work until the complexity of the product that must be modelled increase to a limit where the hardware stuck. Is the moment when most end-user are blaming the "poor software" with so many glitches and unfinished solutions, having no clue that their "poor hardware" is making the troubles.

We experienced that too. First it was in our team development, within the Siemens PLM Training Center, Technical University in Cluj-Napoca, Romania and then at our "beneficiaries" where a lot of work and test were run mostly to identify the problem and occasionally also in helping solving it.

That bring some experience we like to share with the ones decided to try this ways. In the following examples are displayed hardware solution tested or used within our Center, hardware that confirms a proper functionality adequate for wide applications.

The hardware destination is not for gaming or to create animation; in gaming applications the requirements are totally different. Here is about design, modelling, virtual products simulation and process management.

For the CAD applications is considered in the first place workstations. In a company as standard it is a single software platform shared in parallel by different kind of specialists that can be located even in different countries. Regardless the approach point of view is about professionals that are creating a reliable and innovative products using interactive a lot of advanced graphics capabilities. Even the sale department is in the loop using a lot of views, movies and simulations based directly on the on-line client data management. Therefore workstations advocate as internal the most reliable solution.

The computing power, the graphical representation, the unique data base access, swift communication, intelligent input devices; reliability, security and safety are the unavoidable demands from the hardware configurations.

Therefore we are starting with the data processing components the CPU.s, the Central Processing Units, with their requirements in memory and graphics data communication management, all centralised on the proper motherboard. In our examples is not forget about the obsolesce risk and the upgrading necessities offering a mid to long term solution where the availability and the affordability was also considered.

For the development of lifecycle management and related applications are considered in the first place also the desktop workstations solutions, but with the actual trends and the technological developments the mobile workstations will do in this type of applications.

The hardware presentation includes the desktop workstations; the graphical displays for 3D solid models or for the windows oriented hierarchical structured data and lifecycle features management; keyboard for the configuration settings and data inputs; specialised devices for 3D modelling and representations; graphical content input and windows manipulation; finalizing with the most recent solution of mobile workstations that supports the PLM applications.

18.2.1 Desktop Workstation

Models	T7600	T5600	T3600
Processor	Up to sixteen Core HT 2 × Intel® Xeon® E5-2687 W, 150 W 20 MB Intel Smart Cache 3.1 GHz Turbo	Eight Core HT 1 × Intel® Xeon®	Intel Xeon series E5-1600 or E5-2600
ECC RMT RAM Memory	ECC—512 GB	ECC—128 GB wth 4 channels	ECC 64 Gb 1600 MHz
Graphics NVIDIA Maximus simultaneous: visualisation, rendering, simulation PCI Express 3.0/4.0, 30 GB/s	NVIDIA Quadro 6000 or 2 × NVIDIA Tesla C2075 processors Full graphic slots × 16 600 W power consumption	NVIDIA Quadro 5000 or 1 × NVIDIA Tesla C2075 processor	1 × NVIDIA Quadro 6000 or 2 × NVIDIA Quadro 5000 or 1 × NVIDIA Tesla C2075 processor
Communication security Intel vPro™	Lockable ports and TPM	Lockable ports and TPM	Lockable ports and TPM
Power/Energy	1300 W, 90% efficiency	825 or 635 W	825 or 635 W

A good workstation is useless without a performant display that works perfectly with the nominated graphic cards and the application software and therefor is to be considered solutions as the following models based on a balance between engineering and management part of the application.

As long as all examples offer a full compatibility for the lifecycle management specialized software, the final decision will consider a multitude of added criteria as: price, delivery time, service and maintenance, warranty conditions, extra options like no of ports, ergonomic feature, curvature, multiple monitors connection, number and types of video ports, viewing angle, PiP, and of course dimensions.

Display Model	Resolution	GtG test timing	Contrast ratio
AH-IPS LED Philips 34″ BDM3490UC/00  [Courtesy Philips]	3440 × 1440	1 ms	1000:1
Samsung U28E590D 28″  [Courtesy Samsung]	3840 × 2160	5 ms	1000:1

(continued)

(continued)

Display Model	Resolution	GtG test timing	Contrast ratio
Samsung LU32R590CWUXEN, 31.5″, Ultra HD 4 K	3840 × 2160	4 ms	2500:1

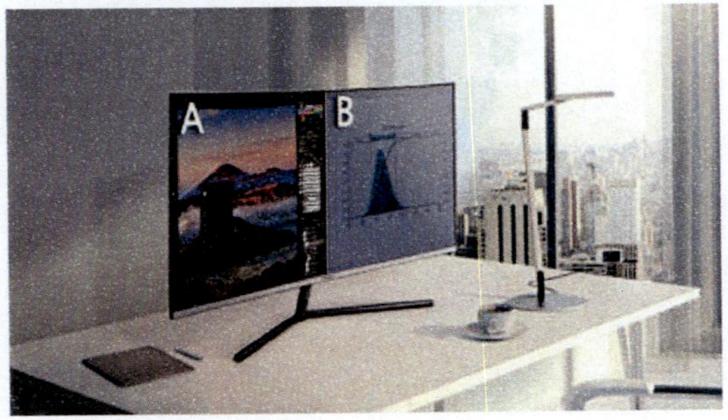

[Courtesy Samsung]

Dell U3417W, 34″	3440 × 1440	5 ms	1000:1

[Courtesy Dell]

Going from the output terminals to the input terminals one of the indispensable is the keyboard, necessary for starting sand setting the standard and specialized software but also as data input and commands device.

It is a large selection of compatible devices of this type. The selection is based on a pretty long list of criteria, some of them being selected as primary decision making characteristics but more can be added (colour, key lighting, etc.):

- Price;
- Weight;
- Security;
- Durability;
- Affordability;
- Dimensions;
- Operating temperatures;
- Mechanical characteristics;
- Number of dynamic extra keys;
- Silent typing;
- Rate of errors in pressing (typing);
- Wireless versions;
- Transceiver frequency;
- No of channels
- Signals rate;
- Protection to eavesdrop cut out the vulnerabilities of password reading.
- Power consumption;
- Mouse combo solutions;
- Internal mouse resolution;

Some examples, corresponding to this range of requirements are the following:

Model	Technology	Keyboard type	Weight	OS

Fujitsu LX410	2,4 GHz 16 channels	105 key layout	900 g	Win 10, Win 8, Win7

Fujitsu LX960	2,4 GHz 16 channels	105 key layout + 11 multimedia keys	1300 g	Win 10, Win 8, Win7

(continued)

(continued)

Model	Technology	Keyboard type	Weight	OS

Logitech K400 PLUS Black 920-007145	Logitech 2.4 GHz Wire-less Technol-ogy	117 key layout	390 g	Win 10, Win 8, Win7 Android 5.0 Chrome OS

DELL PREMIER KM717 US INTERNATIONAL	2.4 GHz wireless or Bluetooth LE 4.0	104 key layout	494 g	Windows® 8/8.1/10 Android

Beside the "standard mouse" that can be directly purchased with the keyboard (or separately) there are some extremely useful features input or features manipulation devices. Like 3D mouse.

Both types of systems are used within the design phase but also for visualisation, geometric representation and windows manipulations management.

The CAD Mouse 3D Connexion, is lowering the application costs by increasing the development rates by 20%, enabling the engagement of both hands. On one hand is the standard mouse, for traditional selections, and in the other the CAD Mouse for the model positioning doing the commands selection when navigating within the assembly and drawings.

More clicks and more accomplishments with less hand off.

Actually is a navigation device that reduces with at least 50% the need of other type of interventions and the chance of disruptions and errors, by optimal positioning of the inspecting views, easier access to the application functions (including the capacity to personalize the functions or to create smart functions).

The use of the CAD Mouse made not obsolete the keyboard, the key modifiers like Alt, Ctrl, Esc or Shift can be used also in combination with the mouse functions

Fig. 18.2 Representation of the CAD mouse (*Courtesy* 3dconnexion.eu)

(in the case that somebody is very used with them) or they simply can be transferred to the CAD Mouse functionality, Fig. 18.2.

The "drag handle" functionality is very handy, repositioning components and features, and also for applying imposed constrains together with dimensions, annotations or symbols (including key symbols for direct and quick access) [1, 2].

All presented functionalities are possible due to the technical characteristics, and the implemented software management modules, like 7 mechanical keys, 1 scrolling wheel, laser motion detection, 8200 DPI motion resolution and USB interface.

The next level of external device is the 3Dconnexion's patented 6-Degrees-of-Freedom (6DoF) Space Mouse. That can be used (rarely) as standalone device or in connection with a standard mouse, but more performant in connection with the CAD Mouse, the left side in Fig. 18.3.

Fig. 18.3 The combination space mouse and CAD mouse (*Courtesy* 3dconnexion.eu)

The Space Mouse, easy input the digital data content in industry leading applications. The 6 axes sensor system enabling performant intuitive 3D motion control as tilting, twisting, rotating, pan and zooming with one hand that can be done in parallel with other functions like editing, selection or content generation.

With an incredible low foot print and its own radial concept menu, together with a separately settings menu, maximize increase of performance and the user ergonomics, transforming the Space Mouse device from a hardly trusted "new thing" in an unbelievable indispensable assistant.

The pressure-sensitive handle that can manipulates models through different environments is compatible with over 100 performant applications like the ones that we like to mention: NX, Solid Edge, SolidWorks, Maya, Virtual Earth 3D, Blender, Autodesk Inventor and 360° Fusion, Second Life, NASA World Wind, T-FLEX CAD, etc., etc. etc., (3dconnexion, 2020).

The ultimate solution is the pricy CAD Space Mouse, the device with a total of 31 programmable keys. The CAD Space Mouse represents the top of the features manipulating device extending the functionality of the 6-Degrees-of-Freedom (6DoF) sensor and pressure-sensitive cap for instant access to Standard and Customized Views.

The CAD Space Mouse supports up to 12 favorite from the dedicated applications, automatically updatable when the application environment is changed.

Including the most advanced ergonomically design for the long hour's users, with les hand travel for the keyboard access, having a complete set of keyboard modifiers and a color LCD for visualizing function key assignment, 3 custom view keys and full-size, soft-coated hand rest.

The CAD Space Mouse in combination with the CAD Mouse, creates a synergy that represents the ultimate two handed power, see Fig. 18.4.

Fig. 18.4 The PLM platform (*Courtesy* 3dconnexion.eu)

Connected through a single twin-port USB, a comfortable single driver solution and a powerful user interface, results the capability to customize and optimize each product development for peak performance.

Easy tailoring of each key and settings using the intuitive interface, the system is running well under the Microsoft Operating systems Windows 10, Windows 8.1 and Windows 7 SP1.

18.2.2 Mobile Workstation

As a lately trend that combines the work from home and the mobility with the hardware compatible solutions, powerful mobile workstations came to enrich the desktop workstations alternatives, Fig. 18.5.

Fig. 18.5 Dell precision mobile workstation (*Courtesy* Dell)

Not forgetting the variety of lifecycle applications, from the product range of mobile workstations the selections stops at the 17″ monitor Dell Precision 7740 in the configuration with the 6 core Intel® Xeon E-2276M at:

- Max Turbo frequency of 4.7 GHz;
- 128 GB Memory
- 12 MB smart cache;
- Memory band width of 41.8 GB/s
- BUS speed, 8GT/s
- Graphic Processor Intel® UHD Graphics P630
- A base frequency of 350 MHz and
- 4096 × 2304@30 Hz HDI resolution
- Together with 7 levels of security systems
- 1 TB HD or SSD.

In the same good manner as the desktop workstations, the mobile workstation can use engineering and management solution for mastering complex cognitive technologies in robotics; AI (Artificial Intelligence); DL (Deep Learning); Economics and Financial services, in connection with ISVs (Independent Software Vendors), banking, financial analysis or trading floors; ML (Machine Learning), or to improve the products to market design; Healthcare and telemedicine together with virtual and augmented reality content creation.

18.3 The Managerial Connections

Beside the powerful Kernel and special hardware configuration, of most importance are the physical and logical connections together with the management algorithms, included under the name of "managerial connections". The managerial connections within the PLM platform are about the process customization for a product manufacturing.

All start form an "innovative idea", "the concept", a dream that must become true. The overall concept and the "concept manipulation" is considered within the module or Department "STYLING".

The "STAYLING" content things like shape, size, color etc. that must be transferred to the designer. The designer will put the idea in a standard technical understandable language: the DESIGN.

And here starts the dialog. The designer receives the entire list of "wishes" and put them in the design standards and representation rules. In the same time the designer send the feed-back in form of observations ("impossibilities"), suggestions (versions or variants derived from the designer experience), and request of clarifications ("indeterminations").

After the designer search in reuse old projects, files or designed parts, finally the dialog ends with the design approval, as "accepted representation" of the emitted innovative idea.

As “virtual object” at this stage the product is a “perfect product. On the paper, or screen, but is nothing sure about the functionality, and therefore within PLM Platform the next module is “SIMULATION”.

In this stage powerful digital tools like FEA (Finite Elements Analysis), kinematics or dynamics module are gradually used in order to demonstrates the resistance at stress functionality, fractures and the overall behaviour, in different imposed conditions with an vigilant eye on conformity standards, considered materials, interactions compatibilities, risks and cracks, errors states and interventions possibilities not ignoring things like ergonomics, manufacturing conditions, environmental issues, recycling and nevertheless the costs.

The SIMULATION MODULE reach the “inventor wishes” regarding the environment and the desired functionality and from the STAYLING module the structure and the functionality conditions that must be simulated.

In return the SIMULATION MODULE offers the functionality capabilities for the STYLING MODULE and the improvement conditions or functionality demands from the DESIGN MODULE.

In this way, the 2 way feedback (STYLING—DESIGN) is transformed in the 3 way feedback (STYLING—DESIGN; STYLING—SIMULATION and DESIGN—SIMULATION).

Now it is pretty sure that the product got a functional shape. The product fits with all the other components within a larger assembly and the material and all other considered features are known. At this stage start the problem about processing the row materials until the finish product is obtain the “TOOLING”. All tools and associated technologies must be identified, purchased and provided in order to ensure the processing.

In some situations special tools must be designed and manufactured like in the case of international company Gühring KG that with about 8,000 employees is designing and manufacturing customized high-precision cutting tools and modular tooling systems, especially for automotive industry.

With the digitalization and Industry 4.0 the conventional tooling is undergoing profound transformations, connecting the industrial product with information technology and data communication. In this way appears a transparent machining analyse in relation with the tools provider. The connection is not limited anymore the catalogue with the pallet of products presentation, consultancy and acquisition advices is followed by the price negotiations.

The link is much deeper, now the products presentation being supported by the “customized cutting forces calculator App” that can be used also latter within the manufacturing process. Also is the ‘Tool Management Software” that can be implemented in the customer fabrication process, and the “Training Academy” that offers a special customized training.

At this stage in the PLM platform exists all the meanings and the procedures required by the product fabrication, already in connection with the materials, tools and auxiliaries suppliers and a draft of the logistic chain and the early defined BOM (Bill of Materials).

The TOOLING MODULE reach from the STYLING MODULE the concept restriction "what to not happened!" with the product, form the DESIGN MODULE the exact dimensions and materials specifications, and from the SIMULATION MODULE the technological restrictions in processing the part.

In return the TOOLING MODULE offers to STYLING the "touch" of potential final finishing and aspects characteristics, to DESIGN MODULE the requirements in respect with some geometry, tolerances or material changings, dictated by the tooling processing capabilities and to SIMULATION MODULE the new processing requirements in order to be checked and approved.

This is the moment to consider the company endowment (machine-tools, devices, conveyors, etc.) that can be used by the module "MACHINING".

In this stage general technologies and the proposed tooling list must be adapted at the existing machine-tools. Firstly is determined if the actual endowment may ensure every step of the process or, some investments must be made. Secondly is to adapt the entire group of processes, down to every detail, at the company structure.

Now in the PLM platform enter the characteristics, features and requirement of every manufacturing workplace. That creates the capability of managing the entire group of electronic documents, the staff allocation and the machining management and control.

The MACHINNG MODULE, in its turn, is connected with the entire above modules. From STLING receive the overall dimensions and components that must form at the end the product, from DESIGN al the geometric and finishing characteristics, form SIMULATION the limits that must be considered for the process intensity and from TOOLING the meanings, technologies and the order of operations that must be materialized.

In return the MACHINING offers the shape and the "look" of the final part to be accepted by the STYLING, and potential requirements for geometry changes, dictated by the machining capabilities. For SIMULATION the machine-tools limits to be consider in parts processing and functionality simulation, and for the TOOLING the capacity of tools handling and tools powering.

Once ready with the MACHINING MODULE, the parts may be processed and we have to consider another module existing only in some cases, namely the "ASSEMBLY" one.

In many manufacturing fields once a product is manufactured the process is continuing by testing, packaging and shipping. In other situations the process is about manufacturing a group of parts that later must be assembled in order to obtain a complex product, the actually final product, followed by testing, packaging and delivery.

Considering the specificity of the two situations, within the PLM platform can be implemented the ASSEMBLY module like a combination of the "assembly activity" plus "the packaging activity". As long as in the assembly case the packaging may impose also disassembly activities, in case of a complex product, for wrapping and positioning the components in a specific order and after a well-studied plan.

From this description results two cases:

Case 1: Single product.

Case 2: Complex Product.

By listing all the processes involved after the manufacturing stage result:

- Parts gathering—P1
- Components assembly—P2
- Product testing—P3
- Product disassembly—P4
- Components wrapping—P5
- Product packaging—P6
- Product storing for delivery—P7
- Delivery—P8.

Connecting these processes with the two cases results the following situations.

Case 1: Single product

From P1—P2—P3—P4—P5—P6—P7—P8 =>

⇨ P1—P3—P5—P6—P7—P8

Case 2: Complex Product

P1—P2—P3—P4—P5—P6—P7—P8

It can be observed that Case 1 is just a simplified form of Case 2, and the P2 and P4 activities is something that can be planned and executed separately as A specific subprogram of the general main one.

In both cases the ASSEMBLY MODULE is connected with every single previous module. The STYLING regulates the extensions and the impact, of any kind, of the assembled product.

The DESIGN MODULE design is offering every aspect about the components geometry, components connections and fitting.

The SIMULATION MODULE shows the final functionality of the assembly, the connections behavior, the order of assembly and the components capability to be assembled.

The TOOLING defines the assembly methodologies and technologies that can be used and the MACHINING MODULE reveals the tolerances aspects that can made easier or harder the assembly activity and the storing necessities before and during the assembly process.

In return the ASSEMBLY offers for STYLING the overall considerations of the assembled product physical capabilities.

For DESIGN, the ASSEMBLY MODULE specifies the ergonomic conditions, possibilities and requirements for the product to be correctly assembled, in the normed time and without accidents and faults.

For the SIMULATION MODULE is offered the conformity with the simulation results and for the TOOLING MODULE how adequate is the list of tooling and their application.

For the MACHINING, is a useful feedback regarding the grade of manufacturing if the adopted tolerances admit an easy and correct assembly and the assembly functionality and if "generosity", in adopting increased tolerances, loss the foreseen functionality quality of the final product.

In a similar situation with the ASSEMBLY MODULE is the ROBOTICS MODULE. It is again something that not exists in every industry or in every company. This is a sort of activity with an increasing role and presence within digital activities but is not necessary to coexist with every considered process.

On one side is more and more pressure to robotize the activities and on the other side exists the logistic systems that must ensure the company handling. The large diversity of situations, that now are solved by humans, to be converted for the robots introduction needed the standards implementation, protocols, rules and specifications with a thigh monitoring of every aspect in the company activity.

This rigorous approach is not coming for free; it requires feasibility study costs, acquisitions costs, implementation costs, reconversion costs, maintenance costs and recycling (or replacement) costs.

All the specified costs and the details about their generation are subjects for the PLM platform. The final result is "the Decision" to switch or not for robotic applications within TOOLING, MACHINING and ASSEMBLY activities.

The ROBOTICS MODULE starts from the STYLING MODULE requirements, regarding the amount, destination, manufacturing location and surface material of the innovative part and the part number of components.

From DESIGN MODULE are required all the geometric data, material characteristics and the payload.

From SIMULATION is defined the surfaces contamination possibilities (the robot may contaminate the surface or the part may contaminate the robotic joints or components, i.e. high temperature or corrosive materials), the places and the points together with the way in which the part can be grabbed and manipulated or how a robotised process may be applied on the parts surface.

The TOOLING specifies if the robots will be used in the tool handling and tool storing activities.

The MACHINING MODULE specifies how can robots load and unload the machine-tools and/or the pallets with row material, unfinished or finished parts. In the same time specifies the robots involvement in the part processing, part inspection or part storage.

The ASSEMBLY MODULE specifications can be very detailed, if is decided to fully robotize the assembly process.

Vice versa, the ROBOTICS MODULE will offer feedback data about the advantages to use a robotic approach to the STYLING MODULE, details about the griping or manufacturing restrictions to the part DESIGN MODULE, geometric, kinematic and dynamic characteristics to the SIMULATION MODULE, payload and geometric characteristics for the tools and tool holder selection. From the SIMULATION MODULE may be also generated the code source for the robotic activities.

For the MACHINING MODULE, the feedback consists in the robot dimensions and gauge, dexterity, motion velocity and acceleration, payload and access requirements including the processing tact.

For the ASSEMBLY MODULE the motion capacity, the manoeuvrability, gripping and manipulating parameters are guiding the new assembly specifications.

The Process is one but the Production is a different matter. Once the tooling, machining, assembling and the robotized processes is clarified the next important module is dedicated to the workshop management, the PLANT MODULE.

A company may have one location, one plant, or several locations with different groups of employees, different machine-tools and different robots, in a word with different production capacities.

The contractual form is dedicated to the company not to the plants and therefore is the company duty to split the tasks and resources in order to fulfil all the contractual engagements.

The Plant consist in all the human and technical capacities, offering a production capacity relied on technologies, tooling and human skills, on the supply chain, inner logistics and resources assurance.

The virtualization, monitoring and management of the Plant is one of the most dedicated and important tasks within the PLM platform. The considered "innovative idea" must be representative for the Plant, all the characteristics must be sent form the STYLING MODULE and fit with the Plant profile.

The DESIGN is considering the Plant technological culture, in achieving the final product.

The SIMULATION MODULE must "simulate" also the every Plant functionality form the access capabilities, internal flows, processing capacity, materials transfer, data generation, storing and processing.

The Plant tools capacity, tools reserve and tools requirements is in direct dialog with the TOOLING module and totally based on the MACHINING module, together creating an auxiliary supply chain for tools, the technologies related with these tools, conditions to be created for the tool storage, tools manipulation, tools use and tools recycling.

The Plant assembly spaces and assembly capacities are correlated with the ASSEMBLY module; nevertheless the robots placement and robots interconnection are completely defined within the PLANT module. The placement mapping of every machining, processing, storing or transferring device is made for every new product considering the low or high re-configurability capacity of each Plant.

So looks the "manufacturing map" of the plant. Based on the mapping are creating the "logistics paths" as a different layer overlapping the "manufacturing map" where the manufacturing workplaces are considered as "fixed elements of the manufacturing".

Within this environment are generated, monitored and managed all the internal fluxes, in correlation with the contractual forms, external fluxes and resources allocation.

The fluxes monitoring and management is the new group of feedback connections that is created within the PLM platform, having a major role for the completely description of the production capacity.

The Plant must fit the STYLING as a representative Plant for the STYLING, promoting the new idea and on other hand capable to offer the adequate technological and managerial resources.

So the Plants require the representation and offer the capability, in relation with the STYLING.

The DESIGN must fit with the technological culture of the Plant. The drawings and the specifications from the DESIGN MODULE must be complete accepted by the Plant structure and the Plant facilities and must be capable to change if a new design specification appears during the products process.

The SIMULATION MODULE must offer the exact behavioural aspects of the Plant specificity guiding the changes, if necessary, considering every aspect.

The TOOLING MODULE defines the entire list of tools, the applied technologies and specifications, and this must be conform with the Plant capabilities in applying this technologies, including the necessities for the staff qualification and training.

In relation with this module may appear negations, regarding the capability or willing in using the proposed tooling list. Actually, all is coming from the confrontation between the Plant technological culture and the one that generates the part design and the one that is generating the tooling. In most cases, in the end, the confrontation may lead in changing the Plant technological culture.

MACHINING MODULE considers the Plant endowment and the machine-tools placement. That may lead to new investments in machine-tools, if is desired to change the profile and the capacity of the Plant or negotiations with the DESIGN and TOOLING modules in order to profit and maximize the existing capacity of the Plant.

The ASSEMBLY MODULE will have an important influence in the Plant spaces, the logistic fluxes and the resources distribution allocation. The Plant specific restrictions may lead to reconsider the assembly process.

The ROBOTICS MODULE plays an important role and a special attention must be paid to the ROBOTICS MODULE, due to the increase of the robotics influence on different activities within the product and product processing lifecycle.

The initial application of robotics was the "handling" for loading and unloading the machine-tools, activity that was extended to the logistic system for loading and unloading the storage spaces and the pelletizing activities. The next major extension was for the assembly and lately with the Cobots implementations the entire manufacturing and production activities including the quality control, the planning, the management and the service within warranty and post warranty period that is more and more ensured by the robots or the Corobots.

The robotics implementation is reconfiguring the position and the space allocation for the storing capacities, creates a different logistics approach, the auxiliary services for the machining and tooling management. The new workplaces and the assembly reconfiguration, together with the auxiliary services and energy supply are made based on the requirements of the robotic systems.

Once produces, the parts must be conform to the specifications. The monitoring and management of the quality control is made within the PLM platform with the help of the QUALITY MODULE.

This module is monitoring every qualitative aspect related with the part production. In the same manner the QUALITY MODULE is dealing interactive and interacts with all the other modules, starting from STYLING.

In accordance with the envisaged potential clients, the STYLING MODULE specifies the quality demands to be performed. In return the quality control offers a permanent monitoring of the products quality evolution and conformity.

The DESIGN MODULE specifies the dimensions and the tolerances levels that must be measured and monitored at the end of the machining, after the assembly process and at the end of the production line, continuously or in a batch system.

In return the QUALITY MODULE specifies the actual level of the conformity with the design but in the same time identifies the potential errors generators that may came for the considered design or can be eliminated by changing the design.

The SIMULATION process may identify the adequate technology that must be applied within the metrological process in the QUALITY module and even creates intervention scenarios for different faulty situations.

The QUALITY MODULE offers to the SIMULATION MODULE a permanent feedback regarding actual situations that may allow the weighting of the simulation process for the improving of the simulation algorithms.

The TOOLINGG MODULE offers the theoretical shape of the practical aspects regarding the end quality assumptions. The direct connection with the QUALITY MODULE may identify the correct administration of the selected tools or requirements for a different tooling approach.

The MACHINING MODULE, beside the programing facilities in the final product quality achievement, incudes also the tools-using-methodology in connection with the processed materials and the external conditions imposed by the working environment or the timing cycles (intensity of the process). Different machining conditions may lead to different quality distribution over the processed parts.

In return the QUALITY MODULE may identify the "machining spots" that are leading to quality nonconformities or even indicates the hierarchical approach in changing the machining parameters for achieving the imposed quality level.

The ASSEMBLY MODULE is acting in a similar way as the MACHINING MODULE. The difference is made in how much is robotized this activity. A low robotized process is leading to a QUALITY MODULE feedback over the working staff and managing personnel, generally a difficult situation hard to change that requires more resources, effort and longer implementation times.

A highly robotized process is leading to a QUALITY MODULE feedback directly to the robots programming or programming parameters or indirectly via the TOOLING MODULE to the robotic activities.

Now, QUALITY and ROBOTICS modules are interactive. That was already observed from the previous modules, but is also a direct interaction function. This direct interaction is considering the performances of the robots in use, the maintenance the service and the programming capabilities, in a very strong relation with the existence of qualified staff and allocated resources. All this data and capabilities are "offered" by the ROBOTICS MODULE and the QUALITY MODULE identifies and address to one or more of this capabilities in order to regulate the situation, but just in case of nonconformity.

In the case that when no capabilities parameter can be changed all the processes are transferred up to the beginning, in a backwards activity, until the desired quality level is obtained.

The PLANT-QUALITY modules relation is related to the succession of the operations or the storing-logistic activities, all having the potential to harm the level of final product quality.

The PLANT MODULE preserve all the essential manufacturing details data that QUALITY MODULE may consider in the errors generation detection and the QUALITY MODULE may address for changings in the plant distribution, for the quality level assurance.

The next considered module is the PRODUCTION one. This module connects the contractual forms (quantity, quality, delivering time and delivering conditions) with the manufacturing capacity.

The STYLING MODULE define the clients distribution and somehow the production rhythm.

The DESIGN MODULE gives the technical specifications that must be corroborated with the contractual forms.

The TOOLING MODULE offers the data about type, quantity, costs and the supply chain characteristics for the entire production process.

The machining specify the manufacturing capacity, the ASSEMBLY MODULE specify the assembly capacity, the ROBOTICS MODULE specify the variation range within the logistic activity the plant specify the rerouting capabilities for the production flexibility and the QUALITY MODULE specify the testing capacity.

All the specified capacities are registered, analysed and superposed over the contractual conditions. The bottlenecks, the rhythm changes, the supply chain, the staff allocation and the rerouting capabilities are considered over the entire production activity based on continuous or discrete monitoring.

In return the PRODUCTION MODULE offers data about the production evolution and may emit requests for the other modules.

For STYLING MODULE are emitted data about the products delivery and the contractual conformity for the client satisfaction.

For the DESING MODULE is offered feedback about the PRODUCTION capacity in respect with the design specifications.

For SIMULATION MODULE, are delivered data for the validation or the improving of the simulation algorithms.

For the TOOLING module feedback about the real demand on tooling based on the real demand resulted from cracking tools, the necessary of twin tools or about the delay in tools availability or tools conformity.

MACHINING MODULE may receive data regarding the necessity of re-programming or re-setting the machine-tools and the operational staff necessities.

For ASSEMBLY are given data about conformity or changes requirements for the assembly process and the staff necessities.

For the ROBOTICS MODULE may be sent data regarding the necessity of re-programming or re-setting the robotic systems.

For the PLANT MODULE are requests regarding the manufacturing locations, the logistic path between location and the acquisition of new manufacturing equipment.

For the QUALITY MODULE, request about different timing for quality detection or changes in the quality testing methodologies.

The next module the SUPPLY MODULE is defined the management of all suppliers for al modules that are requiring internal or external supply. This module is administrating in parallel different supply chains, generally predetermined and lately modular structured, flexible enough and capable to be reconfigured based on actual situation. Al the supply chains are provided with KPI's as triggering factors for changes within the supply chain for: materials, energy, resources, tooling or staff.

The STYLING MODULE defines some restrictions about the potential providers.

The DESIGN MODULE offers the specifications and technical data which the suppliers must ensure the conformity.

The SIMULATION MODULE may define the level of risk for a certain variety of suppliers.

The TOOLING MODULE defines the list of potential suppliers in according with the dedicated tooling set.

The MACHINING MODULE defines potential dependencies of some suppliers.

The ASSEMBLY MODULE defines the required delivery timing for the suppliers, and the ROBOTICS module may offer specifications about the batch quantities, packages and the delivery conditions.

The PLANT MODULE is connected with the SUPPLY module through the logistics, defining the delivering place and time, together with the unloading conditions and the merchandise manipulation and storing.

The QUALITY MODULE defines the merchandise (materials, tools, auxiliary materials etc.) reception by their quality depending also the final product quality.

The PRODUCTION MODULE deals impose the quantity, the quality and the delivering schedule based on the contractual production needs.

In the other sense the SUPPLY may introduce some changes in the STYLING MODULE for the concept revision; in the DESIGN MODULE for changes imposed by the existing and available configuration of suppliers; the SIMULATION MODULE get the real data about the suppliers configuration, delays and nonconformities management; the TOOLING MODULE is using the data for generating the dimensioning for the backup tools storage, the MACHINING MODULE may regulate the process based on the supply activities; the ASSEMBLY MODULE may regulate the activities and the staff necessities based on the SUPPLY MODULE data; the ROBOTICS MODULE may display the flexibility facilities in reprogramming the robotic tasks; the PLANT MODULE may reconfigure the manufacturing and storing devices configuration; the QUALITY MODULE may impose alternative regulations regarding the testing and the acceptability margin; and the PRODUCTION MODULE may benefit from the modularization, reconfiguration and flexibility of the company in order to fulfil the contractual specifications and duties.

Not only once the product is finished manufactured but also during the manufacturing process, it came into discussion the SUPPORT MODULE.

In the SUPPORT MODULE there is a duality that denote from the existence of two branches: an internal one, responsible for the process maintenance; and an external one dealing directly with the customers, for the product implementation and the service during and after warranty time.

Dual aspects are to be considered also regarding the branches interdependences in the connection between the SUPPORT MODULE and the other PLM platform modules.

The STYLING MODULE must consider that the new idea, the new concept, will not be successful only base on the novelty of the product, the quality of the product or of the affordability of the product. A large share of success will be owed to the service activity and to the service management, as was depicted in Chap. 17 “INDUSTRIAL PROTOTYPES OF DIGITALLY NETWORKED PLM”. Therefore the STYLING MODULE defines the philosophy regarding the maintenance activities, required to ensure a later cost efficient service to the customer.

The DESIGN MODULE will provide the maintenance necessities and the service methodology and SUPPORT MODULE offering the feedback regarding the functionality and the faults statistics within the clients environment that in some cases (like automotive industry) may result in call back of millions of products in order to remedy some errors or to ensure a conformity that just was supposed to be fulfilled.

The SIMULATION MODULE offers procedures and instructions regarding the maintenance and the service activities. These days the VR and a lot of new gadgets end devices offers a real support in the training and the assistance of the people that are ensuring the product support. In return the SUPPORT MODULE may offer new sets of data that are leading to better algorithms for the SIMULATION MODULE.

The TOOLING MODULE offers also the tools for the management and service activities but in the same time is using the SUPPORT MODULE feedback about the as appropriate are the actual tooling sets for these activities or in extension about stress that leads to fractures or other unconformities due to bad tooling for machining or assembly.

The tool connection is extended to MACHINING MODULE that may have strong influences regarding the maintenance of the machine-tools, different intensities of machining leading to different maintenance requirements. Also the service can be influenced the machining programing having a large variety of variables to deal with, depending on the part geometry, part tolerances, part material and available tools. All mentioned parameters are influencing the machining and finally the quality of the product and the services necessities.

Similar things happened also in relation with the ASSEMBLY MODULE. The difference is how much is automatized the assembly process. On that depends the maintenance procedures and the service intervention methodology and also on that depends how the feedback form the SUPPORT MODULE will be processed and applied within the assembly activity and how is to be insert and reflected within the ASSEMBLY MODULE.

The ROBOTICS MODULE may ease the way for a proactive maintenance process management, like predictive maintenance, or remote maintenance. More data about maintenance activity are given in Chap. 21 "THE DIGITAL FACTORY".

Once with the Cobots extension we may assume also a bigger role in using robots for the service activities. The SUPPORT MODULE offers data about the necessity of reprogramming the robots for different task during the manufacturing process or details about the necessities and capabilities of the service robots integration.

The PLANT MODULE allocates space, devices and resources for the maintenance activities in accordance with the Plant map, the required data for allocation being generated within the SUPPORT MODULE.

The QUALITY MODULE defines already the manufacturing process quality and the quality of the delivered final products. The maintenance and service activity are validating first the data from the QUALITY MODULE and second if the validation is denied, offers supplementary data to redesign and manage the quality management algorithms within the QUALITY MODULE.

The PRODUCTION MODULE defines the maintenance necessities that must be managed by the SUPPORT MODULE and considers the feedback data from the Service activities, dedicated to the Production process improvement. This is usually a key element in reduction, not necessary the production cost but the general cost with the profit increase, and better define the value adding activities.

The SUPPLY MODULE ensure also the supply for the maintenance and service activities and therefore store valuable data to carry out this activities.

As feedback the SUPPORT MODULE update the necessities customized at each type of maintenance and each type of service that must be provided.

It is important to notice that the SUUPORT MODULE manage also the recycling activity considering as a service. In this context we may have different situation, depending on the cultural technology in the company and the customer relationship:

- Ensure the bay back facility;
- Recycle the consisted materials in company own facilities, with potential integration on new products or selling to other companies;
- Deliver the used product or "waste materials" to specialised companies in return of some benefices.

At this moment practically are defined all the modules consisting data for the entire lifecycle of the product. The last one, the PLANNING MODULE, is used for the managerial implementation strategies by connecting the intentions with the resources, with the contractual conditions and the capacities.

For the PLANNINGG MODULE the STYLING MODULE describes practically not the product concept but the destination of the product, identifying the potential beneficiaries and clients that will be contacted or addressed. The advertising and media connection is included in the planning activities for clients' number rising or promoting the presale activity.

The DESIGN MODULE is connected with the PLANNING MODULE through the parts and technologies description that is divided in planning on operation and tasks.

The SIMULATION MODULE is offering a source and an environment for scenarios generation so the planning activity may be oriented on different decisional criteria to define the selected process to be managed for the products manufacturing.

The TOOLING ensures the connections with the technologies and tools management and allocation strategies.

The MACHINIING offers the real adapted technologies with time and resources estimation.

Similar the ASSEMBLY MODULE offers the data for the tasks-resources allocation.

The ROBOTICS MODULE ensures the data for some processing but mostly for the logistic operation planning.

The PLANT MODULE offers the data for initial tasks distribution, task rerouting and triggering factors of that.

The QUALITY MODULE defines the planning possibilities in connection with the contracted forms and the company existing status (technological culture).

The PDUCTION MODULE is the main module of interaction with the PLANNING being the dynamic factor that must ensure the contacted obligations.

The PRODUCTION MODULE is offering the desired input data and the PLANNING MODULE must ensure the correlation between this data.

The SUPPLY MODULE "describes" the supply capabilities and the instant statues of this activity the PLANNING module ensuring the desired start position, the evolution of the supply chain and the possibilities to adapt it at potential changes during the production process.

The PLANNING MODULE manage the maintenance activity, even more if is decided that the maintenance activity to be a predictive one. About service the PLANNING MODULE describes task to locate the parts that must be in service, the allocated resources and the design of the service network distribution.

In Fig. 18.6 are visualized all the specified logical interconnections within the PLM platform.

The PLM platform modules benefit from a single data base, considering all the data mentioned to be generated or transferred within the PLM platform. The physical connection is made over the bus/servers/internet system, but the potential logical connections are made as represented in Fig. 18.6.

In conclusion, the representation of the type of data transferred between the PLM platform modules and the logical representation of the intermodal connections, can be better understand how the PLM platform is so important for economic applications and why is so successful.

In the last years and especially with the digitalization progress, customization and decentralization made from customers the only factor that defines the production process pace.

The "must", for delivering the desired product, at the desired quality, at desired time and in desired place, together with the criteria that ask about being environmental

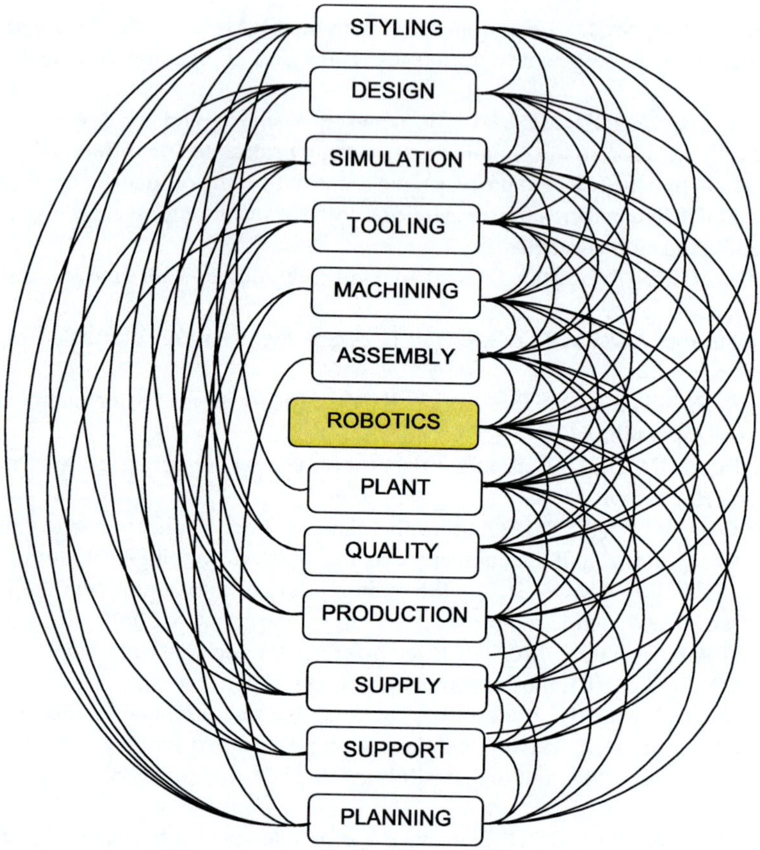

Fig. 18.6 Total logical connections within the PLM platform

friendly, cost effective and safety, requires from the production planning to be made in very tight conditions, more complex and complicated, beyond the TTM (Time to Market) conditions.

From this perspective we have to look at the companies that beyond the internal administration needs to manage also the clients and the suppliers that represents practically a multiple stages PI system (Production-Inventories) that can be monitored and managed with the PLM platform.

References

1. EPB: The Economic Payback of 3D Mice White Paper. https://oneplm.com/wp-content/uploads/3Dx-white-paper-summary.pdf (2008)
2. Reducing Physical Discomfort and Pain among 3D Computer Users, VSI Risk Management and Economics. http://www.vsi-consulting.com/files/129693397.pdf (2005)

Chapter 19
Teamcenter Data Management

Success is when I add value to myself. Significance is when I add value to others.
John C. Maxwell

Paraphrasing Maxwell [1].

> Success is when Digitalization adds value to production. Significance is when Digitalization adds value to the company.

The management control is about leadership. Within the Siemens PLM software exists a specialized module dedicate to the management dimension of the company, named the "Teamcenter Digital Platform". This digital platform has the role to share the DATA within the organization, based on the company optimized logic, enabling the communication between all the organization departments.

Capturing knowledge and engineering processes are crucial for improving success, starting with the early stages of the product lifecycle and therefore we have to know:

What is Doing the Teamcenter?

Exchange Components

1. Documents
2. CAD Data
3. Product Data
4. Workflow Data

J. Niemann and A. Pisla, *Life-Cycle Management of Machines and Mechanisms*, Mechanisms and Machine Science 90,
https://doi.org/10.1007/978-3-030-56449-0_19

The result is the Bottom-line benefit by reducing the product development costs, integrating all the designs created using the company assets within the product's engineering content and in the end within the company value chain.

How to Approach Teamcenter?

By the Team Center Implementation

1. Start an application
2. Search work data
3. Data processing and optimization.

That means the synchronization of the engineering processes enabling all the value chain participants to access all required existing data, necessary to each for doing its job properly and effectively. The result is the Bottom-line benefit by compressing the product development cycles, ensuring the qualitative level and reducing the errors and waste in the manufacturing.

The Teamcenter is Addressed to What?

Known and Unknown

1. Organization structure
2. Process (fabrication) planning
3. Production planning.

Regardless of the locations or number of members, Teamcenter enables the product related teams to work together efficiently and effectively. The result is the Bottom-line benefit by leveraging the company global engineering and management resources.

On What is Teamcenter Useful?

Development Stages

1. Product Engineering
2. Product Management.

Establishing the management and tracking of the product configurations as well as across multiple products versions is ensuring the assets re-use over the entire product lifecycle. The result is the Bottom-line benefit by slashing the development costs.

Which Are the Engineering and Management Components that Are Used by the Teamcenter?

Parallel—Syncron Engineering

1. CAD components (Parts, Assemblies, Sketches)
2. CAE restrictions conformity and technology elaboration
3. CAM/CNC
4. FEA—stress analysis for optimization
5. Cost Calculation
6. Service Documentation
7. ERP—Enterprise Resources identification monitoring and Planning
8. PPS—Production Planning System
9. Packaging
10. Logistics

Enabling a seamless collaboration between development teams, manufacturing teams, suppliers and customers, accelerates the product delivery to market.

The result is the Bottom-line benefit by placing the innovative products to be first to market, increase the annually number of released products increasing significantly the revenue form the product portfolio.

Which Are the Engineering Components of the Teamcenter?

Product Engineering

1. **Conceptualization**
   - Initial concepts, sketches, ideas.
2. **Engineering Specifications**
   - Requirements regarding the parts or the assembly, including material types, strengths, corrosion, coatings, tolerances, or surface quality.
3. **Design Studies and Analysis**
   - Analyses, systems engineering and interactions, which made the design works better, more aesthetically or pleasing, etc.
   - Determinations like strength, strain, elasticity, reliability, durability. This can be utilized to help make determinations on cost vs. durability.
   - In industries like aerospace or automotive, that may be life and death determination.
4. **Design Reviews**
   - Evaluate the current design, relative to milestones, design criteria, conformity, product functionality targets

- Lately, a great deal of decision making and evolution of the product happens in the form of parallel engineering quality checking and cost/functionality supervision.

5. **Drawings**
   - Are an important part of the product development processes, suffering an intense process of digitalization.
6. **Electrical Engineering Integration**
   - Mechatronics is the today's product development. There is more electronics integrated than ever before. Not only the cables, wires, and harnesses, but there is consideration of integrating chips and firmware like never before. This aspect of the product development process has often been neglected in the past, and the technology and particularly the integration thereof is still considerably behind the mechanical CAD.
7. **Simulation**
   - Useful to simulate the motions and the interaction of products or systems
   - Crucial for a better understanding of functionality and design quality
   - A great asset in the manufacturability and assembly of a product, including robotics, machining, and human factors interface.
8. **Technical Illustrations**
   - Specific type of documentation that shows detailed information about a product and how it is to be assembled or manufactured, useful for maintenance, service or marketing.
9. **Work Instructions**
   - specific documents explaining how to perform a specific task, a work area, or a technological process.
10. **Installation Plans**
    - Similar to Work Instructions but targeted on the product, on the assembly or the manufacturing process of the product. This may include technical instructions (how to bolt something) or managerial aspects (run a cable from one area to another).
11. **Quality Assurance**
    - Inspection procedures must be doubled by appropriate tools and trained people for the entire processes
    - Must cover all aspects from materials purity to the final shape, tolerances, fittings and functions, but also including the calibration of the machine tools and robots, together with the imposed climate control for the manufacturing process and the storing facilities

Teamcenter as the leader in multisite collaboration allows geographically dispersed product teams to collaborate on common design within an automated engineering and manufacturing process.

By knowledge sharing enables to federate the database into a single logical system.

Which Are the Management Components of the Teamcenter?

Product Management

1. **Initial Requirements**
   - voice of customer, previous product information.
2. **Comparison**
   - studies Version 1 versus Version N
   - analyses
   - cost versus function
   - market studies, which products to take forward.
3. **BOM Configuration**
   - Bill of Materials considerations, how to structure the sub-sections and group of tasks:
   - design
   - larger assemblies
   - manufacturing
   - assembly
   - production
   - packaging
   - distribution
   - service and repair
   - etc.
4. **BOM Conciliation**
   - as a product design matures, the BOM may change and morph depending on configurations, or more importantly which group/discipline is leveraging the BOM
   - it may need to be configured differently for manufacturing than for design and even differently for maintenance, packaging, etc.
   - important that this BOM to be maintained with the correct components in the assembly when the product gets delivered.
5. **Configuration Management**
   - is considering the modular or switchable or optional elements that can be used to way to configure a product, creating offer options to the customer

for different styles, options, features and avoiding the demands that find the company unprepared.

- in many industries is a practice to make the most of existing designs in offering a more vast and/or niche array of products and derivatives, i.e. automotive industry, robotics or CNC control panels.

6. **Quality Management**

- specific quality policy and procedures must be defined in order to deliver a robust product
- must be implemented from the raw material selection to the finished components.

7. **Documents Inspection**

- are the documents utilized to track the evolution of the products and processes based on automatic or human driven inspections of the materials, tooling and components
- Documented Data Processing (DDP) and Reporting deviation that may need attention.

8. **Engineering Change**

- for any change in design, due to marketing demands, reconceptualization, optimization manufacturing constrains or failures it is a need for quick actions as followers of the quick and accurate communication system that inform the engineering organization in order to ensure the continuity of the process implementing on line the manufacturing transition from the initial version to the optimized design at the lowest possible costs.

9. **Supply Chain Management**

- continuously the updated Data are shared with all the categories implied in material, parts, tools, sub-assemblies, energy, consumables and auxiliary components supply
- this is a bidirectional process
- ensure the receiving of the quality parts, on time, and at a competitive price
- the global environment implies different strategies for the SCM: JIT, RFQ, etc.

10. **Regulatory Documentation and Documentation Compliance**

- are documents required by regulatory organizations
- is a portion of product development often overlooked in the manufacturer view, and can cause a great deal of delay, pain and penalties if it is not kept concurrent.
- all required or specific documentation must be generated and stored in a planned manner and standard offering a direct tracking for all products and processes evolution for evaluation and for standards conformity and certification.

- the documents can be internally developed or imported from a large industry driven or even governmental organization such as EU-OSHA (European Agency for Safety and health at Work, as equivalent of OSHA-Occupational Safety and Health Administration) or EEA (European Environment Agency, equivalent of EPA-United States Environmental Protection Agency).

11. **Service**
    - documentation on when and how to repair a particular product or sub-assembly
    - extremely valuable to leverage CAD/Engineering information in the service area.
12. **Maintenance**
    - is a preventative activity to eliminate accidents and unplanned stops of the process
    - is an opportunity to inform the customer when they need maintenance on their product
    - ensure better customer satisfaction and relationships with a longer lasting, higher quality experience
    - is opening up an opportunity for service revenue as well as predictive service/maintenance revenue.
13. **Packaging**
    - a marketing and management subject that dictate about how a product is broken into sub components and what design, how and what materials will increase our ability to package it in specific configuration.
    - it means a great deal of managing the supply chain, inventory, and shipping costs.
14. **ERP Integration**
    - PLM is including all the ERP components
    - the manufacturing organization will be able to leverage product data in business decisions
    - fill the gap on what "traditionally" is the disconnection between Product Data and Business Data
    - as unbelievable it is, manufacturing organizations can make financial plans, product direction and sales/marketing decisions without leveraging their own product data in consequent, robust and/or concurrent fashion, a chasm that PLM alleviate.

Teamcenter engineering provides a multi-CAD environment for all the members on the market teams (NX, SolidEdge, Por/Engineering, Catia, SolidWorks, AutoCAD). By integrated visualisation improves communication and collaboration between all members and others lifecycle teams that can view and understand the virtual product without having knowledge how to use a CAD system.

The innovative content is "injected" in the virtual product with all definition data, properly aligned and full concurrent, that can automated expedite downstream product lifecycle process with a continuous design validation and product lifecycle management, overcoming the communication barriers among OEMs suppliers and their intrinsic allied partners.

Capable to manage MDE (Multiple Discipline Engineering) and PDM (product Data Management) modules the Teamcenter facilitate: SEM (Systems Engineering Methodologies), Design for Six Sigma and Lean Design.

The SEM allows considering the product as a whole. The holistic approach systematically account the product change, evaluates and optimize the trade-offs that affect the product and the interlaced processes across the product lifecycle.

By enabling to consider all the negotiated customer requirements by building the customer requirement in the manner that can be measured, optimised, tracked and verified along the entire lifecycle it means that they are fulfilled all the requirements to apply the Six Sigma set of techniques and tools for the process improvement.

The capability to eliminate all aspects of waste and by the knowledge management and process automation is actually improving the effectiveness, the efficiency and the speed of the development processes that means nothing else but on a long-term continuous improvement by systematically incremental little changes, actually the design and the application of a Lean Management style.

19.1 Data Management—Data Continuity

From the title is already indicated that Teamcenter is a company monitoring device and a decision making assistant. Teamcenter challenge the IT department for the data acquisition, data structuring and most of all the Teamcenter Digital Platform administration. That means that the company strategical decision to purchase the Teamcenter Digital Platform brings a lot of trouble, but the right implementation will lead at the expected success, the "Guess" decisional process will be transformed in "Informed" process deployment.

The Teamcenter is supported by any common platform and leads the company to become a PLM beneficiary as a PLM user.

The Teamcenter is designed in a matrix form having Teamcenter levels of applications (the destination), represented on the matrix lines, and a different number of specialized modules (the application), represented on the columns.

The task of the lifecycle manager is to identify each data that must be placed in any specialized module, by whom and under whose supervision and certification the data can be accessed. Inserted or changed. In parallel with the data identification must be defined the connections between data cells, constructing a systems-level, behavioral logical defined and functional that must perform as intended.

The Teamcenter will assist from the beginning the PLM implementation. In any application Teamcenter becomes useful offering the capability to identify the company status in respect with products, processes, persons and assets.

The beginning starts with the Level 1. All groups of data: mechanical, electronical, documents, BOM, regulatory requirements, environmental impact, cost and quality, software and simulation specifications are gathered to be standardized, centralized, managed and shared.

In one sense this is a basic activity that implies the most and the hardest part of the implementation work: putting the Lifecycle Manger in control.

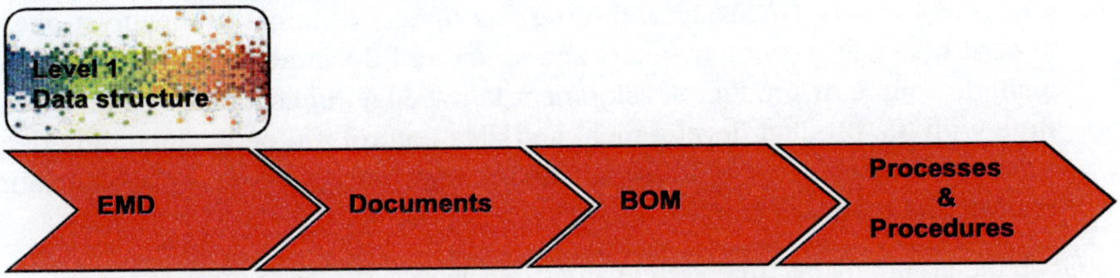

The 1st Level establishes the following abilities to the Lifecycle Manager:

1. Cut the processes managing costs;
2. Identify and streamline the manufacturing operations;
3. Creates the capability to manage cross functional teams;
4. Reduce the manual effort in documenting and updating the operations data;
5. Enable design collaboration with the design re-use, design completeness and quality validation, company assets intellectual property protection;
6. Implement "best-practice" models for shortening the gap between design/development and manufacturing, and between planning and manufacturing;
7. BOM accuracy across organization, with a complete common source connecting the company internal demands with the suppliers profile and end-user interaction;
8. Proven content, for re-use in new products, meeting the desired quality standard, extend the variety of products offer and exceed actual market demands for complex and varied products;
9. Manage the multi CAD, multi domain, multi applications mechatronic parts generation and simulation data, dealing with virtual products, in a single environment as a unique product, updating and synchronisation being made instantly;
10. Eliminates the CMP (Change Management Profile) as a costly, fuzzy, error-prone and slow solution; by delivering consistent, right and updated data, at the right time, at the right location, to the right person with the proper authority to do the workshop activities;
11. Accurate BOM definition for stakeholders, with no "common situation" of incorrect or out-of-date data. Create the image of special demands determined by: changings in the product lifecycles, the company organization evolvement, the dynamics of the supply chains or the new products complexity;
12. Generate a full overview about tasks, status not just implementing processes step by step, linking the disconnection between planning and execution, by reducing the large amount of time spent by managers (not Lifecycle Managers) for statuses/deliverables tracking, or rerouting the work in execution systems;

13. Eliminates the different BOM representations within different company departments with the inherited representations synchronisations that lead only to delays and costly mistakes. The single part product or the million parts product with hundreds of configurations will be treated in the same managerial way, taking advantage of the Teamcenter flexible and scalable advanced tools;
14. Documents management and authoring catalogues, technical publications, operational manuals, service manuals and structured documents production, from a single source of product development knowledge, in accordance and in real time with the product development and data control protocols, including data links, printing controls, product data and change management, version/revision control and their workflow.

Once the company became visible and manageable, the gaze turns to the related domain and processes. In this categories are the contractors and suppliers for related domains with the company activities, but also to the connection between the engineering and development department with the manufacturing area and the maintenance and service area, as a related process to the company main activities. The purpose is to extend the value of the company assets across the product lifecycle where PLM becomes not only a necessity but also a capacity to extend the value. And these will be the activities to be managed in the Level 2 of the Teamcenter.

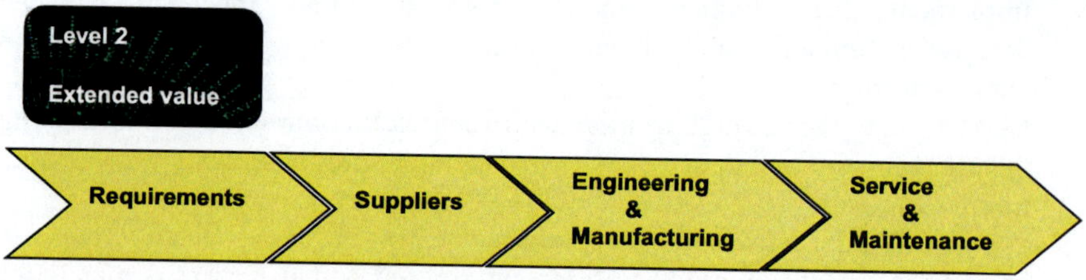

The 2nd Level defines the following abilities to the Lifecycle Manager:

1. Communication platform with the OEM and suppliers, for accurate and up-to-date data, bringing visibility over the supplier performances, reduce the back-office costs, implies the suppliers in the PLM process, and includes the collaboration of the OEM's with first time suppliers ensuring the protection of intellectual property to the first one;
2. Validate a globally distributed supply chain;
3. Ensure the compatibility between the suppliers with the internal CAD environment and manufacturing processes;
4. Enables concurrent engineering simultaneously with the product design optimization and synchronized with the manufacturing deliverables;
5. Maintain the product quality targets managing across multiple domains the lifecycle costs;
6. Avoid the pressures of resource constraints and competition when shifting market demand;
7. Enables lean service operations for scheduling the service and delivery tasks;

8. Generates and use a single knowledge data base;
9. Generates Parts catalogues, Products documentation, Operation and Service Manuals;
10. Avoiding ordering wrong parts due to lack of asset configuration;
11. Monitoring repeated service events and low first-time fix rates;
12. Monitor and consider all contractual aspects. Introduction of the "voice of the customer" into the product by tracking the customer requirements, the costumer contractual compliance, the costumer requirement documents, that some time challenge the identification capability or are out of date;
13. Generates the linkage between the customer requirements and their physical, functional and logical implementation, over the entire product lifecycle.

The Level 3 of Teamcenter is defined under the "Iron triangle" philosophy.

Is the moment when the new technological culture will become active, generating a system-driven approach to the product development, having the Iron triangle regulatory elements: the cost, the quality and the sustainability.

To transform the actual solutions in profound business impact ones, the initiative belongs to the Lifecycle Manger, generating a business leading direction in the company profile.

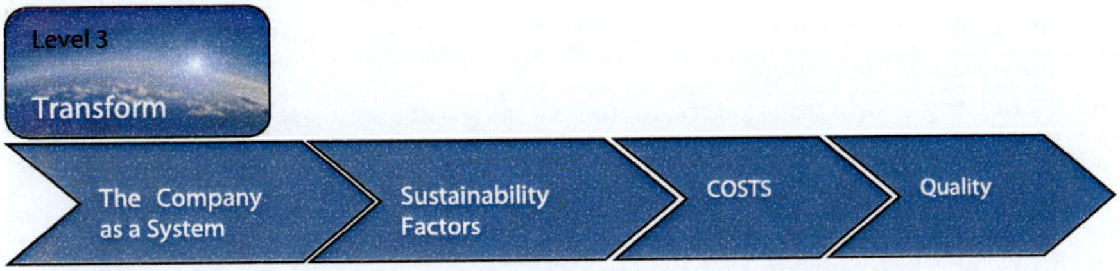

The 3rd Level offers the following abilities to the Lifecycle Manager:

1. Build the brand value;
2. Maintain assets in operation;
3. Shortening commissioning times;
4. Creates centralized trusted records;
5. Track all parts and processes sequences;
6. Establish a long run information management;
7. Supporting connection with emerging markets;

8. Provide support for decommissioning at end of life;
9. Assessing and management of requirements and risks;
10. Running up by deploy cost effective and fast the PDM;
11. Beside visibility and control of processes, offer also predictability;
12. Keep updated and in real time the procurement records – documents handling;
13. Reduce the capital spending by postponing or expel purchases of new equipment;
14. Compensate the rapid obsolesce and product growing complexity with re-use activity;
15. Define the shorter time/cost solution for the production roadmap;
16. Fully integration of the product demands: design-development-planning-manufacturing;
17. Manage design complexity, mitigating the system failures due to subsystems integration issues;
18. Support the energy consumption, aligning the cost management with product development;
19. Intricate supply chains to produce components for the manufactured product, mitigating the manufacturing issues;
20. Risk noncompliance reduction by greater certainty in products development, enabling compliance grading and reporting;
21. Manage everyday tasks with preconfigured or adaptable workflows for engineering change and product release, eliminating the interface discrepancies;
22. Productivity improvement, by minimizing the manufacturing errors and re-work and automating the supplier material and substance declaration request process;
23. Managing the "retirement" term within the product lifecycle by maintenance, refurbishing and service activities, eliminating the lack of product data re-use;
24. Proven intellectual property in a collaborative platform, keep a lifetime's worth of specifications with digital signatures, together with the PCH (Project Change Histories).
25. Initiate company initiatives for upgrading and expansion, implement proactive, forward-looking strategies, assessing the design and material choices impact on downstream processes;
26. Balancing conflicting factors, including trade-offs between cost, time-to-market and meeting design specifications, with mainly impact on additional expenses or impact the quality while increasing the risk of repeat issues;
27. Holistic cost modelling. Reduce standard cost, mitigate the unforeseen costs from labour, materials, technology or tools, and create value, making cost analysis with included impact of purchase and sale, anticipating the profitability of a program;
28. Overcome challenges raised by global fiercer competition, limited resources, rising costs, regulations, short time to market, exceed the inability to assess the impact of design or requirement changes and being forced to make substantial compromises late in the process;
29. "Going green"—as a must, with legislative environmental compliance monitoring the every material and substance (quantity and concentration) being

reported, avoiding in any way fines for failures to complain with regulations, the damage of the brand or to be banished from markets.

Finally the Teamcenter looks like an operational tool of the Lifecycle Manager, offering him the driving levels for the business activities.

In the end the Teamcenter is the he overall matrix, the dashboard for monitoring, and the control panel for the activities.

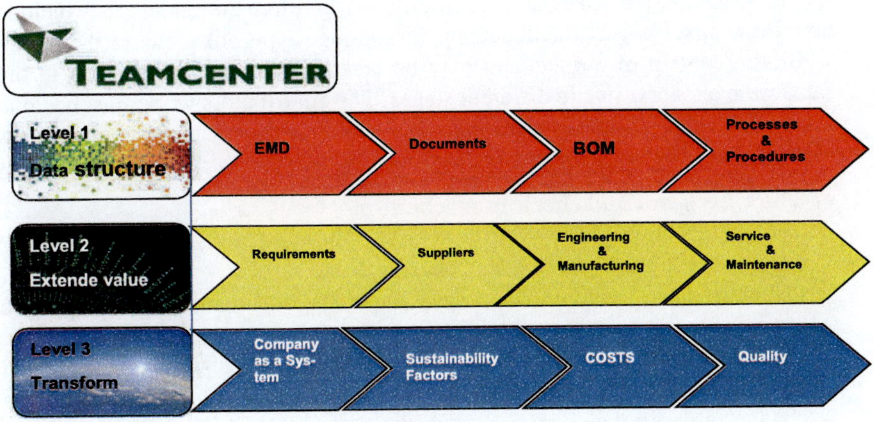

Teamcenter, as extensible platform, is growing up with the company, having also the deployment flexibility to be deployed at a single location or simply at the global scale with all the clients, partners and suppliers.

As the most advanced solution for Defense (including ITAR—International Traffic in Arms Regulations) and Aerospace, is offering the Best-in-class security, using the Teamcenter Mobility™ that provides secure and easy access to PLM anywhere, anytime for remote or mobile workers and in the same time reliable in powerful visualization and sharing the 3D product data by using the JT™ data format.

The document management processes and systems bring value to the organization by ensuring the accountability and transparency of documents creation process and how particular data, as information objects, come to be.

A manufacturing company is dealing with components and processes. Both elements are "represented" in documents, documents that must be tracked, processed and managed.

The Teamcenter deals with all documents managing the entire process from inception through completion, enabling the efficient automation of key task regarding like approval, assembly, quality assurance or packaging process.

The accountability and transparency of documents is ensured at every step in the process for the entire lifecycle. Actually there are 7 standard steps to be considered:

1. **The Document generation**
   In the initial stage of the PLM—Teamcenter implementation, for each type of document is created a Template, in accordance with the company profile. Then each time a document must be created is made based on the blank Template.
2. **The Drafting**
   Each document type has certain specificity and therefore the content may include text, drawings, images, the document formatting, hyperlinks, etc. In the meanwhile the content of a document may be created by more contributors, in the same time or successive in different stages. The contributors work must be done in parallel and sometimes synchronous but without overwriting each other work. There are several mechanisms that are ensuring this protective behavior:
   - **The Check-out and check-in mechanism**
     The EDMS allows a single user to check out the document, allowing other users to read it but not make any changes to it.
     Once the user is done making any changes, the document is checked in and is available for another user to check out.
     Every time the document is checked in with changes, a new version is created so everyone involved can see what has changed between each version. If a change is made that is not desired, the document can be rolled back to a previous version.
   - **The Co-authoring mechanism**
     The EDMS allows multiple users to work on a document simultaneously, but does internal locking within the document.
     For example, in SharePoint, Word document is locked at the paragraph level.
3. **Review**
   A different person that the one that generates the document is assigned to review the document for its overall content and also including things like the accuracy of tables and images, document flow, grammar, spelling, etc.
   The EDMS (Electronic Data Management System) may assign automatically a review to a particular individual or role (based on the implemented business rules)
   The EDMS also is monitoring and ensure that the review is complete before a deadline and before the document can move on in the initiated process.
4. **Revision**
   After the Review completion the document draft return to the creator(s) to insert or process any necessary change.
   Like in the document generation process, if any change occurs a new version is generated.
5. **Assembly**
   This is a step demanded only by complex documents, like contracts with multiple terms and conditions.

Based on the business rules, the EDMS ensure the inclusion in the correct form of the terms and conditions into the document and in connection with the work is to be performed.

6. **Approval**

   The approval process is well defined in Teamcenter, most of the documents have to pass such a process in one or more steps, ending with the confirmation signature of the responsible person.

   Final representation of a product, final form of a project or the contractual form is the trivial examples of the documents of this sort.

   The approved document represents the formal form that may be published and ready to use by all designated people.

7. **Storage**

   Represent the last processing step of a document, once the document is completed. The document is usually stored in a repository together with indication about the information it contains, the authorization conditions for access and the list with authorized users.

During the product lifecycle the standard steps are applied of all sort of documents generated by different departments of the company like 2D or 3D representations or models, tests and simulation results, product specifications, variations analysis, technical or managerial reports, images and product views, technical publications etc. all must be accurate and in due time finalized under the pressure of readiness and in getting shorter time to market.

This every day stressful working environment became critical for the sustainability and the success of a company. Therefore, by managing simultaneously the documentation, the technical publication and the user access Teamcenter offers a simultaneously evolution of the product for all the company departments, keep updated all the products features and the documentation evolution, contributing to a better quality of the final product, reduce the documentation, the number of revision, documents and product generation time and costs.

At the global level Microsoft and Microsoft Office applications are much spread is important to mention that Teamcenter integrates organic all the Microsoft Office Applications within the entire Siemens PLM applications portfolio, offering the entire capability to create, modify or update the documents using the Office environment.

This enhanced compatibility enables a swift integration of the new incomes, in a familiar environment for documents editing, documents check-in and check-out, sending and receiving of the document along the working and processing fluxes using the Word®, reviews and sign-off using the Outlook®, or creating reporting spread sheets or considered BOM with Excel®, with no necessary allocation of learning time or being dependable of a learning curve or certification.

In this way the effort to generate documents form any PLM application, once reduced, the working based on Templates became valid for the entire company. The insertion of related documents with different products structures can be electronically checked, approved and signed, and is not created only a control of data and documents but in the same time increas every department productivity.

Advanced Products are coming with a diversity of data already from the conception phase. Advanced products require also advanced publication methods in relation with clients, suppliers, stakeholders or media. The Teamcenter offers the chance of publishing using XML, the metalanguage which allows users to define their own customized mark-up languages, for annotating a document in a way that is syntactically distinguishable from the text, especially in order to display documents on the Internet.

Additional, Teamcenter enables the use of S1000D and DITA standards. The S1000D standard contains XML specification for preparing, managing, and publishing technical information for a product, as international specification for the procurement and production of technical publications that is now free of charge and can be downloaded from the S1000D website.

DITA (Darwin Information Typing Architecture) is a standard that promote the documents creation and management specification that builds reuse content into the authoring process. The latest version of DITA was approved as an OASIS (Open Standards Opens Source) standard. This is also considered and used by the Teamcenter through integration with Cortona 3D RapidAuthor. The powerful cost-effective authoring suite offers facilities in reuse the existing CAD data to generate the insertion of the interactive 3D representation, instead of multiple text pages, and produce support documentation like: Parts Catalogue, Operation Manuals, Maintenance Manuals, Training Manuals or Work Instructions.

In conclusion Teamcenter is the most important managerial tool within the PLM integrating all the other aspects. The preparation and training for the PLM system engineering, the implementation of the Teamcenter and especially the operation of the Teamcenter are special dedicated activities, actually the big challenge for every company. Therefore small companies cannot afford to sustain such an activity, specialists are hard to train and people that dedicate themselves for such an activity are hard to find but the existing ones are incredible well paid.

Reference

1. Maxwell, J. C.: https://en.wikipedia.org/wiki/John_C._Maxwell (2020)

Chapter 20
Documents Handling

"Just as oil was likened t black gold, data takes on a new importance and value in the digital age. data is the new gold—let's start mining it"
Former EU Commissioner Neelie KROES.

20.1 Requirements and Fulfilment

As long as most of the discussions are around the single data source and the fact that data must be stored, updated, synchronized and available within the entire company, seems obviously for PLM that documents handling is one of the most considered activity.

Since 1995 has been formulated a demand for the way to judge if a repository is properly acting. The founders, stakeholders, administrators and the user data must know if they are in a safe situation as long as one side must know if the investment worth and the other side that they pose their valuable digital encoded data that may contain the entire accumulated company intellectual capital Worthing billions of Euro, will not catastrophically vanish over the night.

The PLM is coming with the rigorous application of an ISO standard 14721, the OAIS (Open Archival Information System), the most modern digital preservation initiatives, widely adopted by diverse types of digital preservation communities (over 66%) and also for organizations, to inform their implementations of the new preservation system. Published in 2005, the OAIS standard is considered as the optimum standard to create, maintain and preserve a data base over a long period of time. The reference model is reproduced in Fig. 20.1 [1].

Chronological there are 3 standards created by members of PTAB (Primary Trustworthy Digital Repository Authorisation Body), in establishing an internationally recognized and certified set of trustworthy digital repositories:

2012

J. Niemann and A. Pisla, *Life-Cycle Management of Machines and Mechanisms*, Mechanisms and Machine Science 90,
https://doi.org/10.1007/978-3-030-56449-0_20

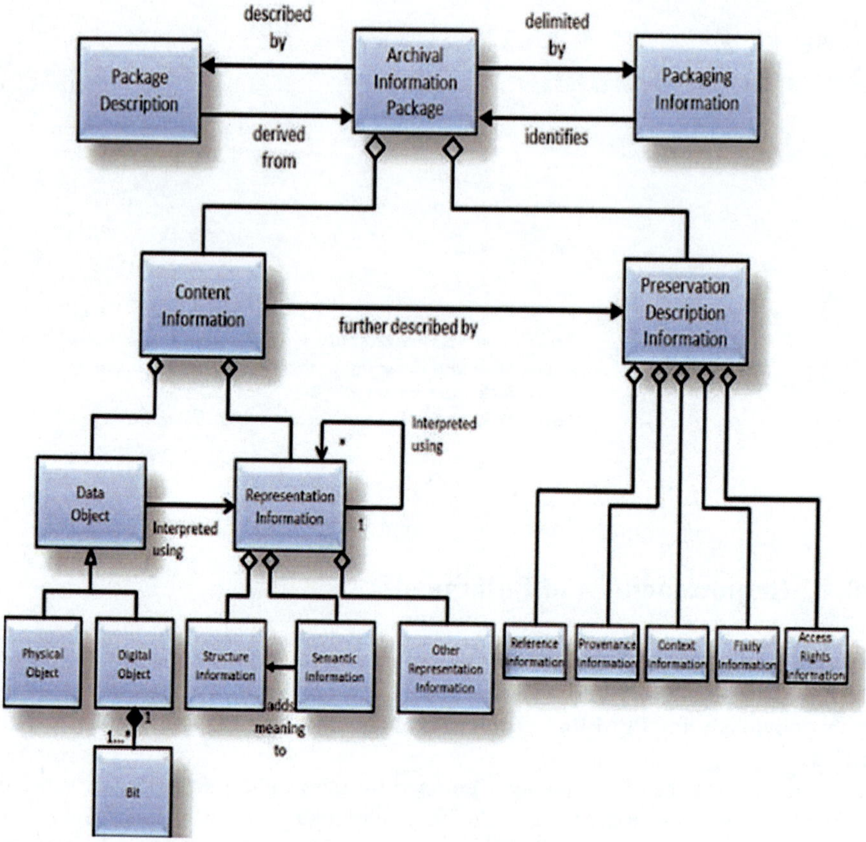

Fig. 20.1 The reference model of the ISO 14721 standard (the OAIS)

The ISO 14721, the OAIS standard (CCSDS 650.0-M-2), the reference model for: *What is required for an archive to provide long-term preservation of digital information?*

2013

The ISO 16363, (CCSDS 652.0-M-1), based on OAIS sets out comprehensive metrics for *what an archive must do, providing Audit and certification of trustworthy digital repositories.*

2014

The ISO 16919, (CCSDS 652.1-M-2), specifies the competencies on auditing bodies, defining the requirements for bodies providing audit and certification of candidate trustworthy digital repositories.

Where the CCSDS is the Consultative Committee for Space Data Systems, and both the CCSDS and the ISO adopted the OAIS Reference Model as an organization of people-and-systems that has accepted the "*responsibility to preserve information and make it available*" to a DC (Designated Community) defining the set of responsibilities an OAIS archive must fulfil and make it distinguished from other uses of the term archive.

Regarding the Siemens PLM system, it is more stable as is assumed from the conformity with ISO standard 14721, the OAIS, as a DPS (Digital Preservation System—how is called and considered), by having an older origin in the DAM (Digital Assets Management) generally designed to ensure long-term accessibility of digital assets.

The DAM actually acts insides as well at it acts outside for its customers. Probably people are generally familiar with the term of "captive client" regarding the customers that are "captured" by a product or vendor because of many reasons: simply the price, the audience trend or the market, made the customer technology to be adapted (with the customer investments) to a specific product. The customer preferences are not mastered by the competitors.

No, the customer commodity and habits breaks the change and the long-time using of the product is not "disturbed" by other forces. The customer is reluctant to invest in a new product or product suppliers, to let him being "accustomed" to a different product, or simply is not willing to lose the advantages of the actual product.

A captive customer (monopole situation) is a great competitive advantage for any company, playing around this situation with different business model tools like: adding new features to the product, discounts, loyalty programs and various alternative offers. The captive customer is an asset, imagines only the logo "British Royal Warrants" as supplier of the royal house.

In some cases also for the captive customers are some advantages like: reliability from good knowledge of the product, controlled risks, well done potential estimations for their products and little or no investments in the search for alternatives. The connection with its only suppliers could be also considered as an asset in some cases. Imagine another logo "We are using only OEM Philips parts".

In this way we may create from a "no alternative" an asset. Starting from that is not surprisingly that people are not having the idea of "captive customer" when they are relate with big companies, usually family companies (Bosch, BMW, Siemens, ThyssenKrupp, etc.).

How come?

Imagine all the towns and provinces around the world that are built around such corporations. Remember localities where the street names are connected to the company, the founder, or perhaps his favorite child or pet [2]. Generation of families is these companies DNA, after decades of working at, and for these companies. So the companies News become practically the employees' family album.

But a "family album" is also the knowledge accumulated inside the company from all these working people. And now the DAM and what happen without it? It may be lost all the company assets documents, knowledge, drawings, images, records, the entire company lifetime heritage, everything can get lost.

In our case, the system is the company using PLM, capturing the customers internal & external staff, suppliers and clients all "captured" in a single information source and data management system. The assets must be preserved exploit and capable to be used also by the next generation. The digitizing activity and digital preservation implying activities, policies and strategies especially designed for the company assets, using the DAM basic functionalities:

- Single Source of Truth, cited as a key benefit of DAM.
- Digitization.
- Asset ingest.
- Ensure multiple format support.
- Long-term archiving.
- Search.
- Disaster recovery.
- Rights management.
- Metadata management. Many DAM automatically extract rich (technical) metadata from the media files and are flexible in customization of metadata fields. This allows for the implementation of any metadata schema standard, but unlike a Digital Preservation System, DAM do not provide any built-in templates of metadata schemas such as EAD (Encoded Archival Descriptor), MODS, or Dublin Core.
- File formats: define the internal structure and encoding of digital objects. They become obsolete when we lose the ability to interpret and render them in human-accessible manner.
- File formats play an important role in digital preservation as they allow characterization of digital assets and the assessment of DPS (Digital Preservation Risks).
- DPSs make use of authoritative file format registries and include functionalities such as format identification and validation. They also support the workflows related to format migration.
- DAM can interpret common media file types, provide access to, and use this information for sorting, browsing and searching digital assets. DAM also support transcoding, which involves decoding and re-encoding already encoded audio or video files, changing frame size, bit-rate, codec, or audio signal.
- A checksum: is a calculated string of fixed length associated with a piece of stored or transmitted digital data, as the result of running a hash algorithm.
- Comparing checksums before and after data transmission or at regular intervals is a common method to detect error and ensure bit-level data integrity in digital preservation systems.
- Checksums can be generated automatically or manually in many DAM but the main use case seems to be de-duplicating files, rather than ensuring ongoing data integrity.

Digital Preservation Strategy [3] it is practically a holistic, technologically and environmentally neutral approach using metadata without any focus on specific standards or procedures. The strategy requires context and curation, and is supplemented with digital preservation procedures and recommendations that will invariably change with time.

The reason to impose a DPS is about the advantages on long run in helping saving money, time and staff by reusing existing documents and projects, avoid ethical and more important legal problems or bad publicity.

In this way by the assets digital preservation, means that ethically, financially and legally the assets will remain accessible and intelligible regardless of technology with an accurate, authentic and reliable status.

Within its functionalities the DAM includes also dynamic tools to concentrate a higher degree of integration allowing the "assets managers" to access their stored content and content storage infrastructure for re-use purposes.

Butch Lazorchak, Information Technology Project Manager at the Library of Congress, mentioned in "The Signal" [4], five shared incentives that not surprising DAM is already considering.

- Open Infrastructures
  (Openness in general can lower costs and expand possibilities).
- A "National Collaboration Engine"
  (An IT platform that covers all the company departments forcing them to work together like a single team).
- Happy Data, Happy Users
  (For social media services, the more control users have of their own data the happier they'll be as participants).
- Self-Interest Paradox
  (By acting selfishly for yourself, you help others accidentally).
- The Karmic Wheel
  (What comes around goes around).

At a great importance in the implementation is the PRISM (Publishing Requirements for Industry Standard Metadata) initiated in 1999 that "*advances core technology to develop standards and best practices to enhance efficiency and speed information across the end-to-end digital media supply chain.*" and that defines a set of XML metadata vocabularies (XML—Extensible Markup Language, described in Chap. 19).

The DAM and actually the OAIS bring the digital object in an unique perspective regarding the evaluation of any digital asset and digital work in developing the technical and social infrastructure that facilitate the preservation and the use of digital object over time with appropriate access.

Managing digital assets became the major task of PLM in the digital lifecycle. In different search activities we found the term of "nostalgia marketing" in relation with DAM [5]. An exciting title but the fact is that modern industry and especially the manufacturing brands are in a continuous change and starving for the change management. Their lifecycle is a continuous humping road in order to adapt to the market and customer expectations change.

Once transformed in digital markets, place the companies in the position to use the "nostalgia markets" that represents all accumulated assets during the entire company lifetime to creates a modern technological structure that is doing the assets management as an investment, for the future, in revalorization of existing solutions and a source of inspiration for the new designed products.

The existence of expert knowledge enables to achieve organized, presented and selected online content. Curate content, like a helicopter view over the fingertips accessible data, tagged with metadata and accurately described. In this way not only the data can be identified but also the data history can be tracked about: When? Where? and By whom? was created or/and used in which projects? Wat was and what is not yet done? and even: What is working and what don't. Everything is made with less effort and higher efficiency, if the right investment is made.

The documents handling is a process that brings value to the company, ensuring accountability to the documents that are recording the data conducting the business.

As expected a great deal of these documents are to be managed more formally representing legally entities that serves as legally accepted evidences for company obligations, decisions and transactions. All these type of documents are records that must be preserved and managed in the way they safeguard the evidentiary weigh, that is even more vital that in a long-term view of the process, with periodically refresh and ensuring the long-term accessibility through the RMA (Records Management Applications).

A special attention is given in cases when records comprise multiple items, that evidence the company activities, directly according to the value of the records, encompassing data like authorities statements, MoU (Memorandum of Understanding), claims forms, contract approval, contractual specifications, drawings, photographs, purchase orders, the sending and receipt of emails, sketches, statements from witnesses, X-rays, etc.

To these records unique identifiers are allocated, providing safeguards against unauthorized changes, creating an unbreakable audit trail for reasons of accountability and eDiscovery.

Once declared as "a record", these records are not subject of any change anymore, so no further changes are expected or allowed. If the current activities requires and implies some changes as additional details to be exhibit, a new record is created no alteration of the existing one being allowed. The new record is considered the only available version in its own right.

> The Audit trails guarantee an enforceable chain of custody by making it possible to know what a record said at a particular point in time, how its content evolved to that point, and who was involved with it.
>
> This is key to preserving the link between thc record and the process or event it describes, and for being able to demonstrate exactly who made what changes and when.
>
> AIIM—The Association for Intelligent Information Management.

The documents handling it is defined in ISO standard 15489: 2001 as the RM (Records Management) responsible for the records efficient and systematic control of the:

- Creation;
- Receipt;
- Disposition;
- Use;
- Maintenance;
- Capturing evidences about business activities and transactions;
- Maintaining evidences about business activities and transactions;
- Capturing information about business activities and transactions;
- Maintaining information about business activities and transactions;
- ERM (Electronic Records Management) ensures needed records when they are needed.

The records control management processes and systems is made by performing a set of activities based on the associated key capabilities:

1. **Retention Rules Implementation**
   Records show different requirements depending of the record type. Depending on the contents, each record type represents a value for the company. The value could be administrative, fiscal, legal, or historical and that will define for how long the record must be kept. Therefore the EDMS (Electronic Data Management System) will assign retention rules to each record based on the record type.
2. **Access Controls**
   Only authorized users will be able to access, read, or retrieve a record. Even so they are not allowed to make any changes to it. In special cases may be a reason to allow changes to the metadata, associated with the record.
3. **Declaration and Registration**
   The record is placed in a repository. A unique identifier is assigned to it for management throughout its entire lifecycle.
4. **Declaration and Registration**
   At the end of the records lifecycle, they lost all their business value with no further involvement in any sort of matter including legal audits. At this moment the records will be simply destroyed or transferred y to a controlling legal authority such as a corporate library or state archives.
5. **Audit Trails**
   Are the final documentation about the considered records were managed from declaration to disposition. In some cases, companies or organizations, the audit trails are themselves records that need to be managed.

At this stage we have to consider how Teamcenter is dealing with the data, documents and processes synchronization.

So we are back to the ideas displayed in Chap. 19 and it is again to consider the duality engineering-management specialization.

From the Engineering point of view we have the PDM frame and in the middle the CAQ as a module that must encompasses the entire functionality of the product as the product Quality.

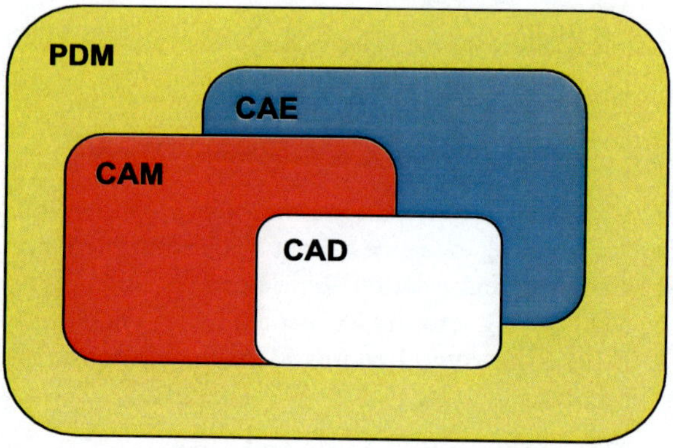

Fig. 20.2 The engineering dimension of the Teamcenter imbrication

The product results after following a manufacturing process as a result of a CAM activity, but CAM is applied after the CAE simulate and correlate all together the product characteristics.

To apply CAE we need the product shape and parameters that are defined within the CAD process.

All this engineering processes are and must be documented, especially as documents regarding the part (item) attributes and the part development with all the functional and qualitative aspects. Therefore are documents directly related with a specific frame but also inter-frames documents that ensure the continuity, the responsibility and the testimonials for the actual or future activities. In the end the documents will be generated and stored based on the company framing, as it looks like in Fig. 20.2.

From the Managerial point of view are considered the processes and the work flows that reflect the company-specific business rules concerning the related processes.

Again every process and processes interrelations are documented for the same reasons to that ensure the processes continuity, the staff responsibility and as testimonials for the actual or future activities.

In this case we have a different model of imbrication for the company activity stability and sustainability. The documents are very well located in their frame generator (i.e. CAD—Parts), but all of them will belong also to the next frame along with the frame dedicated new generated ones. In the frame "Requirements-Documents", where are specified the contractual required and the certification procedures and documents that will validate the requirements fulfilment, are also specifically highlighted marked the 'CAD-Parts" documents related to the contract. And so on until the "Project Schedule" frame, where all the Project related documents are logged and catalogue as briefly depicted in Fig. 20.3.

In conclusion the DHD (Documents Handling Department) perform a separate activity, benefit from the IT support, assisting all the legal and managerial activities of a company corresponding with the specified frames. For each frame the DHD applies first the standards in the industrial company's profile, each engineering part (Design, Manufacturing, Assembly, Quality control, etc.) supposing to have the.

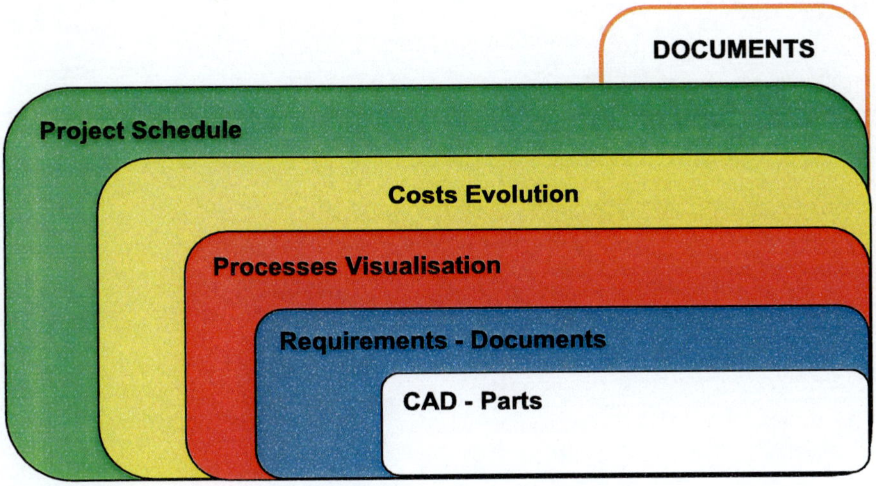

Fig. 20.3 The management dimension of the Teamcenter imbrication

1. The (Design, Manufacturing, Quality ....) Manual
2. The (Design, Manufacturing, Quality ....) Policy
3. The (Design, Manufacturing, Quality ....) Objectives
4. The (Design, Manufacturing, Quality ....) Records

And 6 procedures to control the documents as the 2008 ISO standard updated as ISO 9001:2015, ending with the part ISO 9001:2018 "Documentation Requirements".

1. Control of Documents;
2. Control of Records;
3. Internal Audit;
4. Control of Nonconforming Product;
5. Corrective Action;
6. Preventive Actions.

Regardless of the novelty of the business or the company, usually the start is with a mix data environment that must be integrated in the company policy for documents handling and PLM implementation.

Based on the managerial wish in the end all the "business image", materialized in documents must be converted to an electronic form.

We have to admit that is not within the reach of any company, being an expensive and time-consuming task for many SME's businesses.

References

1. https://www.oais.info/ (2019)
2. Real Story Group: https://www.realstorygroup.com/Blog/should-you-rely-dam-your-digital-preservation-needs (2013)
3. Kolvitz, E.: Digital Preservation for DAMs, Technology Journal. https://www.insight.com/en_US/content-and-resources/tech-journal (2015)
4. Lazorchak, B.: DAMs Vs. LAMs: It's On!, The Signal (2012)
5. Emily Kolvitz, B.: Using DAM for Digital Preservation and Nostalgia Marketing. Amsterdam (2018)

Chapter 21
The Digital Factory

"Life is a gambling what harms in making also profit on that?"

Digital Factory and Industry 4.0 are not really totally new paradigms but the globalization pressure implies a quick response to extended product responsibility: new; better, faster, cheaper.

The roots of Digital Factory (DF) are already old, having a correspondence in the USA and Japan, as Factory of the future or Smart manufacturing, but let's face the fact that without I.40 the concept never touches the global market. The similar initiatives formed as Consortium, IIC (Industrial Internet Consortium) in the USA started in 2014 with about 200 members in 2016, IVI (Industrial Value-Chain Initiative) in Japan, IDF (Industrie du futur) in France and obviously the China initiative remains as "local" at national level initiatives.

Historically we may start with the CNC (Computer Numerical Control) for the machine-tools programming and control. The CAD evolution made also the CAM evolution to join and be grouped in a larger complex the CIM (Computer Integrated Manufacturing) in the early '90, for using numerical control for the entire production processes within a company, with the extension for allowing the data exchange between processes with influence on the flexibilization, logistics and robotisation and here we have now the ISO 9001 for Manufacturing. Aiming the processes control, data integrity and the suppliers integration, together with the development of industrial communication protocols like PROFIBUS standard for fieldbus communication in automation technology (used by Siemens) or PROFINET standard for Industrial Ethernet.

21.1 The CIM

The CIM: *Use* the Lean manufacturing technology; *In combination* with Robotics, FMS (Flexible Manufacturing Systems), ASRS (Automated Storage and Retrieval Systems) and AGVs (Automated Guided Vehicles); *Based on* the CNC, DNC (Direct Numerical Control), PLC (Programmable Logic Controllers), HMI (Human Machine

J. Niemann and A. Pisla, *Life-Cycle Management of Machines and Mechanisms*, Mechanisms and Machine Science 90,
https://doi.org/10.1007/978-3-030-56449-0_21

Interfaces) and digital networks, *Providing* pretty much of the PLM platform a common Database, the CAD, the CAE, the CAM, the CAQ (Computer Aided Quality), CAPP (Computer Aided Processes Planning), PPC (Production Planning and Control), the ERP (Enterprise Resource Planning) and a sort of IBS (Integrated Business System).

Back to closer times Digital Factory is perceived like a virtual company with virtual manufacturing where every detail, every sensor or data can be seen on a holographic screen base on simple command, vocal if possible. This perception is encouraged by any web representation of the Digital Factory.

The reality is somewhere between the harsh start and the science fiction (someday maybe achievable) visions.

The Digital Factory is dealing with the Smart manufacturing containing optimizable processes, very much focused these days on additive manufacturing, the Industrial Internet of Things that enables the implementation of Artificial Intelligence and especially Machine Learning an nevertheless Robotics and the Data centralized storing, analytics and management.

Practically is a production facility, where all the production processes are sharing data within all stages in a holistic real time activity. The managers real time access to operational data facilitate to overcome bottlenecks and indecision blockages, avoiding in this way the business inefficiencies.

The smart sensors and IIoT enables the implementation of Bots, as a software automated application that performs automated task in different "production corners".

The Bots will be used by the AI software modules in the processes behaviors simulation within the decision making process. The Machine Learning as an integrated module within AI that is a self-adjusting component when is exposed to more amount of data, exactly the situation created by Bots and IIoT. The Machine Learning module is an human independent module not reliant on human expertise and therefore it must be very well consider where to be applied.

The data transition and "transaction" follows the trend of the blockchain technology for safety and structural reasons. But finally everything relay on how will be implemented and operated the system: The Digital Factory.

In the end the Digital Factory is not implemented because we can, but because we can't otherwise respond to extended product responsibility, the Iron triangle is not covering anymore the managerial demands and the multicriterial decision is to slow to enables managers to offer a products that is: new; better, faster and cheaper.

The digital factory is not fast and certainly not cheap but offers a viable solution in the asked quality and the activities synchronization within the complexity increases companies.

The Digital Factory offers different layers of data. By overlapping these layers, results an instant solution for the on run activities and scenarios generation.

In the Digital Factory must exist the capacity to define and monitor the behavioral fluxes and ensure the activity discipline on all managerial levels during the entire lifecycle, ensuring the framework for the manufacturing and logistic functional structures.

In the same time the Digital Factory implementation ensure the interfaces for the intercompany activities (and/or the subsidiaries) and the interfaces for the training and the development activity as a single tool.

Complex systems are hard to manage but the companies increasing complexity is not an arbitrary result but a reaction at the intrinsic or extrinsic flexibility demand, arriving from the dynamic market changes. Demand that is differently in different geo-political situations or national and supra national legislations (i.e. EU).

The increase of companies complexity have also a technological reason, defined by the need for diversity and larger functionality, reliability and low prices, dedicated to more and more sophisticated customers.

The complexity and the growing in complexity of a company is not the result of a decision, *but a capability*, resulted from a process based on the inter-individual performance. The number and the type of developed tasks, the structure of the labor division, the monitoring and coordination mode, the functional aspects and the dysfunctions, human resources pool and training facilities, etc., all together combined with the managerial abilities to cope with actual and future challenges are simply defining-elements for the company complexity.

The expected result is a company model, that reflects the "state of the art" that may be represented in a digital form, and by the devices integration, divisions representations, disciplines and methods integration the digital form may lead to the Digital Factory.

The digital era, introduced by the IT development, touch not only the factory (the Plant) with production and service lines, but also the private consumer and the education systems. The result: a more informed partner on the market dialogue, more and diverse alternatives, stronger influences from the surrounding environment, more players and a lot of more suppliers.

The time decisions, and the trust created in time, starts to become history in face of certification and the novelty that the new business models bring with them.

Observed first in fashion and electronic industry, this trend moves to automotive, machine-tools and robots, aerospace, energy, constructions engineering, furniture, pharmaceutical, food and even leisure industry, so everybody will face it.

This dynamics attract around each type of mentioned industry complementary activities for logistics, data management and storing, economic and legislative services, emerging together with the regulatory activity.

In the same time must be observed that all this activities must follow the same paradigm: *the company evolution is based on the sustainable development and durability facts that must follow the local legislations, environment impact rules and social requirements* [1].

Having different locations around the world, a company that was created based on initial idea or philosophy may function only by managing unitary the entire data sets within the "business". Nowadays that is possible only having a performant informatics system at the company level that leads to the Digital Factory.

21.2 The TLCM

The dynamic of changes and the more interconnected and interdependent fluxes, actions and responsibilities justify the creation and development of the Digital Factory that becomes from the first concept as a single self-standing organization, a part of a supply value creation chain network.

Such a complex development with so many interlinked components is justified and supported by the desire to remain sustainable within the global puzzle. In this context is integrated another relatively new paradigm: the Total Life Cycle Management (TLCM). A natural question appears, in what is different TLCM from the LCM (Lifecycle management)?

The TLCM is not considering anymore the pressure to offer different (new) and better value through innovative products as a compromise (gaming) between the company internal and external factors. All the factors that are influencing the change: economical, environmental, legislative, political, social or technological are treated as a TCD (Total Customer Demand).

Customer and not client, because all the influence factors are representing a permanence for the company activity, being more or less demanding but exists on a regular base. The clients will be the beneficiary of a contract with the company; the contractual conditions are additional components in the customers list.

This paradigm enables modelling the TLCM in a compact mathematical form as a change management model that accept any form of variable form digital to complex or fuzzy. Any customer is coming with a set of variables and a list of applicable rules that just extend (or shorten) an elaborated mathematical model. The advantage of this approach is to offer a unique compact mathematical form.

If a company is oriented to follow this approach will have the benefit to be aware of a particular strong influence in case of rapid changes on "any customer". This could be more important in industrial cases with many suppliers, with many manufacturing locations in different regions with different legislative regulations. Within the globalized market that may represent a large variety of product variants, accompanied by corresponding services and maintenance activities, and a large number of distributors and clients.

Beside the early and reliable warns, sustainability and sustainable development, force the company to encapsulate in the TLCM Model the capacity to increase not only the efficiency but also the effectiveness of resources consumption, emissions reduction, hazardous risks and the capacity of providing recycling.

Within the LCM this aspects may be considered but are not encapsulated, being more (or less) a reaction at the accountability, or EU directives, or markets trends, regional legal or social pressures, or simply for gaining competitive advantages.

For example, a new paradigm in the company TLCM approach is what was described in the development and supply of product-accompanying service models [2], as a Services functional sales. This is an example on how a customer enters in the TLCM model.

Whether or not considered, products are going along with the client. The manufacturer extends its responsibility (and costs) for the product in order to ensure a greater sustainability. This is showing the need and utility of considering the TLCM as an integrated tool for the company adaptability and not as a new management challenge.

TLCM appear not as an invention, is more as an innovation, a reaction to the companies' inner structures and external developments. The new generation of people are considering the task based performance and the inter-individual labour for integrating the performance management. The cooperation between the human resources and the institutional functionalities makes the TLCM contribution for a sustainable development and the related research lifecycle disciplines (engineering & management) to promote the integration of uncover interdependencies between disciplines and methods.

The TLCM is not on the shelf methodology it still needs to be integrated in some considered divided area of expertise, like the "standard" LCM the DFE (Design For the Environment) [3–6], or like LCA (Life Cycle Assessment) ISO 14040–ISO 14043; maintenance and/or spear parts life cycle; LCP (Lifecycle Planning) [7–9] the lifecycle (closed-loop) supply chain management [10–12], and the EOF (End Of Life) management [13].

The capacity for TLCM is given by the company technological culture that is defining the:

- structure (location);
- the activities (tools and devices available);
- and the behavior (the type of people training).

21.3 The Manufacturing Line and the Automation

Having a Digital Factory is not an on the shelf decision. A digital Factory is a complex system built in time, consisting in the IT hardware-software structure able to manage all the datasets that are representing the real resources. Thus a Digital Factory must be adapted to the companies' structures, activities, and behavior, considering also the external challenges and changes. A necessary following the management forms implies by the company leading board. The management solutions define the company reaction (digital or not) to the external developments and to the internal changes.

Without Automation the Digital Factory simply can't exists. That is done by using the web server-based machines control by time and attendance control, remote settings, remote inquiring, correct communication errors and remote maintenance, within secure access permission.

As long as the technical aspects are very detailed in the specialized literature or within dedicated training structures it will be considered only the engineering and management aspects with a little focus on the environmental conformities integration within the Digital Factory aspects less considered within presentations.

21.4 Manage the Digital Factory

Within the Digital factory the life cycle management (LCM) is a way for monitoring and control a manufacturing process considering also the challenges, the trends and the predictions.

For the lifecycle monitoring and control is necessary to define the Lifecycle objectives that usually are not totally independent and may present even a hierarchical structure.

The objectives are related to products. Ones defined the list of products and objectives the phases of the products manufacturing may be detailed, a procedure that is following any product technology from concept to recycling.

The phase's decomposition is defining the activities that can be scheduled and represented by "resources necessities". This approach offers the elements to close the loop of product-phases-activities by adding the supply chain and the rules of the supply chain management (including documents, contracts and specifications).

One of the defined these is the optimization process that may run using the Alternative Scenarios Generation (ASG) for the life cycle design and engineering management.

In the Digital Factory, the ASG is including the company structure, company construction, resources in machine tools and robots. The LCA (Life Cycle Assessment or Life Cycle Analysis) is compiling the inventory of relevant energy and material inputs and is evaluating the potential impacts associated with identified inputs, offering support in interpreting the results to help make a more informed decision in accordance with the environmental releases [13], upgraded in August 2010 and [14].

The LCA is dealing with tooling, manufacturing flexible variants, end product transformation. The LCA procedures are part of the ISO 14000 environmental management standards: ISO 14040/2006 and 14044/2006. (ISO 14044 replaced earlier versions of ISO 14041-14043). The GHG (Greenhouse Gas protocol) envisaged that product life cycle assessments can also comply with standards such as PAS 2050 (Publicly Available Specification—specification for the assessment of the life cycle greenhouse gas emissions of goods and services) and the GHG Protocol Life Cycle Accounting and Reporting Standard [15–17].

Considering that we are living on a small planet with little resources, the companies strive to use different available resources. Adapting or change the internal technology is understandable, and the development of the recycling-society is obviously.

What is not obviously is the commitment of the manufacturer to allocate responsibility for no pollution and recycling. That is imposed more and more by the social dimension through the EU legislation like the Restriction of Hazardous Substances Directive 2002/95/EC.

RoHS 1, short for Directive on the restriction of the use certain hazardous substances in electrical and electronic equipment, adopted in February 2003 and took effect on 1 July 2006. This directive is required to be enforced and become law in each EU member state. The directive restricts (with some strong justified exceptions) the use of six hazardous materials in the manufacture of various types of electronic and electrical equipment. In the same time the RoHS 1 is closely linked with the WEEE—*Waste Electrical and Electronic Equipment Directive* (2002/96/EC) which sets collection, recycling and recovery targets for electrical goods also as a part of a legislative initiative to solve the problem of huge amounts of toxic *e-waste*.

Legislation and directives like that implies also a change in business and is to be mentioned the PASM – Product Accompanying Service Model. The PASM is easy to be observed in the printer cartridge suppliers like (HP) that creates in parallel with the distribution network also a recycling network. The PASM implies for the manufacturers to carry the responsibility for their products to a greater extent.

In conclusion the Digital Factory is a "comfort zone" for specialist, as depicted [Courtesy Digital Twin SRL], allowing them to use the everyday new functionalities for an efficient and effective manufacturing offering a quick response to extended product responsibility: new; better, faster, cheaper.

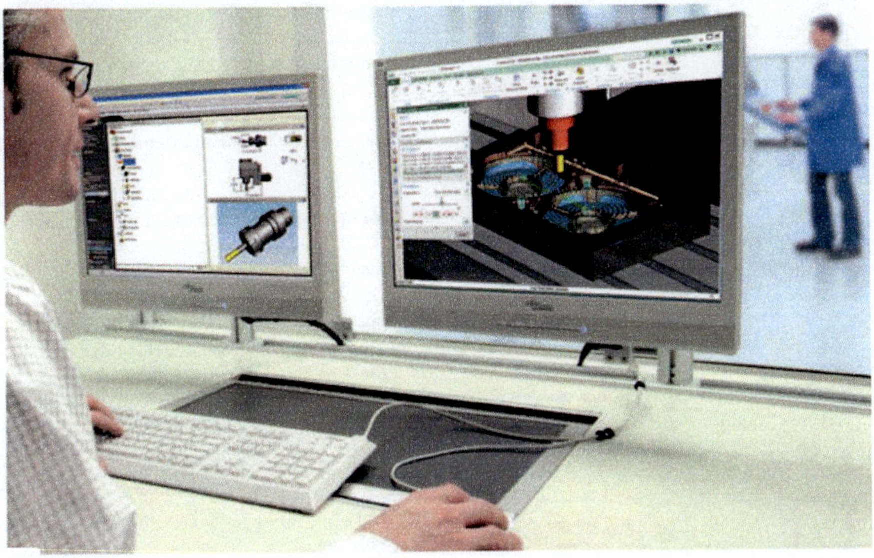

References

1. Herrmann, C., Bergmann, L., Thiede, S., Halubek, P.: Total life cycle management—an integrated approach towards sustainability. In: Proceedings of the 3rd International Conference on Life Cycle Management, Irchel, August 27–29, University of Zurich, (2007)
2. Niemann, J., Fussenecker, C., Schlösser, M.: Measuring the economic impact of life cycle management and service performance. In: International Conference on Competitive Manufacturing, Stellenbosch, South Africa (2016)
3. Seliger, G., et al.: Global sustainability—a future scenario. In: Proceedings Global Conference on Sustainable Product Development and Life Cycle Engineering, Berlin, pp. 29–35 (2004)

4. Westkämper, E., Alting, L., Arndt, G.: Life Cycle Management and Assessment: Approaches and Visions Towards Sustainable Manufacturing. Annals of the CIRP, Vol. 49/2/2000, pp. 501–522 (2000)
5. Niemann, J., Bullinger, H.-J., Warnecke, H.J., Westkämper: Life cycle management—Das Paradigma der ganzheitlichen Produktlebenslaufbetrachtung, Neue Organisationsformen im Unternehmen. Springer, Berlin, pp. 813–826 (2003)
6. Saur, K., Ginluca, D., et al.: Final Report of the LCM Definition Study. UNEP/SETAC Life Cycle Initiative, Version 3.6 November 17 (2003)
7. Herrmann, C., Ohlendorf, M., Hesselbach, J.: Planning WEEE Disassembly—State of the Art and Research Developments, CIRP Seminar on Life Cycle Engineering, Copenhagen, Denmark (2003)
8. Wübbenhorst, K.: Konzept der Lebenszykluskosten. Grundlagen, Problemstellungen und technologische Zusammenhänge. Verlag für Fachliteratur Darmstadt (1984)
9. Spengler, T., Herrmann, C.: Stoffstrombasiertes Supply Chain Management in der Elektronikindustrie zur Schließung von Materialkreisläufen - Projekt StreaM. Fortschritt-Berichte, VDI (2004)
10. Bloemhof-Ruwaard, J., Krikke, H., Van Wassenhove, L.N.: OR Models for ECO-eco Closed Loop Supply Chain Optimization. Springer, Berlin (2004)
11. Dekker, R., Fleischmann, M., Inderfurth, K., Van Wassenhove, L.N.: Logistics: Quantitative Models for Closed-Loop Supply Chains. Springer, Berlin, pp. 357–379 (2004)
12. DIN Deutsches Institut für Normung Der Recyclingpass: Übermittlung recyclingrelevanter Produktinformationen zwischen Herstellern und Recyclingunternehmen, 1049, PAS. Beuth Verlag, Berlin (2004)
13. Curran, M.A.: Life Cycle Assessment: Principles and Practice. U.S. Environmental Protection Agency, Cincinnati, Ohio (2006)
14. Guinée, J.B., Heijungs, R., Huppes, G., Zamagni, A., Masoni, P., Buonamici, R., Ekvall, T., Rydberg, T.: Life Cycle Assessment: Past, Present, and Future. Environment Science & Technology, pp. 90–96 (2011)
15. Ecometrica: GHG Product Life Cycle Assessments. https://ecometrica.com/technology#lca (2013)
16. PAS, 2050:2, Specification for the Assessment of the Life Cycle Greenhouse Gas Emissions of Goods and Services. https://shop.bsigroup.com/forms/PASs/PAS-2050/, BSI (2013)
17. WRI World Resources Institute—GHG Protocol: Product Life Cycle Accounting and Reporting Standard. https://www.wri.org/publication/greenhouse-gas-protocol-product-life-cycle-accounting-and-reporting-standard (2013)

Chapter 22
Applications Modeling

"All time, everywhere for better products by better decisions"
Siemens PLM Motto.

22.1 Life Cycle Management Approaches

Once defined the first conclusion that PLM is a need for globally used products and processes, results that LCM (Lifecycle Manager) will become omnipresent to avoid the exponential problems rise for the next complex and performant products.

It is necessary to stress out again that the implementation is not a standard process that can be simply based on management style presentations but a new way of thinking and therefore you must to prepare yourself for PLM. To get everything needed, must turn up to the last screw of the PLM implementation, and therefore becomes more and more obviously that in a company is a need for:

(a) PLM Department;
(b) PLM Director;
(c) PLM Manager(s);
(d) PLM Project manager(s);
(e) PDM Managers;
(f) Technical specialists for all existing modules SE, NX, NX CAM, Tecnomatix, Team Center, SAP, etc.;
(g) And nevertheless the Lifecycle Manager(s).

The Lifecycle Manager, classify corresponding life cycle objectives and activities considering a subgroup of specialists and their responsibilities, necessary to deal with the DSPD (Data Sources, Processing and Distribution) IT Platforms.

The IIoT and the data management and processing facilities are grouped around the specialized hardware, software or a combination of them that involves the implementation of the so called "virtual machines".

J. Niemann and A. Pisla, *Life-Cycle Management of Machines and Mechanisms*, Mechanisms and Machine Science 90,
https://doi.org/10.1007/978-3-030-56449-0_22

The Virtual Machine (VM) emulates the entire functionality of physical computer architecture, grouped in three main structures, based on their functionality: the Process VM (PVM), the Architecture VM (AVM) and the System VM (SVM).

The PVM, are designed and destined to execute dedicated computer programs in a platform-independent environment (hardware neutral).

The AVM, are designed and destined to emulate different hardware architectures for enabling the execution of an operating systems over the software applications written for different hardware architectures and processing units (CPU).

The AVM are using the existing resources (sharing and partitioning) but are not universally interchangeable. An often given example is the QEMU (Quick EMUlator), an open source for the hardware virtualization.

The SVM, are designed and destined to execute the entire functionality of an operating system, managing multiple hardware resources remotely placed and isolated one form the other.

The SVM is playing the role of the "hypervisor" being the Virtual Machine Monitor (VMM) and in the modern versions the VMM benefit of the dedicated hardware assisted virtualization.

The duty of the Lifecycle Manager is to designate the 4 target groups that are dealing with the VM's (IT Platform Administrators; Approvers; IT Staff and Users) enabling the Administrators to track and control the company virtual machines through a consistent approval process throughout the entire lifecycle.

In this process, to each specific group are destined specific tasks to be followed, considered as key features:

The IT Platform Administrator

Creates a guiding catalog for the considered standard IT services.

The catalog consist in the pre-defined "virtual machine templates" (VMT). Depends on the end-user needs, he may select from the catalog the most adequate Template based on the VM offer:

- machines siz;
- memory size;
- storage capacity;
- offered services;
- etc.

Approver

The IT Administrator is responsible for the complete approval of the users request but it is assisted in the approval process by an Approver.

The Approver has to approve or reject first the requests, after detailed information about the purpose, necessities and technical implications of the request.

After the Approver accept or deny the request, the IT Administrator connect the request with economical, human resources and managerial implications for the complete request approval.

IT Staff

The members of the IT staff represent the support specialists for the IT Administrator and for the Users. The tasks are about the Users requests that once with the complete approval from the Administrator, are transformed in a decision about the "intelligent placement" of the VM within the company environment.

The User

The users represent also an IT Platform Administrator responsibility. The users must be known, identified and monitored as long they have access to the IT Portal with the company standard IT services through the Catalogues from where they may select Virtual Machine Template (VMT). The Users select a VMT as the best considered for their needs and applications and define the filled template as a "Request status".

Once approved, the Users may access the VM through the basic VM Controls (VMC). The controls are powering, snapshots (i.e. add or change a disk in the VM, etc.), revert to snapshot (i.e. the added disk at the snapshot point is removed), shutting down or even suspend the VM.

Despite the consideration that is the IT Administrator duty, is the Lifecycle Manager duty to interfere and determine the infrastructure optimization, the establishment of security policies and the policy-based management.

The infrastructure optimization refers to optimize the resource pools for the virtual machine deployments; avoid scarce, redundant or unnecessary resources regarding the VM folders; and the definition of the applied chargeback metrics.

For the Lifecycle Manager, security policies, refers in how are assigned the groups and subgroups of users to the specific VM resources.

Actually the policy-based management (PBM) is a technology dedicated to simplify the complex task of the networks and distributed systems managing process, where the developed dedicated VM are included, by deploying a convenient set of policies, governing the VM networks and distributed systems characteristics in a flexible and simplified manner.

The PBM is setting up key criteria, for user to be specified in the requests, criteria used for the automatically configuration or placement of the VM in the company environment.

Within the manufacturing process a good example for the key criteria is the automatically allocation of the storage locations and the activation of the Distributed Resource Scheduler (DRS) that is balancing the computing workloads with available resources in a virtualized environment, allocating the physical resources among the VM's, based on the user defined rules.

To summarize the **IT Platform Administrator** setup and defines the VMT catalogue and what types of approvals are required prior to virtual machine deployment.

The users can view and select the VMT that helps them to determine the characteristics of available virtual machines (machine size, memory, storage, backup services, etc.) and can log and make a request for a VM. During the request process, the user enters information to help in selecting the specific resources that best support the request and can log back any time to check on the request status, the Requests being visible only to the user who made them.

At the request submission the end user must specify also the end date, as a checking element for the Lifecycle Manager decommissioning activity. This date represents that length of time the user will need the resource (VM).

The Approver receives a submitted request.

IF

The request is approved and the IT staff has mapped the user-defined criteria on the existing computing resources (e.g. high performance VM for manufacturing is mapped in the virtualized environment with available highest performing server, network and storage resources).

THEN

The complete approval from the Lifecycle Manager is obtained and the VM is deployed automatically.

After all the tasks, for what the VM was requested, are fulfilled the Lifecycle Manager duty is to decommission the VM, by archiving and deleting the VM for ensuring that resources come back into the resource pool for future use. In this way is provided better resource utilization.

The Lifecycle Manager enables IT Platform Administrator to track and control VM's through a consistent approval process throughout the entire lifecycle. It is creating the Catalogue for the standard IT services for which the IT Platform Administrator defines the VMT's that allows him about what types of resources are deployed into the IT environment after the users select from the pre-defined VM's.

The Lifecycle Manager enables IT staff to associate chargeback metrics to specific virtual machine deployments and resource pools, chargeback metrics that can be assigned to specific business groups, or tie in to existing financial systems.

IT staff will know exactly when the request is made, approved or denied and when the VM is deployed and how long it has been in operation.

As deployments grow, automation becomes a critical factor in helping IT staff do more with less. Lifecycle Manager may automates each step in the VM lifecycle based on predefined policies.

The Lifecycle Manager establishes a consistent and scalable mechanism to route and approve all requests for VM, ensuring the compliance with the internal policies, streamlining the entire request and approval process. He is tracking who owns VM's in a virtual environment, keeping record when a VM is created, deployed with a web-based request log or decommissioned. Manually eliminate repetitive and error-prone tasks.

The Lifecycle Manager uses a web interface for managing the interactions among everyone involved in the lifecycle of a VM, web-page that integrates some of existing management tools such as change management, asset management, networking or storage.

Everything must be synergistically incorporated within the company operations, frameworks, disciplines, and methods [1].

22.2 Life Cycle Implementation

The lifecycle implementation is considering the PLM implementation having in mind the actually lifecycle phases, the lifecycle engineering side, the lifecycle managerial side, the new specificities of the lifecycle end of life, the capacity to evaluate the status of the implementation evolution and the modelling of the company functionality.

22.2.1 Life Cycle Phases

In Fig. 13.1 is depicted the PLM platform in the short form with 13 modules and in Fig. 22.1 is depicted the PLM platform in the extended form with 15 modules.

Over the PLM platform are superposed the 5 phases of the product lifecycle:

- Product idea;
- Product definition;
- Product manufacturing;
- Product support;
- Product end-of-life.

The economic environment, the client specificity and contractual conditions, the legislation and the company inner structure, time restrictions and logistics all are captured within the 5 phases in the product specifications. From here on, the Product is the King and therefore the PLM vision and PLM strategy represents a holistic approach including process, people, and information, in the end the PLM strategy it is leading to success.

22.2.2 Life Cycle Engineering

An interesting observation is that Lifecycle Engineering starts always from a non PLM solution or strategy but from people with more or less experience in PLM that just identify the need to do PLM and run a company PLM initiative.

Usually a new Lifecycle Manager activity is not related with a start-up as a brand new company where PLM must be implemented. The usual situation is the employment in an existing company and the usual entry place is what in Fig. 22.1 is represented in the lower side of the image as an "empty space". These come from the general consideration that the Lifecycle Manager is planning and then monitoring

Fig. 22.1 The extended form of the PLM platform (*Courtesy* Siemens PLM)

the lifecycle of an existing situation and that because the managing dimension often is ignored or misunderstood.

Planning is one of the last stages of the PLM Implementation; the start is with the Product Data Generation and the PDM (Product Data Management).

The Implementation is considering virtually the management of all data fluxes regarding the technical, administrative, structural, material, social and financial aspects, as was detailed in Chap. 18. Considering the 5 phases of the product lifecycle, in the PLM implementation we must stick to them for the PLM platform dedicated modules integration within the lifecycle phases.

Product Idea

Module 1 The Product conceptualization—**Styling**.

For any product the first step is to define the client specifications (the client business characteristics, market, standards, main technical parameters specifications, the external design concept and main functionalities.

Module 2 The Product technical representation—**Design**.

The conceptualization will be followed by the technical (CAD) design form mechanical, electrical, mechatronic point of view.

Product Definition

Module 2 The Product technical representation—**Design.**

The product design offers the base line for all the other product development stages.

Module 3 Product documentation—**Documents.**

From the product sketches, technical draws and specifications together with the contractual forms, at this stage starts the documentation of the product lifecycle, including the start of BoM.

From this point on the Integration of the development modules with the lifecycle stages is in total cooperation within the digital company products and technical documentation management.

In order to reach the maximum profit from it is absolutely necessary to consider all as an integrated solution with a strong accent on the product lifecycle management, putting into evidence:

- the collaborative way of working;
- the contract (project) management;
- the quality standards application;
- the reuse of the accumulated knowledge.

based on the unique Database and using the Product Data Management (PDM) software solutions.

Module 4 The Product simulation and analyses of the virtual prototype—**Simulation.**

After the prototype is obtained it must be functional analysed from the real environmental conditions point of view, the so called digital validation.

In this phase the motions, internal tensions, displacements, vibrations, thermal and aerodynamically conditions will be simulated, as close as possible to real conditions, the necessary analyses will be made.

The module benefits of the support of CAE (Computer Aided Engineering) software modules.

Product Manufacturing

Module 5 The definition of the product processing operation—**Tooling.**

The manufacturing phase debut of the product lifecycle is made by the definition of the processing operations that lead from the row material to the finite product.

The operations list is completed by the tooling that must be allocated to each processing operation and the added technology in respect with the processed material, ending with the digital validation of the part design, allocated processing operations and tooling.

Module 6 The Digital manufacturing based on the virtual prototype—**Machining.**

After digital validation the product pass to the digital manufacturing. In this phase are considered the machine-tools, resulting the programming of the CNC machine-tools and robots, together with the simulation of the entire manufacturing capacity.

Safety measures and technological manufacturing standards are applied simultaneously with the optimal use and allocation of the existing resources. The module, benefit from the support of Computer Aided Manufacturing (CAM), also known as Digital Manufacturing.

Module 7 The Digital analyses for part components management based on the virtual representation of manufacturing lines—**Part.**

Module 8 The Digital assembly of the part components based on the virtual prototype and assembly lines—**Assembly.**

Module 9 Manufacturing and logistic resources allocation—**Resources.**

What in the short form of the PLM platform was the Robotics module, here is a larger view about the parts and parts components handling together with the transfer lines, logistics and all the auxiliary needs to run the processes. Within a smaller company that can be considered as a single module, in larger ones can be divided in submodules specific for each type of considered resources, even for separate the assets and IPR and legal conditions. Advisable is to have a separate Robotics module.

Module 10 Manufacturing area description and analyses—**Plant.**

As a company may be formed by several Plants the manufacturing digital representation for each is essential to manifest the PLM capabilities.

Module 11 Human resources evidence and management—**Human.**

No manufacturing process is really imaginable without the human resources regardless of the particular categories involved in a company.

Beside the allocation and monitoring management the module may consider also the simulation of the ergonomic aspects, safety conditions, conformities and alternative scenarios analyses.

The module, benefit from the support of Teamcenter and PSH (Process Simulate Human), with Jack and Jill for testing and improving the working conditions but also ergonomically the products and the processes as a Human-centered design tool.

Beside the customization possibilities: energy expenditure, fatigue limits, injury risk, line of sight, components reachability, operator comfort, and of course the anthropomorphic parameters; is the access and direct use of national worker population database, integrating the human factors in the design, validation and planning of the lifecycle.

Module 12 Digital analyses of the quality control—**Quality.**

The module incorporates the procedures and measuring devices in accordance with the company policy concerning the quality control to ensure the parts conformity with the contractual forms. Actually the Quality module ends the Product manufacturing phase, as a collection of total conditions, digitally validate, that may lead to obtain a finish product.

Product Support

Module 13 Digital analyses of the production capabilities—**Production.**

Processing a part is dealing with physical procedures for the material transformation. Manufacturing a part is a technological activity that connects a certain tool with a certain machine tool, part design and part material to generates the final technological steps and activities that leads to the transformation from row material to the finished part.

The production activity is linking the contractual forms regarding quality, quantity and delivering times with the Plant capacity and the allocated resources. All together are expressing a production capability and the production capacity.

Module 14 Digital analyses of the maintenance and service management—**Support**.

In the context the Support represent the company policy and capacity to ensure the internal service and maintenance activities to ensure the production capability and the service and customer support for the product(s) warranty and post warranty period. That includes also the internal activities the customer feedback data required in the analyses on product reliability and for the product and processes optimization.

Module 15 Digital analyses of the and planning of the company activities management—**Planning.**

The Holonic attitude of the company as a self-standing entity with its own needs but connected with the outside world through the tides of the Technological, Economical, Political and Social forces of change made to suffer to the need to continue adapt to change.

The attempt to control the change is made by planning. The manner in which the deviations from planning correction are made is actually the company adopted change management.

In this way how the planning is made is a personal imprint of each company and the Lifecycle Manager is in position to define the KPI (Key Performance Indicators) adequate for the company technological culture profile and obtain the most adequate Planning style.

Product End-of-Life?

Surprisingly, neither in short forms nor in the extended form of PLM platform the Recycling or end-of- life issue is not present. If we cannot see something don't means that does not exist. What is observed in common practice is an avoided topic The end-of-life alternatives may be encountered within the research activities and in extreme situations, when under the pressure of one or more of the forces of change, reparatory measures are taken. Generally there are compensatory measures, mostly in financial forms.

22.3 Life Cycle End of Life Management

From cradle-to-grave, sounds the advertising for the PLM implementations designed to ensure the IT technology and the services to manage the company assets along the entire process lifecycle with special attention on the conceptual phase and plans for retirement.

The multi 3D CAD facilities and support for the assets representation ensure important reduction in the development costs. The practical full access anytime from any location around the world, with security services, hardware and middle ware support, legacy data integration and consulting support for business transformation, including the maintenance activities ensure the evolution management of the process lifecycle.

Things are less defined for the retirement aspects. For any "retirement implementation" is asked for the PLM specialist, but not any specialist can be valid, only the ones with customized experience in the field were the considered company activates.

The well-known sign is defining "a wish" and in some cases "a request", but is far from the status of "a must".

The most usual approach for the lifecycle end of life is the search for new week markets, relocation, and in some cases to find an application for the assets repairing. The real high-tech involvement in a recycling process is still in the baby steps phase. The environmental issues and the resources scarce availability, increase dramatically in the last years, together with the shortcoming in trained and skilled specialists.

With the start of the 4th Industrial revolution concerning the activities digitalization, the deep data collection and the deep data management, appears an interesting asymmetric distribution of the data in the products end-of-life.

22.3.1 The Recycling Trend

What in Germany for decades is a successful story, reminds to the rest of the world about a forgotten activity to exploit the available resources also through the reuse of the objects and artefacts with a special emphasis on remanufacturing. Soon that could be the main activity for a lot of actors.

This "trend" is not hitting just the low tech activities but really the very important high-tech ones. If we are simply search abound new demands and projects in aerospace we will find a huge demand for products recovery and for regaining the possession of the lost value.

To fill the demand is only possible if we may close the information gap by combining the product data and product service data (manufacturing, production, maintenance) within a single oriented architecture based on specific KPI in terms of capability, viability and profitability.

The products lifecycle analysis reveals that financial situation and the necessity for using the products lifetime declines dramatically, especially in the developed countries—very strong oriented on business and production—so the replacement time is short and repairing is going little by little to disappear. In the same time the globalization arrived with higher pressure on new products and in getting shorter manufacturing times. This sort of behaviour expands the conception and manufacturing pressure to the environment ethical issues and resources use, landing on the conservation and reuse issues. That is the moment when the financially necessity incline back in the sense of using and reusing the product lifetime. You may call it crises time, but is the actual period.

In this context the new business driven philosophy and technological capacities are leading to the "remanufacturing". The will provoke new conceptual products or like-new (refurbished) versions in an equivalent quality assurance and with exactly the same functionality, including the documents, drawings and projects.

The new lucrative activity came with a price, but as long as are ensured the capability, viability and profitability, the price will be accepted. Probably you will be surprise, considering that is not a new concept so "Nothing new under the Sun!".

In real life is a deterring in uptake digitalization and remanufacturing in the same sentence and this is the real data and information gap once the products leaves the OEM. The reason is we face a practically lack of data regarding the products usage and actually the product lifecycle. The knowledge regarding the use of the product, the maintenance and the repairing activities, the parts and components replacement is in the hand of the user. The lack of reliable data and reliable information on the product usage lifecycle, leads to miss opportunities and have a bad economic and environmental impact. All these because of a remanufacturing process, the input data is unknown or of unknown quality.

This is about to change when we are considering the PLM, especially major changes in aspects regarding the "Documents handling". Product and production engineering in companies are typically fragmented across different functional units, distributed across companies along the value chain, requiring input from experts from a variety of disciplines by using different methods and tools. At this must be added the trend of "compulsory feedback" form the users. The conformity legislations for usage, storage and recycling will impose the documents existence at the used site. In time the documents will be open entering in the PLM system.

In the initial stage that will lead to a high coordination effort to synergies work and information transfer as well as to sub-optimal decisions and unused knowledge and experiences. But if it will be showed up in research activities and in information session that the actual resulting waste in engineering processes is reflected in unnecessary lengthening of time-to-market and time-to-production of new products and to a loss of competitiveness for European companies the effort worth and the things will start to go better.

PLM will offer a radically new and extensible approach to collaborative engineering, leveraging state-of-the art research on semantics, heuristics and visualization. Cross-disciplinary collaboration and remote consultation enables the creation of an ontology that serves as an -interoperable model and -integrating element for an open engineering system like an open engineering platform based on existing tools and libraries. Research and development tools assist in product and process development, creating a standard environment for analysis and virtual testing, as an optimization based on heuristic methods and simulation that operate on knowledge represented by information which is structured by means of this ontology.

The user centric approach will made that standard defined environment to offer customized solutions, facilitating cross-disciplinary knowledge-sharing and collaboration, accelerating the product and the production engineering by integrated workflows, capturing and reuse knowledge and experiences.

22.3.2 *Life Cycle Management Close-Loop*

The power of the PLM and the correct application is based on the cybernetic close-loop as a feedback cycle. Remembering the Holonic behavior of the company the successful lifecycle management will consider the Holon side of the company as a close system where the interactions occur only between the company components maintain a low level of organization but integrating the Holonic behavior as the interaction with the outside, the "company environment" through the inputs from the environment and the outputs of the company. Being in a higher level organization the PLM must ensure the input/output balance.

In this scenario exits 3 major close loops;

- The Service loop;
- The supply chain loop;
- The manufacturing loop.

In **the Service loop**, the data is generated by itself as a product-service activity along its lifecycle. In a circular economy perspective this requires the product operations monitoring-analytics and control. The first aim of the engineering close-loop regards the preventive and predictive models for maintenance and the second aim is for innovation in the product service and design oriented on manufacturing-recycling and service business economy.

The supply chain loop, is creating in our times the most critical managerial tasks. The data are generated by the customers (companies and clients), distributors, providers, suppliers and retailers. The management is focused on the executive decisions of each data generator in order to ensure the collaboration and the integration within the value chain and in the end effectiveness optimization.

The close-loops engineering regards the synchronization of the collaborative work in an industrial "symbiosis" between manufacturer, consumer and supplier, using the monitored experience as cross-domain product service.

The manufacturing loop, uses the data generated inside the production activity (manufacturing, assembly, automation, internal logistics, robotics and resources), with aims for conformity, safety, diagnosis, efficiency optimization. The close-loops engineering regards the applications modelling and simulation for the production monitoring and control, having the general KPI the energy consumption, HMI and interaction, zero-defect manufacturing, predictions, the waste reduction, work staff monitoring, management and training.

The general opinion considers the approach to this loops in the reverse order based on the chronological monochromic attitude.

Our approaching order is dictated by the disturbance factor that takes the company and the company managers from the comfort zone. The second element, accompany the disturbance, is the process duration that is practically proportional with the intensity of the disturbance like in a birth process: conception, gestation, care.

The parallel is with the: manufacturing, production, service. The manufacturing and the manufacturing managements requires the shortest time of all and once decided the things are easy to plan and achieve. On this stage are focused the most of research and optimization processes. But what importance may have a 30–40% time reduction in the shortest activity?

The production takes more than the sum of all manufacturing processes within. The production is much related with the supply chain issues than the physical processes and every specialist that works in the supply chain logistics and management admit that is a nightmare time.

This is an area of most successful results, but at this time with much less approach and applications than in manufacturing.

The Service area is a relative new approach, with even less attention but as explained with the greatest potential covering the longest period of a product lifetime. The real specialists are the ones that may repair and fix things, a much harder job then manufacturing even for the moment is not really recognised and is worst paid.

22.4 Life Cycle Planning

The Lifecycle Planning deals with innovative strategies to enhance company assets value and performance. That is possible by identifying the value chain, capture the successful processes and replicate them across multiple projects and companies plants.

The core methodology is to improved collaborative knowledge and processes sharing within all company departments. Every company is different so will never find a "Panacea Lifecycle Planning and Manage Template", but as a Lifecycle manager you may benefit from the guidelines in generating and implementing an application.

First is to consider that you are dealing with data. In here people face the first contradiction, de demand is for good and fast. For fast we need little data for a swift decision, for good we need a lot of data for the best informed decision. Seems that we are going back to an "old discussion" in Chap. 17 "INDUSTRIAL DIGITALLY PROTOTYPES", but in the end seems to be a decision matter over the company assets.

Decisions can be made only by intelligent systems and in our case the system is formed by the Humans (with a limited capacity of data processing) and the assisting IT platform (with a larger capacity of data processing).

Understanding in which situation and in which position we are, the decisional process will be oriented on delegating the data engineering to the PLM system and the management of the data processing to the Humans (the Lifecycle Manager).

From this position must know that we have to integrate data from Design to the Back office and this is the foundation to manage the physical and human assets.

What is unknown cannot be controlled!

Problem As humans we tend to reject everything avoiding "consciously" what is not under our control or intend to control us (make decisions about us). The uncontrollable elements generate a non-perceptual reality in opposition with the concrete, material reality. The lack of concrete evidences leads to unforeseen events that cannot be controlled. These unforeseen events generate an abstract reality, and seem that generally people do not like abstract things instead of material ones, as elements that may lead to frustration, anxiety and depression.

Solution Abstract field is the domain of mathematics and therefor is the PLM to be charge with all the abstract modelling and then fill with data that enables evolution scenarios and decisional support.

Therefor the Lifecycle Manager must identify and know the: work processes, the existing applications and the business system within the company, having in mind the all these must be integrated in a collaborative environment.

Integrating data in a collaborative environment means to consider engineering, economic, administrative, legislative and ethical data validity and conformity; data that must be stored, processed and share in a single homogenous way.

In order to reach that goal, the Lifecycle Manger must build up a specific strategy, considering the company assets, aiming the company (or plant(s) of the company) production improvement.

Any "improvement" must be clearly defined and that means a change, a change that represents a planned lifecycle that must be managed.

Before starting the new lifecycle the planning must consider the assessing the impact of "the improvement" over the CUS:

- Conversions;
- Upgrading;
- Shutdown.

Planning is good but the Lifecycle Planning is made with people and for the people in respect with the stakeholders demands. When the lifecycle planning is established, must be consider the planning data dissemination especially the impact on CUS and how every type of specialists may access the assets data they need.

The Lifecycle Manager must plan the increasing of people abilities by training the plant managers, engineers, technical staff, economic staff and administrative staff. The next important decision of the Lifecycle Manager is to decide on what to focus?

- Manufacturing process?
- Costumer needs?
- Contractors and suppliers?

The general indications in the PLM specialized literature is that we need, we must and we can do all 3 in parallel. The reality is a little bit different, parallel processing is possible when the system is stable and it works. Usually the Lifecycle manager is asked to do something in change times and during the chance, when the system is for sure unstable.

Let's return to the origin of the chapter and select the 2 indicators that lead the Life Manager activity:

- Dealing with data;
- Enhance company assets value.

We may take that as goals, or restrictions. Practically in this situation is the same thing, and the results is the Life manager "forced" to continue by streamline the manufacturing process, increasing the assets value, improve the profit margin; eliminate the non-value activities and reduce the operating costs.

In most of cases this is considered untraditional approach of the manufacturing process, being not addressed exclusively to the technical- engineering team but to all groups of specialists. In this way interlaced steaming may be visualized, managed and controlled.

The first step is to remember that the Plant design is connected with the 3D machine-tools, robots, equipment and systems layouts; instrumentation diagrams; processes flow; maintenance planning, programming and management; contractual conditions; procurement necessities; warehouse data; back-office applications and handling of the all data sources.

Once optimized and stabilized the manufacturing process the next focus is to improve the connection and collaboration with contractors, regulators and suppliers to ensure the input into the manufacturing process.

The last step is to focus on the core competences and add new services in order to customize the business for a quick, valid and efficient reaction to the customer needs.

Now can be assumed that the Lifecycle Manager is able to manage the processes and the change but it must rule the system and that is possible by using the PLM facilities of data sharing and updating, creates a 4D representation (spatial connection and timing) of all the assets in association with the schematic logic of change and develop detailed scenarios of processes development and maintenance.

Most of the tools and facilities of the Lifecycle Manager was presented in previous chapters and here will be offered just a recap to see how powerful can be the Lifecycle Manger in his every day activities.

- Work order management;
- Provide consistent technical documentation;
- Lifecycle extension through service value added;
- Work position evaluation on actual process costs data;
- Projects management based on company all assets data;
- Operations without added value identification and classification;
- Lifecycle extension through predictive and preventive maintenance management;
- Availability of assets data for all the company departments and also for reporting and auditing processes;
- 3D simulation of the procedures for alternative maintenance, feasibility tests, safety scenarios, human-robots interaction.
- Customized documentation and training for the first-time and the traditional contractors in the company new PLM philosophy.

The solutions provide the foundation needed to manage physical and human assets more efficiently. They also help you create common access to information to support innovation and decision making across the enterprise.

So the last part of the book will expose a way in which we consider that a company can be modelled in order to be integrated in the digitalization industrial revolution and remain sustainable.

22.5 The Viable System Model

This model is considered to be a famous theory regarding management cybernetics, developed based on 30 years of observation and analyses of various businesses and business institutions.

The author of VSM (Viable System Model) is Anthony Stafford Beer, a British consultant and leadership theorist that try to formulate organizational systems which ticked the idea of cybernetics may model the management, proposing the effect principle that "*organization should maximize the freedom of their participants, within all practical constraints of the requirement of them to fulfil their purpose*".

In Beer vision the VSM consists of five interacting subsystems. The subsystems will form a representation of the invariant structure of a viable system (a company, a subsidiary or an autonomous working team. Additionally, remember that the company environment extends the close system behavior to the open system behavior being represented in the VMS as a necessity to determine the internal interactions with their surrounding environment and the considered actions frame.

Subsystem 1

Contains several primary activities and thus stands for what is primarily done in the company (e.g. product design, product development, product manufacturing, product sale, product services, etc.).

Subsystem 2

Represents the information bodies and the information channels and cross channels (nodes) that allow the primary activities in Subsystem 1 to communicate between each other and stands for how the components in Subsystem 1 are coordinated. Additionally the Subsystem 2 manage the access of Subsystem 3 in the monitor, audit, and co-ordinate the Subsystem 1 activities.

Subsystem 3

Is the subsystem that represents the operational management.

The first aim refers to Subsystem 1 and is to: establish the rules, allocate the rights and responsibilities, distribute the resources, and to control.

The second aim refers to Subsystems 4 and 5 with the role to provide and ensure the interfaces between the Subsystem 1, Subsystem 4 and 5.

Subsystem 4

The strategic management of the company. The bodies forming the Subsystem 4 are responsible for interaction with the outwards interaction with the environment, and the monitor of how the company needs to adapted to remain viable and sustainable (management and change strategy).

Subsystem 5

Stands for the normative management that is responsible for policy decisions within the company as a whole, balance demands from different departments of the company and steer the company as a whole.

As PLM the VSM have less use and applicability in SME. To be applicable the company must be large enough to afford to sustain the implementation and to create 5 main subsystems and the adjacent departments and the products that the company is providing to have the complexity of components and relationships the over cross organizational units and interacts with a large number of personals.

Reaching the end of the book we may consider it as a Coffee Start—We are social enterprise members and if we like to act in the PLM is like a discovery journey an entrepreneurship activity for a company, for a product, creating value for all the representatives in the team.

Internal we have to deal with contracts, expenses, teams' dynamics and potential development. But exist an even deeper internal, the self-management. If you cannot manage yourself how can you manage the company?

To go externally and sell a product, to make money of it is impossible regardless how great the product is with internal stability. And this is what PLM is doing. The PLM.

Product—Process—Project—Progress—Program—Personnel—Proficiency Lifecycle Management.

The end game for the company is sustainability.

Reference

1. Liu, D.T., Xu, X.: A review of web-based product data management systems. Comput. Ind. **44**, 251–262 (2001)

Index

J. Niemann and A. Pisla, *Life-Cycle Management of Machines and Mechanisms*, Mechanisms and Machine Science 90,
https://doi.org/10.1007/978-3-030-56449-0

Q

R

S